Aboriginal Peoples and Forest Lands in Canada

Edited by D.B. Tindall, Ronald L. Trosper, and Pamela Perreault

Aboriginal Peoples and Forest Lands in Canada

UBCPress · Vancouver · Toronto

21 20 19 18 17 16 15 14 13   5 4 3 2

Printed in Canada on FSC-certified ancient-forest-free paper (100% post-consumer recycled) that is processed chlorine- and acid-free.

Library and Archives Canada Cataloguing in Publication

Aboriginal peoples and forest lands in Canada / edited by D.B. Tindall, Ronald Trosper, and Pamela Perreault.

Includes bibliographical references and index.
Also issued in electronic format.
ISBN 978-0-7748-2334-0 (bound); ISBN 978-0-7748-2335-7 (pbk.)

1. Native peoples – Land tenure – Canada. 2. Aboriginal title – Canada. 3. Forest management – Canada. 4. Indian business enterprises – Canada. 5. Native peoples – Ecology – Canada. 6. Traditional ecological knowledge – Canada. 7. Forests and forestry – Canada. 8. Forest policy – Canada. I. Tindall, David II. Trosper, Ronald III. Perreault, Pamela

SD145.A37 2013   333.7508997'071   C2012-903637-4

Canada

UBC Press gratefully acknowledges the financial support for our publishing program of the Government of Canada (through the Canada Book Fund), the Canada Council for the Arts, and the British Columbia Arts Council.

Printed and bound in Canada by Friesens
Set in Stone by Artegraphica Design Co. Ltd.
Copy editor: Judy Phillips
Proofreader: Lana Okerlund
Indexer: Natalie Boon
Cartographer: Eric Leinberger

UBC Press
The University of British Columbia
2029 West Mall
Vancouver, BC V6T 1Z2
www.ubcpress.ca

Contents

List of Illustrations

Figures

Tables

Appendices

Acknowledgments

Like the Nisga'a who travelled by canoe from the Nass River to Victoria in the late nineteenth century, this book manuscript travelled on a long journey before it reached its destination: publication.

Thanks to all the contributors for staying in the canoe!

This book began from a desire to present the voices of a mixture of Aboriginal thinkers and social scientists to those who wished to learn more about Aboriginal forestry and related social, natural resource, and forest land issues.

The production of the book began during a critical time in Canada – and in particular in British Columbia – with regard to Aboriginal–non-Aboriginal relations. Over the past decade and a half, issues concerning Aboriginals and forest lands have often been headline news in Canadian media for a variety of reasons (as we discuss further in the introduction). One reason is that some of the first modern treaties came to fruition during this period, when protests and court challenges forced the forest industry (and other natural resource industries) and governments to rethink their relationships with Aboriginal communities. In addition, a process of Aboriginal self-determination was taking root across Canada.

The seeds of this book were germinated in "Conservation 370: Perspectives on First Nations and Forest Lands," a course in the Faculty of Forestry at the University of British Columbia. David Tindall originally designed and taught the course with the help of Pamela Perreault, among others. Later, Ronald Trosper taught this course.

To paraphrase a saying, we would like to thank the following people – but they should not be held responsible for any errors in the text.

A number of visionary administrators in the UBC Faculty of Forestry including Clark Binkley, John McLean, Jack Saddler, and John Innes facilitated initiatives such as Conservation 370 and other activities focusing on Aboriginal issues.

A key figure in all of these activities was Gordon Prest, the founding First Nation Co-ordinator for the UBC Faculty of Forestry Aboriginal Initiative, which he started in 1994 and retired from in 2004. In addition to helping (with Pamela Perreault) establish part of the network of authors who would eventually contribute to this book, Gordon served as the co-chair of the UBC Faculty of Forestry's First Nations Council of Advisors and was instrumental in helping to get Conservation 370 off the ground. Beverly Bird and Madeleine MacIvor also played important roles in following in Gordon's footsteps in various ways.

Gary Merkel, who served as the Co-Chair of the First Nations Council of Advisors for the UBC Faculty of Forestry, also made a number of contributions through speaking engagements, among other ways.

Paul Tennant and Doug Sanders contributed enormously to Conservation 370 by giving guest lectures and sharing other materials. Their written work, especially Paul Tennant's book, *Aboriginal Peoples and Politics: The Indian Land Question in British Columbia, 1849-1989,* also influenced David Tindall's ideas for this volume.

Some other individuals and institutions who contributed indirectly to the book through their involvement in and/or contribution to Conservation 370 include Renel Mitchell, Jennifer Morrison, John Lewis, Paul Kariya, Grand Chief Ed John (of the Tl'azt'en Nation), Nancy Turner, Bruce Miller, Brian Chisholm, David Pokotylo, Eduardo Jovel, Bill Bourgeois, Chief Joseph Gosnell (former President of the Nisga'a Tribal Council), the Honourable Michael Harcourt (former Premier of British Columbia), Jo-ann Archibald, June McCue, Lloyd Roberts, Richard Vedan, Paul Mitchell-Banks, Tracy St. Claire, Ramvir Singh, Ann Doyle, Don Ryan, Patrick Matakala, Brian Lapointe, Peter Lusztig, Vince Stogan, Wedlidi Speck, Harry Nyce Jr., Eric Schroff, Caroline Findlay, Sonny McHalsie, Paul Mitchell-Banks, Richard McGuigan, George Hoberg, Scott Hinch, Peter Marshall, Denise Sparrow, the BC Treaty Commission, the Xa:ytem Longhouse Interpretive Centre, the UBC First Nations House of Learning, the Xwi7xwa Library, the UBC Museum of Anthropology, and the North Vancouver Outdoor School. (Apologies to anyone we have inadvertently omitted.) Of course a big thanks also needs to be given to the many students who participated in the class over the years, and from whom we learned so much!

Several people made important contributions to earlier versions of the book, including Charles Menzies, Caroline Butler, Annie Booth, Bob Ratner, and Andrew Woolford.

We would also like to thank Gary Bull and Charles Menzies for their helpful feedback on several of the chapters; Howard Harshaw, for assistance with one of the figures; Angela Walker, for her long-time assistance with media analysis; and the Social Science and Humanities Research Council of Canada, which supported some of the research in this book (particularly that of David Tindall).

Thanks also go to the UBC Department of Forest Resources Management for its support, to the Department of Sociology for its financial assistance with the index, and to Sociology Department Head Neil Guppy for his general support related to this project.

A huge thanks goes to UBC Editors Randy Schmidt and Megan Brand who shepherded this project along, and to Judy Phillips for her great work in copy-editing the manuscript. We would also like to thank UBC Press for its assistance more generally, and to acknowledge and thank the anonymous reviewers who helped make this a better book.

Finally, David Tindall would like to thank his partner Diana Tindall, and their children Joshua and Jenny, for their encouragement and patience, and for making the whole enterprise meaningful!

Pamela Perreault would like to thank the many First Nations students she worked with at UBC who were willing to share their concerns and vision for their communities – their insight helped shape the topics covered in this book. She would also like to thank the Garden River First Nation Education Department for their constant support during the production of this book and her PhD studies.

Part 1
Introduction

1
The Social Context of Aboriginal Peoples and Forest Land Issues

D.B. Tindall and Ronald L. Trosper

Logging Protest by Vancouver Island First Nation Disrupts BC Ferry. "A First Nation group staging a protest over logging on its territory on Vancouver Island blocked a BC Ferry for more than an hour in Port McNeill Thursday. Members of the Kwakiutl (kwa-CUE'-tul) First Nation had canoes in the water to keep the ferry at the dock and demonstrators on land to stop cars from loading. A BC Ferry spokesman says the demonstration caused a 75 minute delay in the sailing of a ferry that links Port McNeill and Sointula on nearby Malcolm Island. He says the RCMP kept an eye on the demonstrators but there were no arrests and no property damage. The protest was the second of two staged by the Kwakiutl over two days to show their anger over logging in an area the First Nation claims as its traditional territory. The band says the logging violates its rights under the Douglas Treaty, and is demanding the provincial government stop the harvesting and start negotiating commercial forestry rights with the First Nation." The Canadian Press, August 4, 2011.

Judge Rejects Logging in Favour of Native Rights. "The see-saw battle over Native timber rights in the BC Interior has tipped in the favour of the Okanagan Indian Band. A BC Supreme Court judge has rejected the provincial government's claim on logging a large tract of forested land west of Okanagan Lake's north shore. The band says the decision should trigger negotiations to settle a question that's been outstanding since British Columbia's first Europeans arrived. How can the province have title and authority over the forests without a treaty? 'The province does not own our territory. They use it,' said Chief Byron Louis of the Okanagan Indian Band ... The band sought an injunction, saying the area is the water

> source for its 1,800 members. The watershed shows evidence that Aboriginal people occupied it up to 7,500 years ago, and forestry operations would threaten their water and destroy their sacred land, the band said." *Kamloops Daily News,* October 26, 2011, A5.
>
> *First Nation Signs $30M Treaty.* "A land-claim treaty is one step closer to reality for a First Nation in the Powell River region. The federal and provincial governments and the Tla'amin First Nation have signed a final agreement that will now be put to the 1,000 members of the First Nation for ratification. The agreement needs final approval in Parliament and the BC legislature. The deal gives the Tla'amin about 8,322 hectares of land and almost $30 million over 10 years, along with self-government powers, economic-development funding of almost $7 million, resource revenue sharing and forestry and fishing rights. Chief Clint Williams says the treaty will provide a foundation for his nation's children and grandchildren." *Kamloops Daily News,* A6, October 22, 2011.
>
> *Logging Fight Brewing over Flores Island.* "Iisaak Forest Resources received a permit from the BC government April 1 to build logging roads on Flores Island, prompting accusations from environmental groups that the forestry company has broken an agreement intended to protect intact watersheds in Clayoquot Sound. 'Flores Island supports a large, intact ancient rainforest that was included in the historic 1999 "no logging in intact valleys' agreement,"' states a media release from Tofino-based Friends of Clayoquot Sound (FOCS). That 1999 Memorandum of Understanding (MOU) was a mutual agreement aimed at protection of undeveloped watersheds in Clayoquot Sound. Iisaak denies logging activity on Flores Island will break the MOU, and says the right to determine where to log should be left to the First Nations who own the company." *Cowichan Valley Citizen,* April 15, 2011, 9.

As the above news media accounts suggest, Aboriginal people in British Columbia in particular, and in Canada in general, have been struggling over the past several decades to gain control of their forest lands. Blockades against logging have been only part of a larger strategy to assert Aboriginal rights and establish Aboriginal title. But Aboriginals have also been concerned about forestry itself, and their inability to influence how the land is managed, as well as the fact that they have generally not benefited from resource extraction.

In many instances, action began with blockades or other related forms of protest. Such protest was often intertwined with seeking recognition of

Aboriginal rights and title through the courts. The courts have clarified Aboriginal rights regarding resources and have established what Aboriginal title contains in principle. These events have motivated the provincial and federal governments to restart the treaty process in British Columbia and to establish various types of interim measures, such as co-management agreements. It has also motivated companies to enter into joint ventures with Aboriginal groups. In some instances, Aboriginal communities have also started their own forestry companies. In these various ways Aboriginals are coming to have more say in decision making about forest lands, and to benefit more directly from forest resources.

In recent years there has been increased participation among British Columbia's Aboriginals in the management of forests (and other natural resources). An advance toward Aboriginal self-determination in British Columbia was achieved with the signing of the Nisga'a Treaty in 1998 by the Nisga'a, the federal government, and the provincial government.[1] Similar agreements have started to come to fruition through the BC treaty process. Interim measures agreements will continue to be signed while treaties are being negotiated. There will also be increased Aboriginal participation in resource management and self-government through alternative processes.

These events mean that professionals – in the private sector, public sector, and not-for-profit sector – working in forestry and conservation will need to understand the context of Aboriginal participation in resource management. The objective of this book is to provide the reader with an opportunity to learn more about Aboriginal issues relating to politics, culture, forest resource use, and land ethics, so that they may have a better understanding of the challenges and opportunities that will arise in the coming years from increased Aboriginal self-government and increased Aboriginal land management.

Although conflicts over forests between Aboriginals and non-Aboriginals have generated headlines during the past several decades, issues pertaining to Aboriginals and forest lands are broader than just those revolving around conflict. In recent years there have been numerous activities on forest lands, including research into traditional knowledge and traditional use of land and resources, contemporary Aboriginal management initiatives, co-management initiatives between Aboriginals and non-Aboriginals, and joint ventures between forestry companies and Aboriginal communities. This book focuses to some extent on the activities that take place on forest lands (co-management, joint ventures, community forestry), but just as importantly, it also focuses on the context that underlies both the conflicts and the cooperative ventures. Hence, this book is very much about the *context* of Aboriginal peoples and forest lands.

The terminology for referring to Aboriginal peoples varies by context, time, and space; consistent usage is difficult. Various labels have been used to describe the indigenous peoples of Canada, including Indian, Inuit, and Métis. All three of these are used in federal legislation. The term "Indian" was originally used by Europeans to describe the indigenous peoples of North America (excluding those in the far north), and many regard its use as derogatory, substituting "First Nation." The Inuit are the indigenous people living in the north of Canada (as well as several other countries). The Métis emerged as a result of contact and interaction between Europeans and Indians during the fur trade period. In the title of the book we use the term "Aboriginal" as an inclusive designation for these different categories of indigenous peoples in Canada. We also primarily use this term in this introductory chapter. However, different terms are used by different authors in this book because of the different contexts that they deal with. In British Columbia, for example, Aboriginal groups are now officially known as First Nations. This is also the term used in some of the provincial legislation and policies regarding Aboriginals and forestry in the province. But other terms are used in other jurisdictions and contexts, and accordingly, different terms are used in different chapters of this book.

Main Themes

The news media excerpts that opened this chapter illustrate a number of different aspects of Aboriginal–non-Aboriginal relations in British Columbia and Canada. These include confrontations such as blockades, legal challenges that have been undertaken by Aboriginals regarding rights and title issues, and a shift from conflict to cooperative endeavours, as symbolized by some of the joint ventures that have been undertaken between Aboriginals and forestry companies. In some instances, joint ventures have incorporated traditional ecological knowledge into planning and management, and such cooperative endeavours have provided opportunities for cross-cultural learning. Media coverage has helped shape public opinion, which in turn has changed the context in which Aboriginal peoples, governments, and corporations operate.

Four main interconnected substantive themes integrate the material in this book. These are (1) history: cooperation, conflict, and reconciliation, (2) differing visions, (3) traditional ecological knowledge and use, and (4) collaborative endeavours. Below we elaborate on these themes and briefly describe the contributions made by the authors of this volume. A number of the chapters exemplify several of these themes. Hence, in some instances below, we will discuss particular chapters under one theme while their physical location in the book may be within a different thematic section. In some instances we also talk about particular chapters more than once – to highlight their connection to these multiple themes.

History: Cooperation, Conflict, and Reconciliation

One goal of the book is to give readers who are interested in forestry and natural resource issues an account of the history of Aboriginal–non-Aboriginal relations. This history provides a crucial context for understanding contemporary natural resource issues. As Coates and Carlson note in their chapter in this volume, early relations between Aboriginals and the European newcomers were marked by a good deal of cooperation. Aboriginals vastly outnumbered the European settlers. Aboriginals were central to a variety of activities, including the fur trade, and were integrally involved in endeavours such as logging and mining. European settlers benefited from these relations. Over time in the new colonies, however, Aboriginals were alienated from most of their lands, resources, and rights.

The alienation of Aboriginals from their lands and resources, their substantial exclusion from the workforce, and the consequences of their experiences in residential schools resulted in Aboriginals being among the most disadvantaged groups in Canadian society. High school drop-out rates, substance abuse, legal problems, health problems, and unemployment have persisted in many Aboriginal communities. In this volume Frideres discusses these issues of social inequality. Dale builds on these insights in noting how this history of alienation, marginalization, and resulting social problems has created challenges for Aboriginal communities in terms of participating in collaborative decision making about lands and resources. Dale provides some reflections on the Central Coast Land and Resource Management Plan process in particular.

From the 1970s through the 2000s, Aboriginals in British Columbia progressively gained support from the courts. Also around this time, public opinion about Aboriginals started to change (Ponting 2000), with increasing numbers of the general public in support of their claims. Arguably, to some extent, attitudinal support for Aboriginals was indirectly influenced by the civil rights movement and also the American Indian Movement, which had become active earlier in the United States (Morris 1999). Also, in the 1970s and 1980s, Aboriginals started to engage in various types of protest activities in order to bolster their land claims. These contentious actions garnered a good deal of media attention (Ramos 2006) and likely influenced public opinion. In this volume, Wilkes and Ibrahim examine the practice of direct action against forestry and other enterprises. The influence of direct action as a factor in the development of Aboriginal–non-Aboriginal relations and the evolution of land use planning is also noted in several other chapters.

This constellation of forces – especially increased uncertainty about the legal status of land and resources (Kunin 2001) – motivated the provincial government to finally begin a treaty process in British Columbia in 1993.

Ironically, as Mark Stevenson notes in his chapter, the provincial Liberal Party and its former leader (and then premier) Gordon Campbell were initially adamantly opposed to treaties but later became arguably the biggest boosters of the treaty process and other measures to deal with these unresolved issues. Stevenson provides an overview of the BC treaty process and discusses the limitations of the treaty process from the perspective of Aboriginals.

Differing Visions

There are profoundly different visions about how to proceed to address Aboriginal claims. These visions include analyses of past relations, current relations, and future relations. As Smith and several other authors in this volume observe, past relations have mostly proceeded under the assimilation model whereby Aboriginal peoples were supposed to abandon their language, culture, land, and resources and become assimilated in the larger Canadian society. In her chapter, Smith describes several models of co-management and argues that the types of relationships that the parties have with one another have consequences for the outcomes of such collaborative endeavours.

Passelac-Ross and Smith discuss the issue of land tenure and how the land tenure system needs to be revamped in order to deal with Aboriginal rights and title. They note that the determination of the annual allowable cut has generally not taken Aboriginal treaty rights into consideration and argue that current tenure systems reinforce the assimilation model and result in Aboriginal dependency. They explore two specific cases where alternative land tenure arrangements have been tried: the Nisga'a in British Columbia and the James Bay Cree in northern Quebec.

Marc Stevenson writes about the challenges of incorporating traditional knowledge, and the holders of that knowledge, into sustainable forest management. Traditional ecological knowledge is a body of knowledge and beliefs that is communicated through oral tradition and first-hand observation. Stevenson argues that traditional knowledge exists within particular contexts, and that elements of traditional knowledge cannot be cherry-picked and dropped into a Western scientific framework.

Lewis and Sheppard explore Aboriginals' spiritual perceptions of forested landscapes. They describe, for example, that sacred sites are linked to the identity of the Stó:lō in southwestern British Columbia. Lewis and Sheppard use the relatively novel methodology of visualization coupled with interviews to empirically explore this topic.

In the final chapter, Trosper and Tindall discuss consultation and accommodation. They note that although Aboriginals are constrained in their options by recent legal rulings, consultation and accommodation have the potential of making invisible losses (such as lifestyles) visible and thus hold some potential for providing better outcomes than have typically occurred in the past.

Traditional Ecological Knowledge and Use

Another theme explored in the book is importance of understanding traditional uses of forest and other natural resources, and traditional ecological knowledge. Brian Chisholm and Andrew Mason describe some of the ways in which archaeologists have documented traditional use of land and resources. Archaeological knowledge has been important in the land-claims process in helping to legally establish continuous use of traditional territories. Andrew Mason talks about some of the legislation that exists regarding the protection of cultural artifacts and makes some suggestions for new policy in this area.

Several of the contributors to this volume compare and contrast traditional knowledge with scientific knowledge, including Michael Blackstock, who describes insights that these two perspectives provide into understanding of water, especially in a forest context. Marc Stevenson talks about how "relationships" are central to traditional knowledge. Traditional knowledge is not simply a compendium of information about the land and resources; rather, it is a system of relationships that are crucial for living in harmony with the ecosystem. In their chapter, Lewis and Sheppard also talk about the importance of relationships for traditional knowledge and management. They focus in particular on the spiritual ties that the Cheam First Nation have to the land, and about the challenges to forest managers of managing for sacred spaces.

Traditional use and knowledge are intimately linked to the other three themes discussed here. Most notably, perhaps, it is a central component of collaborative endeavours such as co-management and joint ventures.

Collaborative Endeavours

There have been many types of collaborative efforts to repair the damaged relationships between Aboriginals and non-Aboriginals and at the same time address land and resource issues. One such effort has been to introduce Aboriginal knowledge and topics into forest management and natural resources education. In this volume, Trena Allen and Naomi Krogman explore such efforts in the Canadian context.

Another important avenue for collaboration has been co-management. Co-management agreements have been pursued by Aboriginals, governments, and other parities. In certain instances they are meant to establish some progress in relationship building and better resource management in lieu of completed treaties; in others they are integrated with the treaty process. In their chapter, Mabee, Tindall, Hoberg, and Gladu discuss co-management in Clayoquot Sound and on Haida Gwaii, in British Columbia. They argue that an important benefit of co-management processes is the experience that allows the two parties to learn about each other's cultures and world views. They argue also that such cross-cultural exchange and learning processes occurred in both Clayoquot Sound and Haida Gwaii. Other potential benefits of co-management for Aboriginals include capacity building and increased stability. These benefits can accrue through other forms of collaborative endeavours too, such as joint ventures, as discussed by Pechlaner and Tindall. However, in both co-management scenarios and joint ventures, preexisting community capacity is an issue. The benefits of co-management and joint ventures will be diminished if both parties do not have the capacity to fully participate in forest management and planning processes.

Finally, Forsyth, Hoberg, and Bird discuss the evolution of forest policy in British Columbia in recent years, including issues of consultation and accommodation. In particular, they explore a couple of cases where collaborative governance has recently been undertaken: on the north and central coast of British Columbia – the Great Bear Rainforest and Haida Gwaii.

Conclusion

The book provides an overview of the context of Aboriginal peoples and forest land issues in British Columbia and Canada. In this introduction we have given a brief overview of the main themes of the book: (1) History: Cooperation, Conflict, and Reconciliation, (2) Differing Visions, (3) Traditional Ecological Knowledge and Use, and (4) Collaborative Endeavours. To some extent, we are in a transition period from a situation best described as "assimilation" to one marked by "shared power." As Smith in this volume notes, many Aboriginals desire a further transition from shared power to sovereignty. By exploring the four themes in greater detail, the authors of this book help to provide insights for understanding the current transition and what will be necessary for a future transition to take place. A number of authors in this book argue that the "Indian Land Question" (Tennant 1990) will be resolved only when Aboriginals have their world views respected in land use planning, and have their lands, title, resources, and some measure of self-government returned to them.

Notes

1 In 2000, the Nisga'a Treaty, the first modern treaty in British Columbia, became law.

References

Kunin, Roslyn, ed. 2001. *Prospering Together: The Economic Impact of the Aboriginal Title Settlements in BC Updated.* Vancouver: Laurier Institution.

Morris, Aldon D. 1999. "A Retrospective on the Civil Rights Movement: Political and Intellectual Landmarks." *Annual Review of Sociology* 25:517-39.

Ponting, J. Rick. 2000. "Public Opinion on Canadian Aboriginal Issues, 1976-98: Persistence, Change, and Cohort Analysis." *Canadian Ethnic Studies* 32(3):44-75.

Ramos, Howard. 2006. "What Causes Aboriginal Protest? Examining Resources, Opportunities and Identity, 1951-2000." *Canadian Journal of Sociology* 31(2):211-34.

Tennant, Paul. 1990. *Aboriginal Peoples and Politics: The Indian Land Question in British Columbia, 1849-1989*. Vancouver: UBC Press.

Part 2
History: Cooperation, Conflict, and Reconciliation

2

Different Peoples, Shared Lands

Historical Perspectives on Native–Newcomer Relations Surrounding Resource Use in British Columbia

Ken Coates and Keith Thor Carlson

Newcomers to British Columbia are quickly introduced to the tensions over land, resources, and authority that have divided indigenous peoples and immigrants for more than two hundred years. Conflicts over resources, land rights, commercial and food fisheries, access to timber and recreational lands, demands for attention to Aboriginal self-governance, and debates about federal fiduciary responsibilities to status Indians are only a few of the most publicly debated issues confronting the province. For many years, these struggles were very one-sided, with the colonial, provincial, and dominion governments using the authority and instruments of the state to marginalize indigenous people from their tribal lands and resources. Empowered by a series of major court decisions over the past forty years, and bolstered by statistics showing the reversal of two centuries of demographic decline, Aboriginal peoples have recently secured *de jure* recognition of their rights and title, and growing de facto authority over certain economic, legal, financial, and administrative powers and resources. But as the balance of power shifts tensions are not necessarily dissipating.

An historical evaluation of these relationships demonstrates that there is no linear or singular pattern. History is a complex tapestry, and the history of First Nations–newcomer relations over natural resources in British Columbia is an excellent example of the uneven, occasionally circular, nature of the ongoing contact experience.

History Meets the Present

Many people in the province find the current round of confrontations unnerving. Bitter debates over commercial salmon fishing rights in the lower Fraser River basin occasionally turn violent. Roadblocks in the Okanagan and throughout northern British Columbia spark angry words. The demands of Aboriginal political sovereigntists, like the Lil'wat First Nations, infuriate those British Columbians who believe that there is no room for special status for Aboriginal Canadians. The "common sense" arguments of right-wing political groups advocating "no rights based on race" and "one law for all Canadians" likewise incense Aboriginal leaders who reject attempts to undermine indigenous rights that are derived from history and not skin colour. In Vancouver, the long-standing dispute over residential rents for leased lands on the Musqueam Reserve, resolved in favour of the Musqueam First Nation, elicited demands that the Government of Canada intercede to protect tenants' rights. Even successful negotiations, like the resolution of the Nisga'a Treaty in 1998 and the Tsawwassen settlement in 2006, generated protests about the unique constitutional position of First Nations. And although the prominent acknowledgment of the leaders of the four First Nations whose territories hosted the 2010 Olympic Games seemed to signal the arrival of the provincial government's "new relationship," the associated grassroots Olympic Resistance Network's 2010 campaign slogan of "No Olympics on Stolen Land" reminded the world that the unresolved state of Aboriginal title in British Columbia continues to have social as well as economic implications. The contemporary situation is a logical outgrowth of more than two centuries of struggle between indigenous peoples and newcomers. Moreover, some of the most innovative solutions to these tensions hearken back to earlier expressions of Native–newcomer relations in the era before significant non-Native settlement.

The principal and most public differences and disagreements are so fundamental that they commence with the attempt to understand the initial habitation of the Pacific Northwest. Aboriginal peoples explain the emergence of human beings in this region in terms that are locally anchored. For the lower Fraser River Stó:lō, for example, the commencement of life is explained in terms of both "sky-born" people and the subsequent arrival of the "Transformers." The resulting landscape thus consists quite literally of people's relatives and is animated by the presence of human ancestral spirits who continue to reside within objects non-Native newcomers typically classify as "natural resources" (Jenness 1955; Bierwert 1999; Carlson 2010).

Non-Aboriginal Canadians used words like "myth" and "legend" to describe Aboriginal accounts of origins. In their place, Western scholars point to archaeological evidence of a very different explanation. Recent discoveries, typically set in opposition to Aboriginal accounts, point to two separate migrations of new peoples from Asia (Fladmark 1979). Carbon dating indicates continuous indigenous occupation of the west coast for at least nine thousand years (for an overview, see Schaepe, "Village Arrangements and Settlement Patterns," in Carlson [2001]; Matson et al. 2003), and as Brian Chisholm (this volume) shows, First Nations have long been actively engaged in managing the forest resources. Complementing the scholarly discernment is a popular Judeo-Christian discourse that largely constrains miraculous activities related to human kind's origins to the Middle Eastern Holy Land.

Debates about the nature and extent of Aboriginal occupation reveal a great deal about the changing place of First Nations people within British Columbia. The first generation of European settlers in British Columbia in the mid-nineteenth century felt that they were creating a new society in opposition to existing indigenous occupants. The second generation regarded newcomer society as something new and even tenuous in a place where Aboriginal people still made up the majority of the population. With the third and subsequent generations came the dominant notion that non-Native people are inheriting a European-style society. No longer do most non-Native British Columbians strongly identify with the notion that they are newcomers. What has been inherited, however, is a social, economic, and political system built on the denial of Aboriginal rights and the silencing of their oppositional voices.

Contact Encounters

Even before the first Europeans arrived, Aboriginal people had knowledge of the newcomers. With the opening of regular communication routes to Europe and Asia over four hundred years ago, imported diseases caused havoc on Aboriginal populations. Once introduced, these diseases passed quickly from one group to another. Continental-wide pandemics likely had reached certain regions of the Northwest before the arrival of Europeans (Boyd 1999). Certainly, sometime in the late eighteenth century, a major smallpox epidemic swept across most of coastal British Columbia and much of the Columbia and Fraser Rivers plateau. Although scholars continue to debate details, there is a substantial consensus that the imported diseases exacted a dramatic toll. If, and to what extent, populations may have initially rebounded in the wake of these diseases has yet to be determined. Other groups often moved into resource zones "widowed" by the epidemics, resulting in a substantial precontact resettlement of British Columbia (Harris 1997; Carlson 2010).

The arrival of newcomers along the Pacific Northwest Coast in the 1770s ushered in a substantial reordering. The Spanish, British, and Russians who were drawn to the region by the imperatives of the so-called Age of Exploration laid claim to all unmarked lands. The Americans, meanwhile, sought to expand an already lucrative circum-global trade. The principally Bostonian and English trading ships that came to trade for sea otter pelts for re-exchange in China for silk affected imperial relations and resulted in the identification of important new trading territories. The short-lived Pacific coast sea otter trade actually dwarfed the beaver trade of Rupert's Land in terms of both volume and profits.

The arrival of outsiders created opportunities as well as threats for the First Nations. New technologies, in the form of metal goods, rifles, and other trading items, proved extremely attractive. Initial relations wavered between cooperative trading sessions and violent clashes. Cultural confusion abounded, with the Europeans interpreting the First Nations as a west coast variant of the "heathens" and "barbarians" encountered elsewhere in the New World, and with the Aboriginal people wondering about the motives and ambitions of the newcomers. Both the Spanish and the Nuu-chah-nulth (Nootka) incorrectly assumed the other to be cannibals (Archer 1980). Moreover, in addition to the new diseases, indigenous societies had to cope with issues associated with the sexual advances of lonely sailors toward Aboriginal women.

Aboriginals in an Imperial Order

The advent of imperial rivalries drew First Nations societies into sweeping and complex cross-racial relationships. Beneath this imperial veneer, indigenous politics unfolded. Chiefs Maquinna and Wickaninnish of west Vancouver Island were as quick and keen as any British or American trader to take advantage of all the new opportunities. By the 1790s, both these Native leaders had firmly consolidated economic hegemony over vast expanses of territory. Both had sufficient European-manufactured light and heavy arms to defend their interests against indigenous and European challengers alike.

From the 1770s to 1840s, First Nations people responded as partners and rivals in the changing commercial and political realities. European nations had relatively little presence in the area, and Aboriginal people outnumbered the newcomers by a wide margin. The Spanish, British, and Russians resolved their differences by dividing indigenous territories among themselves, with no Aboriginal input. That the initial establishment of colonial boundaries and authority had little on-the-ground consequence for the First Nations masked the fact that the decisions made in this era would shape Aboriginal–newcomer relations for generations.

Until the middle of the nineteenth century, Aboriginal and newcomer resource use were tightly intertwined. The Europeans counted on the First Nations to harvest or bring in the furs, fish, and trees necessary to sustain the tiny settlements. The Hudson's Bay Company dominated after the simultaneous collapse of the ocean-based sea otter trade and the expansion of the land-based trade. The Hudson's Bay Company moved north after the Oregon boundary settlement of 1846 and established several strings of trading posts along the coast and into the Interior (Mackie 1995). There were significant problems at times, particularly relating to trading relations, social contacts, and introduced diseases, but in general Aboriginal peoples and newcomers worked constructively and cooperatively.

The relatively peaceful relationships of the pre-1850 period masked the profound transformations that had occurred in Aboriginal societies. Metal axes and knives not only cut more effectively than traditional tools, they were easier to sharpen, thereby saving Aboriginal craftsmen hours of labour. However, their introduction also meant that people formerly valued for their stone tool-making abilities found their skills devalued. Opportunities for intergenerational sharing and teaching associated with some traditional activities declined. While harvesting proceeded more expeditiously with the manufactured tools and firearms, the opportunities for individual wealth accumulation threatened the communal nature of west coast tribal society, and the fragile environment. At the same time, the commercialization of certain resources resulted in the rapacious harvesting of coastal sea mammals and inland fur bearers. The newcomers introduced a wide variety of new values and world views, ranging from the production of surpluses for the purpose of accumulating personal wealth and private ownership of land to Christian convictions, particularly those transmitted through the dozens of Catholic and Protestant missionaries who came to the region in pursuit of souls. Alcohol and drugs like opium, meanwhile, tore families apart and destroyed the lives of countless individuals.

Collectively, these new products and ideas created enormous pressures for change within Aboriginal societies. First Nations responded creatively at times, and out of desperation on other occasions. The core values of the indigenous communities appear to have stayed generally intact, with significant adaptations to the new realities (see also Lewis and Sheppard, this volume). In some instances, populations shifted from traditional villages. Inter-tribal warfare accelerated in some regions. Several coastal groups became deeply engaged in commerce. Others, particularly in the Interior, maintained limited contact with the newcomers, preferring their mobile harvesting existence. The Chilcotin are probably the most prominent of those groups that sought to maintain a significant degree of isolation from the newcomer economy (see Lutz 2008).

With the arrival of European traders, cultural activities like the potlatch flourished. Other social practices, meanwhile, sometimes faded into the background. Some scholars have suggested that contact with missionaries precipitated shifts in aspects of Aboriginal epistemology and spirituality. Indigenous people have steadfastly maintained that changes were reflective of responses to the spirit realm. Determining whether the Creator and ancestor spirits provided tools for a rapidly changing world, or if innovative indigenous people adapted to altered circumstances through their own agency, is, perhaps, beyond the scope of what historians can answer. What is certain, however, is that prophet movements swept through much of the province in the early through mid-nineteenth century. Unlike most other areas of North America, the west coast prophets generally did not preach the rejection of European technology and ideas, but rather an inclusive expression of spirituality (Suttles 1957). The resulting spirituality retained many of the older traditions and expressions while embracing such outward signs of Christianity as the sign of the cross and the preeminent position of the Sagalie Tyee (Lord Above) over the other spirit forces of the animal, ancestral, and natural world. Many of the early prophets, especially in the Coast Salish area, helped usher in a syncretic approach to spirituality that paved the way for such religious expressions as the Indian Shaker Church (Barnett 1957; see also Neylan 2003). All told, Aboriginal peoples and communities exercised considerable selectivity in determining their relationship with the newcomers and set many of the terms and conditions of contact – even if they typically found themselves as responding to newcomer initiatives, rather than initiating change on their own terms.

The intrusion of the colonial system, however, soon limited the flexibility and opportunities of the First Nations people in British Columbia. Until the 1850s, British authorities paid little attention to conditions on the west coast. The Oregon boundary dispute and subsequent British fears of American expansionism in the Pacific Northwest caused the Hudson's Bay Company (HBC) to withdraw north of the forty-ninth parallel. The gathering tensions pushed the British Colonial Office to assert its dominance over the northern region. Aboriginal demands that their land and resource rights be respected attracted little attention, save from James Douglas, the HBC chief factor and later governor of the colonies of Vancouver Island and British Columbia. Under Douglas's leadership, the British acceded to a request for a series of small treaties on Vancouver Island, but they rebuffed suggestions for more comprehensive arrangements for other First Nations in the region. The Colonial Office regarded treaties as desirable, but it was willing to endorse them only if the colony paid for them. For the fledgling colony of British Columbia, infrastructure and security took priority over negotiating Aboriginal rights and interests (Tennant 1990). And besides, in European eyes, Aboriginal people were destined for cultural, if not physical, extinction in the face of Western advance.

The absence of treaties on the BC mainland west of the Rocky Mountains became a critical issue in subsequent years, with British, Canadian, and BC officials arguing that the protection of Aboriginal rights articulated in the Royal Proclamation did not apply to the area west of the Rockies (Ray 1999). Aboriginal peoples generally accepted the arrival of newcomers, the establishment of trading posts, commercial fishing and farming activities, and the presence of government officials and military units. At the same time, however, various Native groups made it clear that they wanted either assured access to resources or compensation for lost lands and rights. In 1922, for example, the Upper Fraser Valley Native leader Dennis S. Peters explained:

> Gov. Douglas when he set aside the reserve for the Katzi or the Port Hammond Indians said all inside the lines of the reserve would remain the

> real property of the Indians and all outside would become white mans land. The land taken over by the whites would be like a tree, which should blossom and bear fruit for the Indians meaning that the Indians would share in the benefit of the use of their tribal lands by the whites. Later Gov. Seymour called a great meeting of the Stalo and other Indians at Queen's Borough (now New Westminster). There were very many Indians and whites at this meeting. He said (1) The Queen desired the two races to live together peacefully and neither to harm the other. That they would be as brothers, the whites the elder, and the Indian the younger. (2) When Indians worked they would have the same wages as the whites for the same kind of work. (3) Money or revenue would be coming to the Gov. from the lands outside the reserves. This money or revenue would be as in four: ¼ would go to the Queen or Crown or Gov. ¼ would be used by the BC gov. for the purposes of the country development, road making etc. ¼ would go towards education and ¼ for the benefit of the Indians, assistance to them etc. The Indians in those days believed the words of those big men in authority (or the chiefs of the whites) and they never thought of asking them for written agreements nor their words in writing. (see Carlson, 2001, 185-86)

From a purely legal and political standpoint, however, the so-called Douglas Treaties on Vancouver Island represented, for more than a century, the end and not the start of the treaty process.

Resources, Labour, and Land

British Columbia's relatively cooperative contact ERA ended when newcomers discovered resources of immediate value. The recognition in the mid-1800s that Fort Rupert and Nanaimo sat on impressive seams of coal sparked a mining rush. Although many Aboriginal people worked in the coal mines, the broader interests of Aboriginal people in the area were quickly brushed aside. The sudden expansion of the gold mining frontier north from California spoke to even more dramatic changes. News that the HBC was buying gold collected by Aboriginal people along the lower Fraser River drew more than thirty thousand outsiders from the United States to the region in the summer of 1858. The initial focus of activity around Hill's Bar and Fort Yale steadily spread up the Fraser River, sparking subsequent rushes in the Interior at places like Quesnel and Barkerville in the early 1860s. Gold drew the attention of prospectors, traders, camp followers, and developers to this previously neglected corner of the globe. Newcomers pushing into the distant corners of New Caledonia/Colony of British Columbia discovered fertile farm lands in the southern Interior and, later, rich deposits of silver, lead, and other ores throughout the colony. The gold frontier pushed steadily north to the Cassiar Mountains and beyond into the Yukon by the 1890s.

The resource bonanzas transformed the newly created colony of British Columbia and ushered in a new era in Aboriginal–newcomer relations. Except for the power they held as workers in a colony where non-Aboriginal labour was scarce, First Nations people had little authority. Although most of the gold miners quickly moved on, thousands of settlers followed in their wake and settled permanently in and around Victoria, New Westminster, and the lower Fraser River, and at smaller centres along the Fraser corridor and into the Interior. Those Aboriginal people who wished to participate in the new regime found themselves shouldered aside by anti-Aboriginal legislation directed at their commercial, land, and resource rights. In 1867, the colonial government unilaterally reduced Indian reserves in the Fraser Valley by over 90 percent; in 1871, the provincial government disenfranchised Aboriginal people; in 1884, the potlatch and Tamanawas (Winter Spirit) Dance was banned; in 1885, Native people were denied the right to sell salmon caught in nontidal (river) waters; in 1913, Natives were prohibited from obtaining "independent" fishing licences to participate in the ocean-based commercial salmon fishery; and, in 1918, Indian Agents were empowered to try potlatchers without a judge. The now-dominant Europeans had little space or time for the dwindling Aboriginal population.

British Columbia entered Confederation in 1871. Developers rushed forward with grand schemes for the opening up of the new province. Subsequent railway projects, followed by settlement, criss-crossed the province. By the turn of the nineteenth century, commercial farms operated on occupied Aboriginal lands in the Lower Fraser Valley and throughout the low lands of the Okanagan. Small farming operations developed on Vancouver Island and in selected areas in the northern Interior. Ranchers opened sizable operations inland. Commercial fishers opened large-scale harvesting and processing plants along the coast, competing with Aboriginal peoples for resources, and hiring them as seasonal labourers. Mining operations drew hundreds of hardrock miners into short-lived boomtowns, like Sandon and Idaho Peak. Loggers capitalized on the commercial promise of the large coastal stands of Douglas fir and cedar trees (Newell 1993; Knight 1996; Lutz 2008).

Governments seconded this province-wide effort. Provincial authorities subsidized various railway and related projects, granting licences and land rights to individuals and companies, and restricted the rights of Aboriginal peoples and immigrants from Asia. James Douglas's effort to grant treaties or allow Natives to preempt farm lands off reserve evaporated quickly on his retirement from office in 1864. Subsequent officials clearly saw First Nations as an impediment to settlement.

Aboriginal peoples, however, did not sit idly by as others sought their lands. Harvesting activities continued, and not only along the isolated sections of the Pacific coast and in the rugged interior valleys. When opportunities arose, First Nations people joined in the new economic activities. Just as they had traded furs with coastal traders, Aboriginal people participated actively in the gold rush, commercial farming, and selling of meat and fish to the newcomers, or they worked as loggers. Many others worked seasonally and casually in the transportation and construction industries; on the west coast of Vancouver Island, the entire adult male population of many villages were contracted by Victoria-based schooner captains to work as ship-based hunters in the lucrative four-to-eight-month-long Alaskan pelagic sealing industry. Others sold crafts to the newcomers or provided various services in and around the growing urban areas. By the end of the century, it would have been difficult to find a coastal indigenous person whose family was not seasonally engaged in fishing and processing salmon for the rapidly expanding commercial cannery industry.

Social connections, however, proved more difficult between First Nations and newcomers. The settlers and developers assumed that Aboriginal peoples would remain on reserves and at arm's-length. The churches believed that the First Nations were not "prepared" for survival among nonindigenous peoples and argued that prolonged separation would inculcate Christian teachings and government-sponsored education. At the core, social contact between indigenous peoples and newcomers was conditioned by the racial assumptions and stereotypes of the age (Perry 2001).

The Federal System and the Twentieth Century

The twentieth century saw the solidification and codification of Aboriginal exclusion. Through a variety of measures, governments whittled away at Aboriginal land holdings. Key properties, many near urban and industrial areas, were removed from First Nations' control. The arrival of thousands of Asian migrants following the completion of the construction of the railway further overshadowed the Aboriginal presence in British Columbia (Roy 1989, 2003; Ward 1990). Although Aboriginal peoples remained prominent in the remote regions, they receded further into the background elsewhere – except in the new urban areas' most marginalized and impoverished skid-row neighbourhoods. Jobs they had formerly dominated in the commercial canneries and fishing and sealing fleets were systematically alienated from them. Legal proscriptions on Aboriginal involvement in the economy were less important than informal exclusions and racially discriminatory practices that enjoyed wide acceptance across the province.

Government policies consistently aimed to limit Aboriginal involvement in BC society. The consolidated Indian Act (1876) outlined federal responsibilities to, and restrictions on, status Indians. They sought to restrict Aboriginal peoples to reserves, arguing that more time was needed before integration was possible. Local Indian Agents and missionaries oversaw local political, economic, and social activities. Some managed affairs with a gentle and supportive, if paternalistic, hand; others rigidly imposed a new social order. Together they worked to eliminate Aboriginal spiritual beliefs and to inculcate Christian values. The federal government placed particular faith in the workings of residential schools, hoping that prolonged separation from parents and communities would westernize the newest generation.

The federal government also granted its agents the authority to suppress key cultural and economic practices. It focused, in particular, on the potlatch – the elaborate gift-giving feast-ceremony associated with transfers of hereditary resources and prerogatives. Officials and clergy believed that the potlatch interfered with the development of a properly capitalistic and materialist ethos (Cole and Chaikin 1990); what the ban really did was undermine the ability of indigenous people to participate in inter-reserve self-governance (Carlson 2010). The first arrest under the anti-potlatching law occurred in early 1897 (Carlson 1997, 99). In some settlements, the prohibition succeeded only in pushing the potlatch underground. The law was not repealed until 1951. Other restrictions served the direct economic interests of the non-Aboriginal population. The banning of the in-river commercial fishery had the intended impact of undercutting local self-sufficiency and, not surprisingly, strengthening the economic position of coastal non-Native commercial fishers.

By the second and third decades of the twentieth century, First Nations people no longer figured prominently in the provincial order. There was the sense that time would take care of the "Indian problem" in British Columbia and that the "yellow peril" of Asian immigration represented a far greater threat to the province's future. And although some Aboriginal people, like Hank Pennier, traded so skilfully in the currency of hard work and masculinity that they were able to prosper in the forest industry during the 1930s Depression, most First Nations loggers found the economic downturn only reified existing racial barriers to success in the wage labour economy (Pennier 2007).

For many years, Canadians assumed that the First Nations did little to protest their exclusion or to protect their rights. In fact, Aboriginal people worked assiduously from the late nineteenth century to gain government attention to their land, resource, and cultural rights. Indigenous groups launched several nineteenth- and early-twentieth-century delegations to Ottawa and London. They proved to be skilled in arranging their political affairs, working with lawyers and activists like A.E. O'Meara to place their case before various politicians, government agencies, and public interest groups (see Tennant 1990). The scenario, played out by Maori groups from New Zealand at much the same time, attracted considerable attention from the newspapers and the public but left officials in Victoria, Ottawa, and London unmoved.

What attention government officials paid to indigenous people was directed less at addressing their concerns than at muzzling protest. Fear that the First Nations would press their case to higher and potentially more sympathetic authorities in London led to an aggressive revision of the Indian Act in 1927. The federal government made it illegal for Aboriginal groups to raise or give money for pursuit of a claim against the Crown. Then, in 1936, the minister of Indian affairs made it illegal for Native people to accumulate property that "might" be used for potlatching illegally. Public pressure caused this initiative to be dropped the following year; however, other regulations banned community gatherings called for political purposes. Along the north coast, political rallies quickly converted into faux Christian revival meetings if a police officer, missionary, or Indian Agent passed by. "Onward, Christian Soldiers" emerged as the unofficial anthem of the Aboriginal rights movement in the region. Likewise, as Natives moved into nuclear family housing units, the giant longhouses, where potlatches and spirit dances traditionally occurred, were replaced by inconspicuous barns specifically built for the same purposes.

Aboriginal groups were not readily denied. British Columbia was among the most active provinces in Canada in terms of indigenous activism. A significant number of local and regional political groups sprang into existence. In general, Interior and coastal groups went their separate ways. This cultural and geographic divide also reflected another division – that between Christian denominations. The south coast was predominantly Catholic, whereas the Interior beyond Kamloops was largely the domain of Anglican and Methodist Natives. Some groups, like Sechelt, chose to work with the government and supported substantial integration. Others, including many groups in isolated regions, opted for a more separationist approach. First Nations did not, however, have either the federal or provincial vote and therefore had little opportunity to intervene in the political process. This taught Aboriginal people a valuable lesson. Even today, with the franchise extended to Aboriginal people, many choose to concentrate their political energies at the band or tribal level rather than participate in federal or provincial elections. By the middle of the twentieth century, Aboriginal voices attracted little attention.

The Postwar World

The political dynamics of Aboriginal organizational life in British Columbia changed after the Second World War. Many Canadian First Nations people had signed up for military service during the war, even though technically they were exempt. Many Native veterans returned after the conflict, believing that they had earned the right of full citizenship, only to discover that the old barriers remained in effect. The vast majority of the Aboriginal vets never learned of or received veterans benefits after the Department of Indian Affairs in Ottawa insisted that it administer the program to Native people, but then promptly forgot to train or notify the agents in the field (Sheffield 2004). The postwar era, however, also saw the emergence of considerable national and global interest in the rights of minorities and indigenous peoples. This development coincided with renewed Aboriginal activism, continued encroachment on indigenous territories and lands, and growing non-Aboriginal sympathy for indigenous cultures and traditions. In part in recognition of the role Aboriginals had played in the war effort, the Indian Act was amended in 1951 to drop the ban on the potlatch and tamanawas, as well as the prohibition on giving or soliciting money for pursuit of claims. Within little more than a decade, Aboriginal people secured the right to participate in provincial and federal elections.

Despite growing public support for First Nation claims to land and resource rights, the province refused to alter its stance and the federal government was reluctant to move. In a long and costly series of court trials – *Bob and White, Calder, Guerin, Sparrow,* and *Delgamuukw* – First Nations slowly and methodically cut through the legal entanglements that had entrapped their harvesting, land, and resource rights. Through these cases, First Nations people secured recognition of their right to hunt and fish for food and ceremonial purposes; recognition that such rights are not frozen in time; acknowledgment that the federal and provincial governments were obligated to negotiate land claims agreements; and the ability to hold officials accountable for their mismanagement of Aboriginal interests and trusts. The battles were hard-fought and not always successful (Miller 2009).

Through the court process, Aboriginal people discovered that the federal and provincial governments and the public at large understood the significance of judicial decisions. They also learned that negotiations rarely proceeded without the threat of a court case hanging over the federal AND provincial governments. The provincial authorities, who controlled the land that would be crucial to any land claim settlement, resisted all participation in negotiations until 1990, effectively halting any recognition of Aboriginal land rights. It took another eight years to conclude the Nisga'a Treaty (the negotiations of which took place outside the BC treaty process).

The emergence of Aboriginal land claims and resource rights campaigns met with considerable resistance from the non-Aboriginal population, particularly those in the logging, mining, sport and commercial fishing, and hunting sectors. (For a discussion of how social movement theory can help us understand this process see Wilkes and Ibrahim, this volume.) The perception emerged that Aboriginal rights would come at the expense of other resource users. A bitter debate erupted over the issue of indigenous rights versus demands for "equal" treatment for all Canadians. Many of these conflicts revolved around the question of the extent of Aboriginal rights to resources in territories where they asserted title. Corporations operating under provincial tenures often granted decades earlier wanted unrestricted access to resources. But First Nations argued that such exploitation was improper on lands where Aboriginal title remained unextinguished. At their heart, such debates spoke to the fact that Crown title was asserted without having first needed to be proven to the affected Aboriginal inhabitants, whereas Aboriginal title remained only asserted until proven accepted legally by the newcomer society. As such, it is a question of historical injustice and inequity. These issues came to a head in the battle over the definition of the government's duty to consult, and the extent to which Aboriginal people might legitimately stop resource development where they feel they have not been adequately consulted. The concurrently heard River Tlingit and Haida Supreme Court cases in 2004 emphasized that the consultation process required good faith and reasonableness on the part of the province and Aboriginal people but also specified that the duty of consultation did not provide a veto to Aboriginal people. This strengthened Aboriginal peoples' position, but over the subsequent six years, this Supreme Court decision, like so many before it, has proven more of a moral victory than a practical one. Even the landmark Tsilhqot'in (Chilcotin) case of 2007, which dealt with an indigenous people living in one of British Columbia's most remote regions who had historically asserted their rights against the Crown with military force, found the Tsilhqot'in to have title to only 50 percent of their traditional homeland (though their rights existed throughout their entire homeland). The message to First Nations people remains consistent: newcomer society's openness to recognizing the economic implications of Aboriginal title and rights to resources will be incremental and staggered, but resource extraction by third parties will remain ongoing.

Conclusion

The First Nations–newcomer encounter in British Columbia has been marked by several stages: conflict and compromise, confrontation and domination, incorporation and segregation, and incremental accommodation. The years since the arrival of Europeans exacted a dramatic toll on the indigenous peoples through the intertwined processes of depopulation, occupation of traditional lands, and competition for resources. First Nations also contended with government intervention on the side of newcomers, political and legal challenges to their rights, efforts at integration through education and reserve life, and an ongoing effort by Aboriginal peoples to find a place for themselves within a rapidly changing provincial world. There is not a great deal of difference between the conflicts, rhetoric, and outcomes of the nineteenth century and the conditions occurring in the early years of the twenty-first century. First Nations bore the major share of the dislocations and cultural change, and had to fight to secure both basic and Aboriginal rights. Public opinion polls suggest that the support Native people had garnered to their cause in the late twentieth century may now be evaporating. In November 2003, the *National Post* reported on a poll conducted by Environics Research Group and CROP showing that fully 71 percent of British Columbians were opposed to land claims that would result in preferential access to hunting and fishing by Aboriginal people over other Canadians. The same article reported that "half of all Canadians believe 'few or none' of the hundreds of land claims by aboriginals are valid." Forty-two percent of Canadians were reported as taking the view that "it would be better to do away with aboriginal treaty rights and treat aboriginal people the same as other Canadians" (Curry 2003). Shifting attitudes toward Native rights concern Aboriginal leaders, who know from experience that legal and constitutional protections count for little when confronted by a hostile or apathetic electorate.

Conflicts over fisheries, legal rights, and land and resource entitlements cloud the province's future, as they have since the nineteenth century. Add to this confusion over the meaning and extent of Aboriginal self-government and the social and economic crises facing most indigenous communities in the province, and British Columbia's dilemma becomes clearer. Although there are key forces working for reconciliation, major challenges remain. All sides have increasingly relied on the courts and treaty negotiations to determine the future of Aboriginal land and resource rights in British Columbia, though conciliatory gestures by the former Campbell government (including the appointment of Stó:lō Grand Chief Steven Point as lieutenant-governor) improved the negotiating environment considerably.

Understanding the history of First Nations–newcomer relations in British Columbia is essential to making sense of contemporary relationships and activities. Whereas non-Aboriginal British Columbians, particularly recent non-European immigrants, appear to pay relatively little attention to the history of this complicated relationship, First Nations people understand the past quite well. They know of the patterns of dispossession and government interference. They know about discriminatory regulations and the dislocations associated with residential schools. They know of their traditional land and resource rights and activities. And they know the manner in which non-Aboriginal British Columbians systematically stripped them of their inheritance. Despite this understanding, most First Nations remain committed to finding a mutually acceptable accommodation with British Columbia. What is also clear, however, is that the lessons of the past ring loudly in their ears.

Acknowledgments

The authors appreciate the helpful comments made on earlier editions of this chapter by

J.R. Miller and Charles Menzies. We are grateful of the bibliographic assistance provided by Darren Friesen.

References

Archer, Christon I. 1980. "Cannibalism in the Early History of the Northwest Coast: Enduring Myths and Neglected Realities." *Canadian Historical Review* 61(4):453-79.

Barnett, Homer. 1957. *Indian Shakers: A Messianic Cult of the Pacific Northwest.* Carbondale: Southern Illinois University Press.

Bierwert, Crisca. 1999. *Brushed by Cedar, Living by the River: Coast Salish Figures of Power.* Tuscon: University of Arizona.

Boyd, Robert T. 1999. *The Coming of the Spirit of Pestilence: Introduced Infectious Diseases and Population Decline among Northwest Coast Indians, 1774-1874.* Vancouver: UBC Press.

Carlson, Keith. 1997. "Early Nineteenth Century Stó:lō Social Structures and Government Assimilation Policy." In *You Are Asked to Witness: The Stó:lō in Canada's Pacific Coast History,* edited by Keith Thor Carlson, 87-108. Chilliwack, BC: Stó:lō Heritage Trust.

–, ed. 2001. *A Stó:lō Coast Salish Historical Atlas.* Vancouver: Douglas and McIntyre.

–. 2010. *The Power of Place, the Problem of Time: Aboriginal Identity and Historical Consciousness in the Cauldron of Colonialism.* Toronto: University of Toronto Press.

Centre for Research and Information on Canada. 2003. "Canadians Want Strong Aboriginal Cultures, but Are Divided on Aboriginal Rights." Press release, Centre for Research and Information on Canada, Ottawa, November 26. http://www.cric.ca/.

Cole, Douglas, and Ira Chaikin. 1990. *An Iron Hand upon the People: The Law against the Potlatch on the Northwest Coast.* Seattle: University of Washington Press.

Curry, Bill. "Half of Canadians Disbelieve Land Claims," CanWest News Service. *National Post,* Nov 27, 2003.

Fladmark, Knut. 1979. "Routes: Alternative Migration Corridors for Early Man in North America." *American Antiquity* 44(1):55-69.

Harris, R. Cole. 1997. *The Resettlement of British Columbia: Essays on Colonialism and Geographical Change.* Vancouver: UBC Press.

Jenness, Diamond. 1955. *The Indians of Canada.* Ottawa: E. Cloutier, Queen's Printer.

Knight, Rolf. 1996. *Indians at Work: An Informal History of Native Labour in British Columbia, 1848-1930.* Rev. ed. Vancouver: New Star Books.

Lutz, John Sutton. 2008. *Makuk: A New History of Aboriginal White Relations.* Vancouver: UBC Press.

Mackie, Richard Somerset. 1995. *Trading beyond the Mountains: The British Fur Trade on the Pacific, 1793-1843.* Vancouver: UBC Press.

Matson, R.G., Gary Coupland, and Quentin Mackie, eds. 2003. *Emerging from the Mist: Studies in Northwest Cultural History.* Vancouver: UBC Press.

Miller, James R. 2009. *Compact, Contract, Covenant: Aboriginal Treaty-Making in Canada.* Toronto: University of Toronto Press.

Newell, Dianne. 1993. *Tangled Webs of History: Indians and the Law in Canada's Pacific Coast Fisheries.* Toronto: University of Toronto Press.

Neylan, Susan. 2003. *The Heavens Are Changing: Nineteenth-Century Protestant Missions and Tsimshian Christianity.* Montreal and Kingston: McGill-Queen's University Press.

Pennier, Henry. 2007 *"Call Me Hank": A Stó:lō Man's Reflections on Logging, Living and Growing Old.* Edited by Keith Thor Carlson and Kristina Fagan. Toronto: University of Toronto Press.

Perry, Adel. 2001. *On the Edge of Empire: Gender, Race, and the Making of British Columbia, 1849-1871.* Toronto: University of Toronto Press.

Ray, Arthur J. 1999. "Treaty 8: A British Columbia Anomaly." *BC Studies* 123:5-58.

Roy, Patricia. 1989. *A White Man's Province: British Columbia Politicians and the Chinese and Japanese, 1858-1914.* Vancouver: UBC Press.

–. 2003. *The Oriental Question: Consolidating the White Man's Province.* Vancouver: UBC Press.

Schaepe, "Village Arrangements and Settlement Patterns," in Carlson ed., [2001].

Sheffield, R. Scott. 2004. *The Red Man's on the Warpath: The Image of the "Indian" and the Second World War*. Vancouver: UBC Press.

Suttles, Wayne. 1957. "The Plateau Prophet Dance among the Coast Salish." *Southwestern Journal of Anthropology* 13(4):352-96.

Tennant, Paul. 1990. *Aboriginal Peoples and Politics: The Indian Land Question in British Columbia, 1849-1989*. Vancouver: UBC Press.

Ward, Peter. 1990. *White Canada Forever: Popular Attitudes and Public Policy toward Orientals in British Columbia*. 2nd ed. Montreal and Kingston: McGill-Queen's University Press.

3
Circle of Influence
Social Location of Aboriginals in Canadian Society

James S. Frideres

For much of the history of Canada, Aboriginal peoples have been alienated from ownership of and decision making over natural resource development, including forestry. Relatedly, Aboriginal communities have typically experienced very high levels of unemployment and a whole host of other social disadvantages relative to the broader non-Aboriginal population. This chapter pursues the context of this situation in more detail by considering issues of social inequality in Canada with reference to the social location of Aboriginals in Canadian society. It provides an overview of the social location and involvement in the economic sphere, including forestry, of Aboriginal peoples in Canada.

The theoretical perspective that is reflected in this chapter is best captured by what is known as a colonial settler society model, whereby a group of people have been subjected to annihilation, assimilation, and rejection over the past two centuries by immigrants settling in a new country. Powerful forces such as the Indian Act and treaties have been imposed on this group by central and provincial governments with considerable force and determination. Moreover, these forces are supported by the larger judicial and economic systems within Canada. Many Canadians still believe that Canada was without inhabitants when Europeans arrived and that Aboriginal people had not established any kind of culture or institutional structure that provided them with any rights. Even though thousands of Aboriginals lived in Canada at the time of "discovery" and early settlement, most Canadians accept the argument that the country was *terra nullius* and, thus, settlers had unfettered claims to land, water, and resources. Aboriginal peoples have used many strategies over the years to resist these efforts (see Wilkes and Ibrahim, this volume), and although they have not always been successful, these strategies allowed them to retain elements of their culture even as they entered the twenty-first century (see Figure 3.1). The result of their tenacity has been the entrenchment of their Aboriginal rights in the Canadian constitution, an increased land base, and expanding Aboriginal rights.

Figure 3.1

Terminology

Different names are used variously to identify indigenous people

Aboriginal people: A collective name for original peoples in North America and their descendants. The Canadian Constitution, 1982, recognizes three subgroups of Aboriginal people: Indians, Métis, and Inuit. "Aboriginal" is a term used in the Canadian Constitution.

Indian: Sometimes also "status Indian" or "registered Indian." This is a legal term that describes a group of people who are subject to the Indian Act. They can be treaty or nontreaty and can live either on or off a reserve.

Indigenous people: A broad term used around the world to designate people who originally settled and lived in a geographical area.

Inuit: A legal term describing people in the northern regions of Canada and their descendants. Most live in Yukon, Northwest Territories, Nunavut and northern Labrador, Quebec, and Newfoundland.

First Nation: A term that came into usage in the 1980s to replace "Indian," which some people found offensive. This is not a legal term but rather a social one used to identify Indians.

Métis: People who have mixed Aboriginal and European ancestry.

Nevertheless, the dominant society's actions have taken a heavy toll on Aboriginal people and their culture.

We begin with a brief historical review of Aboriginal Affairs and Northern Development Canada (previously known as Indian and Northern Affairs Canada), as the department that has been the central driving force for assimilation of Aboriginals is now known. We then turn to a discussion of how Aboriginals fit into the forestry activities of the country. The second part of the chapter provides a brief socio-demographic analysis of First Nations in Canada. The information profiles the demography of Aboriginals over time on a number of attributes and compares them with the general Canadian population. The chapter ends with an analysis of the unique treaty process in British Columbia and the outcome of the recent referendum. This chapter provides a context for the current tension between Aboriginals and non-Aboriginals.

Aboriginal Affairs and Northern Development Canada

The current department "overseeing" Indian and Inuit affairs was created in 1966 (although over the past century and a half there have been many federal departments that have been placed "in charge of Indians") and has primary but not exclusive responsibility for carrying out the federal government's constitutional, political, and legal responsibilities to First Nations.

Its mandate is to support Indians and Inuit in developing healthy, sustainable communities and in achieving their economic and social aspirations.

The department carries out its management/control function for the more than 2,700 reserves (612 bands) involving 2.7 million hectares of land and nearly 700,000 people. It is responsible for delivery of services – including education, housing, and infrastructure – although today, over 80 percent of the total expenditures on the delivery of services are managed by First Nations. Finally, the department has a unique relationship with other federal agencies – Health Canada, Indian Oil and Gas Canada, Correctional Service of Canada, the Indian Taxation Advisory Board – and is only one of thirty-four government agencies that provide funding for Aboriginal peoples. In carrying out its mandate, the department's budget in 2008 exceeded $12 billion, and if one added additional funds allocated by the other federal and provincial departments, the total would be near $25 billion.

For over one hundred years, the Government of Canada has taken action to assimilate Aboriginal people into mainstream Canadian society (Clow and Sutton 2001). However, over the past quarter century, Aboriginal Affairs has embarked on a devolution process, or what might be called a decolonization policy. Under this strategy, more involvement by Aboriginal communities is to occur, with the eventual goal of Aboriginal control over their own lives. The policy has been accepted by Aboriginal Affairs, yet its implementation has been slow to occur. Today, the department has chosen to take a holistic approach to enhancing the quality of life for Aboriginal people (Frideres and Gadacz 2012), though it doesn't always include all elements – for instance, environmental management issues. Nevertheless, the department has determined that a multipronged strategy involving access to capital; developing educational qualifications, skills, and experience; increasing access to lands and resources; and building the economic infrastructure are crucial to the social and economic development of First Nations. Several programs, such as community economic development, resource partnerships, and resource access negotiations, have been created to support the more than twenty thousand registered Aboriginal businesses in Canada.

Forestry Development

With regard to forest development in Canada, Canada has a history of denial and exclusion when it comes to Aboriginal people. Sometimes it was explicit and blatant, at other times subtle and seemingly benign. Nevertheless, as Passelac-Ross and Smith point out in this volume, the end result is that Aboriginal people continue to operate in an environment of constraint with regard to natural resource development, including forestry. The boreal forest region of Canada is sparsely populated; less than 10 percent of the Canadian population resides within this area. At the same time, it is estimated that nearly half of the Aboriginal population lives in this region. Today, within this region, there are nearly 250 Aboriginal communities that each have more than a thousand hectares of forest land and over 500 communities that each have over one hundred hectares of forest land. The National Aboriginal Forestry Association (NAFA 2005) claims that over 85 percent of the First Nations in Canada live in productive forest areas. As we scan this vast geographical area, we find that it is encompassed by the major treaties established between Aboriginal people and the Crown. Furthermore, even though much of the area is covered by treaty, provincial and federal governments and Aboriginals have taken very different views of what the terms of the treaty are.

Within the area of forestry management and development, provincial policies vary in comprehensiveness, detail, and content. Under current law, provinces own nearly 90 percent of the forest land, though virtually all the timber harvesting and forest products manufacturing are done by the private sector. Through the forty-two forest tenure arrangements (i.e., the mechanics of who may or may not obtain access to the forests) across the country, few Aboriginal communities have been able to be involved in forestry activities (Notzke 1994). At the federal level (the federal government owns less than 10 percent of the forests), a new federal policy, entitled "A Vision for Canada's Forest: 2008 and Beyond," has been implemented. Here too we find restrictions on Aboriginal involvement in forestry through the Indian Act (section 57) and the Indian Timber Regulations. Even when these restrictions can be surmounted, it would appear that the government's objectives in forestry programs rarely coincide with Aboriginal land use and socioeconomic priorities. However, as Passelac-Ross and Smith (this volume) allude to, complementing that new federal policy are other initiatives created by Aboriginal Affairs, such as the Sustainable Development Strategy and the Environmental Stewardship Strategy. Implementation of the First Nations Land Management Act (1996) (Canada 2009) is a new key strategy that will address activities such as forestry.

The National Aboriginal Forestry Association has been active in involving Aboriginal people in forestry development and management and is working with both levels of government, as well as with the private sector. Specifically, it is working on integrating Aboriginal institutions, knowledge, and values into sustainable forest management; accommodation of Aboriginal tenure and treaty rights; enhancing Aboriginal economic and capacity building; and developing Aboriginal criteria for sustainable forest management (NAFA 2005).

Aboriginals have long argued that the exclusion of Aboriginal treaty rights from provincial and territorial forest policy remains a major barrier for Aboriginal peoples to sustainable forest management. They also argue that the current depletion of forested areas impact on their treaty rights to hunt, fish, and trap, as well as impact on their source of food and medicine, and other economic activities (Ross 2003). Proposals by Aboriginal peoples that seek to develop alternative tenure models and reconcile the rights and title of Aboriginal peoples with industrial forest tenures have been resisted by both federal and provincial officials. Most Aboriginal reserve forests are small, in need of rehabilitation, and incapable of sustaining ongoing harvesting. However, many reserves are surrounded by extensive Crown forest lands that could be tapped for forest harvesting. Unfortunately, these lands are no longer available, as almost all forest tracts have been committed to some form of tenure. For example, in much of British Columbia, 90 percent of the annual allowable cut is allocated to three transnational companies. In addition, through evergreen clauses, these commitments are in force for periods of fifteen to twenty years, with the right of extending the leases beyond the fifteen- to twenty-year contract.

Recently, though, the courts have forced the province to create a new relationship with Aboriginal people in the province and, as such, the province has made thirty-five direct awards of timber to First Nations. It also has signed close to a hundred additional agreements whereby First Nations are receiving approximately $35 million per year and have been granted the opportunity to log up to 33 million cubic metres of timber each year (Natural Resources Canada 2006). These numbers may seem impressive, until it's noted that the BC government collects $1 billion annually in stumpage fees from forest companies logging public lands – the same lands claimed by First Nations. Thus, the financial rewards offered to First Nations are less than 4 percent of the total revenue stream. Moreover, these agreements are different from contracts with large companies in that they are for a period of five years only. Until now, Aboriginal involvement in forest products industries centred mainly on the provision of labour and harvested wood. Although this is a very small concession and for a relatively short period, it will give them tenure and a foothold in the forest economy of British Columbia.

As other authors in this book make clear (see Forsyth, Hoberg, and Bird; Mabee et al.), forestry is not a major enterprise for all Aboriginal people, though some are utilizing forestry activities as the mainstay of their economic activities within their communities. What is of equal, if not more, importance is the attempt by Aboriginal communities to restructure the power and responsibilities with regard to forestry and all natural resources within the scope of Aboriginal people. Aboriginal people want not only access to and their fair share of the resources in question but also the right to participate in the management of these resources.

Demographic Profile

We begin our discussion with a profile of the Aboriginal population. Demographic attributes of any group are important for political, social, and economic reasons. Policies developed by government need to be grounded in the social and economic attributes of the population. Demographic factors such as population size and age distribution are crucial in developing and implementing policies on education, labour market forecasts, and economic development. Understanding the population dynamics (births, deaths) as well as the socioeconomic attributes of any population provides a lens through which policy can be implemented and the behaviour of a people can be better understood.

Aboriginal people in Canada are unevenly distributed throughout Canada (Statistics Canada 2006). When the Aboriginal population is viewed as a percentage of the total Canadian population, we find that overall, Aboriginal people make up less than 4 percent. In most of the Atlantic provinces, Aboriginals make up slightly more than 1 percent of the population; in both Manitoba and Saskatchewan, they make up more than 13 percent of the total population. In Alberta and British Columbia, they make up about 6 percent. Only in Yukon, Northwest Territories, and Nunavut do Aboriginals make up a substantial component of the overall population – 25, 69, and 86 percent respectively.

Another way of looking at the distribution of Aboriginals is to examine how they are dispersed. Today, over one-fifth of the Aboriginal population resides in Ontario. Between 13 and 16 percent is located in each of British Columbia, Alberta, Saskatchewan, and Manitoba; 9 percent live in Quebec, and the remaining 7 percent is distributed throughout the Atlantic provinces and northern territories. (For a more detailed analysis, see Canada 2005.)

Within these larger geographical areas, where do Aboriginals have their residences? Nearly 33 percent of the indigenous population lives in an urban setting. Forty-four percent lives in a rural setting, 18 percent lives in a "special access" area, and only about 5 percent lives in remote areas. This profile is quite different for the general Canadian population: over 80 percent lives in urban areas. Census data also reveal that Aboriginal people are more mobile than their non-Aboriginal counterparts. Their high degree of mobility poses specific challenges in implementing programs in education, social services, housing, and health care, both on the reserve and in urban centres.

Population Dynamics

Over the past century, the Aboriginal population has increased tenfold, whereas the total Canadian population has increased by a factor of six (see Table 3.1). However, the rate of growth (for both Aboriginals and the general Canadian population) in the first half of the century was quite different from the latter half. For example, in the first fifty years of the twentieth century, the Aboriginal population grew by less than 30 percent, while the general Canadian population increased 161 percent. A reversal of this pattern

Table 3.1

Aboriginal population by group, Canada, 1996, 2001, 2017

	Year		
	1996	2001	2017
Indian	648,000	713,100	971,200
Métis	214,200	305,800	380,500
Inuit	42,100	47,600	68,400

Note: The projected figures are based on estimates of annual rate of population growth (2% for Indian, 1.5% for Métis, and 2.47% for Inuit).
Source: Adapted from *Projections of the Aboriginal Populations, Canada, Provinces and Territories* (Ottawa, Statistics Canada, 2005), Gionét (2009).

is evident for the second half of the century, and this trend continues into the twenty-first century. The Aboriginal population in 2011 was 1.2 million, and it is projected to increase to 1.4 million by 2017 (Canada 2005).

How do populations of people increase over time? Several dynamics occur that produce changes in the population size and composition: births, deaths, and migration.

Births: The fertility rate (how many children a woman will have during her childbearing years) and the birth rate (number of live births) are two measures that reflect cultural, economic, and biological factors of a population. Today, the fertility rate of all Canadian women is 1.57 (below the replacement rate of 2.2). For Aboriginal women, the fertility rate is 2.7. The birth rate for Aboriginal females is 24.6 per 1,000 population, compared with 12.3 for the general Canadian population.

As a result of such a high fertility rate, Aboriginal people have a much younger population than non-Aboriginal people. Today, the median age for a registered Indian is 24.6 (compared with the median age of 35.9 for non-Aboriginal), and 42 percent of the Aboriginal population is under nineteen years of age (compared with 25 percent of the non-Aboriginal population). The age structure shows that nearly half the Aboriginal population is under twenty-five, whereas the general Canadian population has only one-third of its population under twenty-five.

Mortality: The First Nations and Inuit Health Branch of Health Canada provides health services to over six hundred Aboriginal communities. Through partnership arrangements with Health Canada, the process of devolution continues, and more Aboriginal communities are signing separate contribution agreements, integrated community-based health service agreements, or transfer agreements to provide health care to their members.

Nevertheless, in Canada over the past decade, one out of every seven deaths of children under one year of age was an Aboriginal child. This infant death rate of 10.6 (per 1,000) is more than twice that of the general population. This rate does, however, represent a decrease in the infant mortality of First Nations over the past three decades. Today, sudden infant death syndrome is a leading cause of Aboriginal infant mortality, accounting for nearly one-third of all infant deaths. This represents a rate six times that found in the general population.

As we move into the twenty-first century, the age standardized mortality rate for Aboriginals is twice that of non-Aboriginals. For example, in British Columbia, it is 103.3 deaths per 100,000, nearly twice the rate for other BC residents. Recent analyses of mortality rates by Aboriginal Affairs show they have remained constant since the early 1970s.

Younger Aboriginals contribute disproportionately to the mortality rate, with nearly 10 percent of the deaths occurring among ages nineteen or younger (compared with less than 2 percent for non-Aboriginals) and only 39 percent from the sixty-five-plus age group (compared with 77 percent for all other Canadians). The major cause of death for Aboriginals is "accidents and violence," which account for nearly 30 percent of all Aboriginal deaths; for other residents of Canada, this accounts for about 10 percent of all deaths. Young Aboriginal people have nearly four times the rate of suicide compared with non-Aboriginal people.

In more general terms, Aboriginal people have poorer health than any other ethnic group in Canada. Moreover, this poor health is endemic throughout their lives, whereas for most Canadians, poor health is felt in the last thirteen years of life. We also find the return of diseases that were devastating to Aboriginal people in the past. For example, over the past two decades, tuberculosis has become a major life-threatening disease. As well, several diseases that were previously rare in First Nation communities are now fast becoming leading causes of death, diabetes and AIDS among them (Tjepkema 2002).

What does this mean in terms of life expectancy? The data reveal that the life expectancy of Aboriginals is still less than for Canadians but that it is slowly closing the gap. Moreover, the data reveal that the number of Aboriginals in the age category of 65+ is increasing. In 1996, status Indians had a life expectancy of 72.8; a non-Aboriginal, 79.4. By 2012, status Indians will have an average life expectancy of 74.7, compared with the overall Canadian average of 80.6.

Migration: In the context of population dynamics, "migration" generally refers to international in and out flow: the population size decreases when substantial numbers of people leave the country (out-migration); the population size increases when large numbers of immigrants come to a country (in-migration). In the case of Aboriginals, "migration" reflects the changing definition of who is an Indian, Métis, or Inuit person. Over the past century, the definition of who is an Indian, Métis, or Inuit has changed several times.

Early definitions focused on cultural and lifestyle attributes, but more recently, genealogy and marriage patterns have taken precedent. For example, Bill C-31, passed in 1985, allowed individuals previously unable to be a registered Indian to become one (discussed in more detail below). As a result, an additional 110,000 individuals became registered Indians.

This change in the definition of who is a registered Indian brought annual growth rates for the Indian population to between 5 and 8 percent for nearly a decade. Since 1995, the annual growth rate has decreased to around 3 percent – though still nearly twice the annual growth rate of the general Canadian population. As a result of this phenomenal annual growth rate, the total registered Indian population nearly doubled over the past twenty years. Similar definitional issues have influenced the number of Inuit and Métis.

Bill C-31 created several different "memberships" for Aboriginal people. Under the new rules, the future population entitled to be registered Indians will be determined by patterns of Indian–non-Indian marriages (outmarriages). There are now several classes of Indians, such as registered and membership in a band, registered but no band membership, and membership in the band but not registered. These internal divisions will have profound social, economic, and political implications for First Nation communities and for the federal government. As a result of such changes, there is a growing number of offspring who live on the reserve but have no membership, nor are they entitled to be registered. This means that an increasing number of "non-Indians" will be eligible to live on the reserve but will not have Indian status. It also means that sex of an individual can determine, depending on their marriage choice, the legal status of their offspring. For example, male status Indians can have offspring who are status Indians while women's children may be non-status (meaning children of the brother are First Nations, while those of the sister are non-status). Overall, if the impact of Bill C-31 achieves its purpose, by 2057 there will be fewer First Nations people than there are today.

Over the next twenty years, the overall First Nations workforce (ages fifteen to sixty-four) will increase by nearly 50 percent, to reach a total Aboriginal workforce of over 600,000. It is estimated that the Aboriginal working age population will grow three to five times faster than its non-Aboriginal counterpart. If these projections become a reality, the majority of the Aboriginal youth will be entering the labour force, while at the same time the majority of the non-Aboriginal population will be entering retirement. The dependency ratios show that the young dependency ratio for the general Canadian population is thirty young dependents (under fifteen years) for every hundred persons in the labour force, whereas for the over-sixty-five age group dependency ratio it is seventeen. When we compare these with Aboriginals, we find the dependency ratio for the young people is sixty-five and for the older age group is nine. Inuit and Métis reveal dependency ratios that also are higher than the general Canadian dependency ratio. These dependency ratios have implications for service delivery of social, economic, and health programs.

The Realities of Aboriginal Life

The social integration of Aboriginal people can be assessed by looking at their educational, labour force, and income levels. Education level is a kind of building block that impacts the extent and nature of participation in the labour force, income, and overall quality of life. Moreover, the educational level of a community provides some indication of how well members of the community will be able to compete in the larger society.

When we look at the distribution of the total population fifteen years or older by highest level of schooling for both Aboriginals and non-Aboriginals, we find revealing differences between the two populations and among age groups. For example, more than four times the number of Aboriginals between the ages of fifteen and twenty-four have less than a grade nine education compared with other Canadians (Ministers National Working Group on Education 2002). An increasing number of students are graduating from high school, yet in 2004 only 39 percent of Aboriginal students completed grade twelve; for non-Aboriginals, the completion rate was 77 percent. Fewer than thirty thousand Aboriginals attend a postsecondary institute, translating to a rate less than half of that of the general Canadian average. Aboriginal communities have noted that females are more likely to have postsecondary educations and that Aboriginal children have few teacher role models in the field of education (Mendelson 2006, 2009).

The vitality of language is an important component in cultural identity as well as self-concept. Using numerous measures, we find that English has become the dominant language, whether it is the mother tongue, the language used at home, or the language used at work. Moreover, younger people have much higher rates of English usage than older Aboriginals. Norris (2008), Elgersma (2001), and Norris and Snider (2008) calculated the viability for Aboriginal languages across Canada and found that Cree, Ojibwa, and Inuktitut are the only three languages that show significant vitality.

In 2008, over 40 percent of First Nations individuals were unemployed, whereas for non-Aboriginal men and women, it was 11 percent. The labour force participation rate for female Aboriginals was 42 percent; for their male counterparts it was 57 percent. Comparable data for non-Aboriginal people was 58 and 71 percent respectively. Data from various sources show that for all ages and educational attainments, the participation rate among Aboriginals is more than 10 percent less compared with non-Aboriginals. We would expect participation to be different among the age groups, but what is puzzling is the difference between First Nations and non-First Nations when age and educational attainment are controlled for. Put another way, we would expect all individuals aged twenty-five to forty-four with a university education to have similar income and labour participation rates. However, data on this reveal that this is not the case when comparing Aboriginals with non-Aboriginals. One plausible explanation is that Aboriginals are experiencing discrimination when trying to enter the labour force (Mendelson 2006). We also find that participation in the labour force differs in that Aboriginals are more likely to hold part-time or seasonal jobs than non-Aboriginals. Moreover, Aboriginal participation in the labour force tends to be in lower-status, lower-paying jobs (Drost 1995; Canada 2003).

Income also can be viewed from an individual perspective or from that of a family unit. In 2006, First Nations individuals made on average $16,000, whereas non-First Nations individuals had an average income of just $26,000. If we add to this nonearned income such as old age or disability payments, stocks and bonds, and social service payments, we find that the overall income increases for Aboriginal people but not for non-Aboriginal people. This indicates that a larger portion of household income for Aboriginal people comes from nonearned income, which is reflected in their low labour force participation rates. Table 3.2 reveals the distribution of household income of First Nations families by educational level. As the data point out, over one-third of First Nations families survive on less than $10,000 per year. With a poverty rate set at $24,000 per year for a family of four, this suggests that over half of the First Nations population lives below the poverty line (Maxim, White, and Beavon 2001; White, Maxim, and Beavon 2003). If this data were compared with that for non-Aboriginal people, we would find that more than twice the number of First Nations households makes less than $20,000 per year compared with non-Aboriginal; the profile is reversed when looking at those who make more than $70,000 per year.

Table 3.2

Annual household income for North American Indians from all sources and highest level of formal education completed, Canada, 2006

Income	%	Education	%
Less than $10,000	35	Less than grade 8	25
$10,000-19,999	19	Some high school	26
$20,000-29,999	11	High school	11
$30,000-39,999	19	Technical/vocational	20
$40,000 plus	16	University	18

Source: Adapted from Sharpe et al. (2009), Gionet (2009), Mendelson (2006), Statistics Canada (2006).

We now turn to events and issues that didn't result from the demography of Aboriginals but nonetheless have an impact on Aboriginal communities. These political and legal structures and events have impacted, and will continue to impact, on the lives of Aboriginal people as they continue to seek solutions and to find ways by which they can become economically self-sufficient and integrate into the larger society.

Self-Government/Governance

The issue of self-government has generated considerable discussion over the last century, during which time the federal government took a position that First Nations did not have any "rights" and self-government was not something it was willing to discuss (the chapter by Mark Stevenson more fully reflects on this issue). However, in 1973, the now famous *Calder v. Attorney General of British Columbia* decision ruled that First Nations indeed had Aboriginal rights. This decision was supported first by *Cardinal v. The Queen* (1974) and then, ten years later, by *Nowegijick v. The Queen* and *Guerin v. The Queen*. When the constitution was patriated in 1982, it was recognized and affirmed that all Aboriginal people had rights that still existed if they had not been extinguished (Constitution Act, 1982, section 35). Two years later, the federal government enacted self-government legislation. The 1997 *Delgamuukw* Supreme Court of Canada decision made it clear that Aboriginal rights could no longer be ignored by provincial or federal officials.

Between 1995 and 2003, more than eighty self-government negotiations took place, including comprehensive land claims, sectoral negotiations, and other issues involving self-government. Of these, over half were in British Columbia. Ongoing negotiations involve 513 First Nation/Inuit communities, of which nearly one-fourth are in British Columbia. To complicate matters, in 1999 the *Corbiere* decision by the Supreme Court held that subsection 27(1) of the Indian Act (which held that only on-reserve residents could vote) was discriminatory under the Canadian Charter of Rights and Freedoms. This decision made it clear that off-reserve members could not be denied the right to vote and participate in decisions that impacted on them. This decision prompted the federal government to rethink this and other electoral provisions in the Indian Act.

The federal government recognizes the inherent right of Indian self-government under section 35 of the Constitution Act, 1982. For example, in 2000, it gave royal assent to the Nisga'a Final Agreement Act, which represented the most up-to-date thinking on Aboriginal self-government, comprehensive land claims, and other land rights, among other issues. This recognition is based on the premise that any Aboriginal government or institution will operate within the framework of the Canadian constitution. Moreover, because of the different types of Aboriginal governments, provincial governments are necessary parties to discussions between the federal government and the Aboriginal group. Finally, the right to self-government, from the federal government's perspective, does *not* include the right of sovereignty in the sense of international law.

In the government's view, the scope of Aboriginal self-government is limited to those activities that are both internal to the group (such as marriage or policing) and an integral component of Aboriginal culture. As such, these activities would be considered essential to the group's operation as a government or institution. At the same time, the federal government has noted that any issue that goes beyond matters that are integral to First Nations culture (or go beyond internal First Nations issues), such as justice issues, gaming, or immigration, would not be part of self-government.

Treaties

We now turn to the issue of treaties and how they relate to the social niche of First Nations. Early "peace and friendship treaties" established between Europeans and indigenous peoples in Canada did not involve the transfer of land, rights, or other attributes of sovereign countries, with the exception of the Royal Proclamation of 1763. However, as colonization continued, Britain forced other forms of treaties on Aboriginal people. Starting with the Robinson Huron Treaty (1850), for the next 150 years, the Government of Canada signed over three hundred treaties with status Indians. Of equal importance is the fact that Canada did not negotiate treaties with a large number of First Nations in British Columbia and in the northern regions of Canada.

In British Columbia, only a few treaties have been established (fourteen Douglas treaties on Vancouver Island, and Treaty 8 covers part of northeastern British Columbia), leaving most of the land and water in British Columbia under current legal dispute. In some cases, reserves were set aside for BC Indians, but even these have become the subject of litigation. In the early twentieth century, a question of their boundaries was raised and the 1916 McKenna-McBride Royal Commission established to determine the extent of lands set aside for Indians. In the end, the commission recommended that some reserves be increased, but the overall recommendation was that valuable farmland not being used by Indians be taken away from land set aside for reserves. Ten years later, an Indian delegation set off to London, England, to object to the loss of the "cut-off" lands and to present a petition to the Queen. One year later, an amendment to the Indian Act would make it illegal "to receive, obtain, solicit or request from any Indian any payment or contribution for the purpose of raising a fund or providing money for the prosecution of any claim" (1926 An Act to Amend the Indian Act) without first having approval of the superintendent general of Indian affairs. This clause of the Indian Act would remain in place until the 1951 amendment repealed it and the pernicious rules outlawing the potlatch and prairie dances.

As a result of the claims being put forth by First Nations, the federal government created processes through which the claims could be resolved. The government's early strategy was to simply ignore the claims. Later it would allow claims to be put forth, but the procedures for resolution were unclear and arbitrary. Prior to the 1970s, a land claim took an average of twenty-seven years to be resolved.

In 1973, the federal government presented its first formal policy on Indian claims. Today, two claims policies – comprehensive and specific – have been developed by the government to resolve various disputes. Comprehensive claims arise when a First Nation asserts Aboriginal rights and title to land. These types of claims are most relevant to northern Aboriginals and those living in British Columbia, where few land treaties were established. To date, fifteen comprehensive claims have been settled. Specific claims come about when a First Nation community alleges that the federal government has not honoured the content of its treaty or legal/fiduciary responsibility. The Indian Specific Claims Commission, which began operating in 1991 and concluded its operation ten years later, was an independent body with authority to hold public inquiries into the First Nation claims that had been rejected by the federal government. Of the nine hundred specific claims that have come before the federal government, only about two hundred have been settled. Given that only twenty claims are settled a year, it will take fifty years to settle the $25 billion worth of claims currently in the system. The importance of this becomes clear when one considers that over 20 percent of reserve land in British Columbia was surrendered back to the federal government between 1886 and 1911.

In British Columbia, a task force was created in an attempt to give order and process to First Nation claims. When the British Columbia Supreme Court ruled that both the Gitxsan and Wet'suwet'en people had unextinguished nonexclusive Aboriginal rights *(Delgamuukw)*, the BC government, for the first time, officially and publicly recognized the inherent rights of Indians to Aboriginal title and self-government. In 1993, the BC Treaty Commission was created – its sole function is to facilitate the negotiation of treaties. During 1993-94, the commission began its negotiation process with First Nations whose statements of intent to negotiate land claims it had accepted. Today, about two-thirds of Aboriginal people in British Columbia are participating in the BC treaty process (BC Treaty Commission 2010).

With the introduction of new legislation concerning both specific and comprehensive land claims, First Nations have gained access to increasing amounts of land. In Canada, approximately 350,000 hectares of land have been added to the inventory of First Nations since 1989. At the same time, the number of bands has increased (from 596 in 1989 to 612 in 2003), as have the number of reserves (from 2,263 to 2,617 in the same time period).

As Ross and Smith (2001) point out, the decisions of the courts as well as the outcome of ongoing treaty negotiations will have significant implications on resource developments, specifically in the area of forestry and economic integration. Aboriginals argue that the solutions to allow Aboriginal people to become involved in the resource sector must first see that forest owners, managers, and tenure holders are willing to modify industrial forestry practices and management planning to include Aboriginal values and land uses. An example of the new treaty process is evident in the Nisga'a agreement concluded outside the BC Treaty Commission process and involving a province-wide referendum. (See Ponting [2006] for a full discussion of the process.)

Why are treaties an issue today when they seem to not have been of concern to the provincial or federal governments for years? Two factors have made them a prominent issue. First, the courts have forced the federal government to address the issue through their decisions. Second, as noted by Mark Stevenson in this volume, there is a belief that the BC economy will remain volatile until land claims and Aboriginal rights of BC First Nations are resolved. Investors for economic development are unwilling to make large investments and long-term commitments if the land issues are not resolved. From the First Nations' perspective, successful resolution of treaties is an essential component in strengthening their culture and economy. For both groups there is a belief that successful treaty negotiations will establish certainty about lands and resources and thus provide a secure climate for investment and economic development (Hodgins et al. 2001).

Conclusion

The sociodemographic profile presented in this chapter indicates that the quality of life for Aboriginals is marginal. To be sure, there have been positive changes in Aboriginal communities, yet all these changes have not reduced the social, economic, or demographic gaps between Aboriginal people and non-Aboriginal people. With regard to the development of natural resources, including forestry, Aboriginal people argue that there is a need to fundamentally rethink Western resource use. They have argued that sustainable resource development can take place only if they have improvement in their access to and participation in resource management and in the implementation of traditional environmental knowledge.

For over a century, the mandate of Aboriginal Affairs and Northern Development Canada was to assimilate Aboriginal people. Only in the last quarter century has that mandate changed. Nevertheless, the department has had nearly 150 years to identify the issues confronting Aboriginals, thoroughly understand the context in which Aboriginals operate, and evaluate their own policies and programs. Its active intervention and benign neglect have resulted in the social location of Aboriginal people today such as it is. At the same time, Parliament and the courts have been supporting actors to this scenario. In the end, considerable time, money, and energy have been devoted to improve the quality of life of Aboriginals. Yet, one could not conclude by any stretch of the imagination that they have been successful. It is true that from an absolute perspective, the length of life for Aboriginals has increased, their incomes have increased, and their mortality rates from some diseases have decreased. But when we compare their quality of life, income, and mortality rates with that of non-Aboriginals, we find that they are falling further behind. One would expect that detailed evaluations of the policies and programs implemented in the past would have taken place; that funds allocated were spent in an effective and efficient manner; that after spending over $12 billion last year, some changes in the quality of life of Aboriginals, even if small, would have occurred.

What is clear is that over the past one hundred years, there has been little change in the relative quality of life of Aboriginals compared with other Canadians. Notwithstanding the commitment to "devolution," the federal government continues to hold the ethos that it knows best and that ethnic groups are incapable of making decisions in their best interests. There is a belief also that Aboriginal people need to assimilate into mainstream society if they are to increase their quality of life. As a result, a huge bureaucratic structure with a large budget has been put in place, and it continues to control the lives of First Nations people.

References

BC Treaty Commission. 2010. "Negotiation Update." http://www.bctreaty.net/.

Canada. 2003. "Basic Departmental Data – 2002." Department of Indian Affairs and Northern Development, First Nations and Northern Statistics Section, catalogue no. R12-7/2002E.

–. 2005. *Projections of the Aboriginal Populations, Canada, Provinces and Territories*. Ottawa: Statistics Canada.

–. 2009. *Canada's National Forest Strategy, 2003-2008—A Sustainable Forest, The Canadian Commitment,* Ottawa.

Clow, R., and I. Sutton. 2001. *Trusteeship in Change*. Denver: University of Colorado Press.

Drost, H. 1995. "The Aboriginal-White Unemployment Gap in Canada's Urban Labour Markets." In *Market Solutions for Native Poverty: Social Policy for the Third Solitude, the Social Policy Challenge,* vol. 11, edited by H. Drost, B. Crowley, and R. Schwindt, 87-102. Toronto: D.D. Howe Institute.

Elgersma, S. 2001. *Aboriginal Women*. Ottawa: Information Management Branch, Indian and Northern Affairs Canada.

Frideres, J., and R. Gadacz. 2012. *Aboriginal Peoples in Canada*. Toronto: Prentice Hall.

Gionet, L. 2009. "First Nations People: Selected Findings of the 2006 Census." *Canadian Social Trends,* May: 52-58.

Hodgins, B., U. Lischke, and D. McNab, eds. 2001. *Blockades and Resistance*. Waterloo: Wilfrid Laurier University Press.

Maxim, P., J. White, and D. Beavon. 2001. "Dispersion and Polarization of Income among Aboriginal and Non-Aboriginal Canadians." *Canadian Review of Sociology and Anthropology* 38:465-76.

Mendelson, M. 2006. *Aboriginal Peoples and Postsecondary Education in Canada*. Ottawa: Caledon Institute of Social Policy.

–. 2009. *Why We Need a First Nations Education Act.* Ottawa: Caledon Institute of Social Policy.

Ministers National Working Group on Education. 2002. *Our Children: Keepers of the Sacred Knowledge.* Final report. Ottawa: Department of Indian Affairs and Northern Development Canada.

NAFA (National Aboriginal Forestry Association). 2005. *A Proposal to a First Nation.* Ottawa: National Aboriginal Forestry Association.

Natural Resources Canada. 2006. "Forest Industry Competitiveness" (Fol-6). *The State of Canada's Forests 2005-2006.* Ottawa, Natural Resources Canada, Canadian Forest Services, 19-24.

Norris, M.J. 2008. "Voices of Aboriginal Youth Today: Keeping Aboriginal Languages Alive for Future Generations." *Horizons* 10:60-68.

Norris, M.J., and M. Snider. 2008. "Endangered Aboriginal Languages in Canada: Trends, Patterns and Prospects in Language Learning." In *Endangered Languages and Language Learning,* edited by T. de Graaf, N. Ostler, and R. Salverda, 123-54. The Netherlands: Foundation for Endangered Languages.

Notzke, C. 1994. *Aboriginal Peoples and Natural Resources in Canada.* Toronto: Captus University Publications.

Ponting, J.R. 2006.*The Nisga'a Treaty*. Peterborough, ON: Broadview Press.

Ross, M. 2003. *Aboriginal Peoples and Resource Development in Northern Alberta.* Paper no. 12, Canadian Institute of Resource Law, University of Calgary.

Ross, M., and P. Smith. 2001. *Accommodation of Aboriginal Rights: The Need for an Aboriginal Forest Tenure.* Synthesis report. Edmonton: Sustainable Forest Management Network.

Sharpe, A., J. Arsenault, S. Lapointe, and F. Cowan. 2009. *The Effect of Increasing Aboriginal Educational Attainment on the Labour Force, Output and the Fiscal Balance.* Ottawa: Centre for the Study of Living Standards.

Statistics Canada. 2006. *Aboriginal Peoples of Canada: A Demographic Profile.* Catalogue no. 96F003XIE2001007. Ottawa: Minister of Industry.

Tjepkema, M. 2002. *The Health of the Off-Reserve Aboriginal Population.* Supplement to Health Reports, vol. 13. Ottawa: Statistics Canada.

White, J., P. Maxim, and D. Beavon. 2003. *Aboriginal Conditions.* Vancouver: UBC Press.

4

Treaty Daze

Reflections on Negotiating Treaty Relationships under the BC Treaty Process

Mark L. Stevenson

In Canada, the historic treaties between First Nations and the Crown were an attempt to resolve early conflicts over land ownership and open up the country for settlement. Modern treaties are also about the reconciliation of different world visions and overcoming the deadlock in the struggle between competing views. The history of double talk and broken promises related to treaties is well documented.[1] In recent times, a number of institutions have been developed to resolve land issues related to the historic treaties and facilitate the negotiation of modern treaties. In particular, the BC Treaty Commission is overseeing attempts to resolve land claims. Although the BC treaty process is a heroic undertaking, the process is in trouble. After close to two decades of negotiations, there are few success stories, and the debt load of First Nations is becoming unmanageable. The failure of the Lheidli T'enneh Final Agreement ratification and the growing support by First Nations and chief negotiators for fundamental changes to the mandates of governments are signs of a need for change.[2] Recent statements by Sophie Pierre, chief commissioner of the BC Treaty Commission, suggesting that the process be shut down if it cannot be improved magnify the need for changes to both the substance and process of treaty negotiations in British Columbia.[3]

Background

Federal Claims Policy: Failure of Imagination

The federal government has the exclusive legislative jurisdiction over "Indians, and Lands reserved for the Indians," pursuant to section 91(24) of the Constitution Act, 1867.[4] This would include the development of policy for claims based on Aboriginal rights and title, what are known as comprehensive claims. Comprehensive claims are based on the fact that there are continuing Aboriginal rights and title to the lands and natural resources within Canada. For the most part, these claims arise in parts of Canada where

Aboriginal title or the land question has not been resolved through historical or modern treaties. Comprehensive claims negotiations encompass a wide range of issues, such as land (including surface and subsurface rights), water, language, culture and heritage, fishing, forestry, wildlife, environmental management, tax and fiscal matters, revenue sharing, social and economic benefits, the administration of justice, and self-government.[5]

In 1996, the Royal Commission on Aboriginal Peoples (RCAP) reviewed the claims policies and process, identifying significant weaknesses.[6] The RCAP report itself and the recommendations in it continue to be relevant. According to RCAP, little or no effective interim measures are in place to protect Aboriginal interests either before or while negotiations are being conducted.[7] In British Columbia, thanks to an evolution in the province's approach to land protection, this concern now applies primarily to those negotiations that have not achieved an agreement in principle. However, during the course of negotiations where there is no agreement on the land package, governments may create, on traditional lands that have been identified as potential settlement lands, new third-party interests, which will then take precedence over rights included in final agreements.[8] This tends to diminish any sense of urgency on the part of non-Aboriginal parties over the settlement of claims. There are also numerous substantive difficulties that arise at the negotiating tables. Most importantly, government mandates often seem too narrow for reaching agreement because the price that Aboriginal peoples must pay for entering into a final agreement is too high. Extinguishment – or to use the more fashionable term, "certainty" – is one of the more difficult issues to reconcile in the treaty or land claims negotiations.[9] Because of the threat of extinguishment, whether perceived or real, the land and cash volumes offered by governments seem miniscule. And, in an economy that is growing each day, the per capita formula that governments bring to the table is diminishing in value compared with the increase in land values and the cost of living. There are other difficulties too, some of which will be addressed in more detail later.

Indeed, the current federal claims policy is outdated and cries out for change. Since it was announced in 1986, it has continued to guide federal land claims negotiations, although some adjustments have been made for the BC treaty process.[10] Since the policy was announced, there has been a dramatic shift in the law. The shift has been of a magnitude that the early claims policy makers could not have foreseen. *Sparrow* had not been decided by the Supreme Court of Canada, and the consequences of section 35 were barely understood.[11] At the time the policy was announced, it was still unclear whether and how Aboriginal rights and title could be extinguished in the post-section 35 era. The concepts of justification and infringement were barely explored, and *Gladstone,* on the existence of a commercial right to harvest herring roe on kelp, had not yet been decided on.[12] More importantly, the claims policy did not contemplate that the fiduciary relationship, first articulated in *Guerin* just a few years earlier, would be incorporated into section 35.[13] *Delgamuukw* had not yet been decided, and Aboriginal title had not been recognized as an interest in land somewhat akin to fee simple ownership and under the exclusive jurisdiction of the federal government pursuant to section 91(24).[14] Also, the courts had not decided that infringements of title would ordinarily require compensation, and it had not yet been determined in *Haida Nation* that the duty to consult and accommodate arises prior to proof of a right.[15] *Sappier* had not determined that harvesting timber for domestic purposes is a section 35 right,[16] and *Mikisew Cree* had not yet determined that the honour of the Crown is always engaged in addressing Aboriginal issues and that breach of that honour is an independent cause of action.[17] And, as regard the Métis, the historic decision in *Powley* had not yet confirmed that the Métis have Aboriginal rights with the same constitutional status as the section 35 rights of the Indians and the Inuit.[18] Yet, astounding as it may seem, Canada's policy continues to be driven by the 1987 Comprehensive Land Claims Policy,[19] and key federal policy initiatives continue to be driven by the same old twentieth-century mindset.

BC Treaty Commission

Although the federal government has the exclusive jurisdiction to make laws for "Indians, and Lands reserved for the Indians," the provinces have jurisdiction over property and civil rights, pursuant to section 92(13) of the Constitution Act, 1867. The provinces also have jurisdiction over nonrenewable natural resources, forestry resources, and electric energy, pursuant to section 92A. In addition, the provinces have jurisdiction over education, the administration of justice, hospitals, the management and sale of public lands, local works and undertakings, and a host of other matters that are an intimate aspect of a self-governing regime.[20] A successful resolution of land claims in British Columbia will necessarily involve the province.

In 1990, Canada, British Columbia, and the BC First Nations Summit created the British Columbia Claims Task Force to address the land question. In 1991, the *Report of the British Columbia Claims Task Force* made nineteen recommendations on how to negotiate and resolve the settlement of comprehensive claims in British Columbia.[21] Recommendation 3 of the report was to establish the BC Treaty Commission in order to facilitate treaty negotiations. In 1992, Canada, British Columbia, and the BC First Nations Summit reached the BC Treaty Commission Agreement, which established the BC Treaty Commission and supported the process of treaty negotiations recommended in the task force's report.[22] The purpose of the commission as set out in both the task force's report and the BC Treaty Commission Agreement is to facilitate the negotiation of treaties among one or more First Nations, Canada, and British Columbia. As part of its role, the BC Treaty

Commission determines whether the parties are ready to commence negotiations, allocates funding for negotiations, encourages timely negotiations, attempts to facilitate the resolution of disputes, provides the public with status reports on negotiations, and performs other related duties.[23]

Treaty Daze

When originally announced, the *Report of the British Columbia Claims Task Force* and the BC Treaty Commission Agreement were met with much fanfare. At the outset, the BC Treaty Commission process eliminated a number of the problem areas in other claims. The use of an independent commission and a chief commissioner agreed to by all three parties gives the process more legitimacy. Making the commission responsible for funding eliminates some of the blatantly disproportionate discretion exercisable by the federal government. The use of a statement of intent of the First Nation to be filed with the commission, and the requirement that within forty-five days of the filing the commission convene a meeting of the parties, eliminates costly and time-consuming research to establish exclusive use and occupancy and long delays before the first meeting. In addition, the language of the recommendations with respect to interim measures and the agreement to allow any of the parties to put any agenda item on the table are refreshing.[24] But other realities of the BC Treaty Commission process have proven to be very different.

More than fifty First Nations are participating in the BC Treaty Commission process. Most of these are labouring on substantive negotiations in Stage 4 (Negotiation of an Agreement in Principle), though some are still in Stage 2 (Readiness to Negotiate) and Stage 3 (Negotiation of a Framework Agreement).[25] A number of First Nations have negotiated and ratified their agreements in principle, and several have negotiated their final agreement (Stage 5).[26] But many negotiations are stalled at the table, with no prospects of success in the foreseeable future.

After years of dithering, even the commission has said that the BC Treaty Commission process and the current treaty model require fundamental changes.[27] The complexities of the negotiations are beyond what was originally anticipated, and the negotiations themselves are taking much longer than was expected.[28] In 1998, the commission advised the governments of the "serious consequences" of this for First Nations and predicted that the situation would likely worsen.[29] It has now been more than fourteen years since these concerns were raised and still not much has changed.

Likewise, with the exception of modest breakthroughs accomplished at the final agreement tables, there have been no fundamental changes to treaty policies.[30] In fact, over the years, the federal government has narrowed its mandates and treaty policies rather than broadened them. Canada's policies are outdated and inflexible. There is very little dialogue. Canada tables positions and leaves it to First Nations to either agree or disagree. For instance, a fisheries agreement package similar to that negotiated by the Nisga'a is no longer on the table. Elements of what was negotiated in chapter 11 (fisheries) of the Lheidli T'enneh Final Agreement are no longer on the table. For the most part, Canada is not negotiating fisheries matters and is awaiting the results of the Cohen Commission to determine whether to return to fisheries negotiations. Canada's fiscal policy is also narrowing, so that the negotiation of the fiscal chapter and fiscal side agreements is a frustrating experience. As in fisheries negotiations, Canada's fiscal negotiators come to the table with a take-it-or-leave-it mentality. This is not good faith negotiation by almost any standards.

Just as troubling is a growing doubt about the capacity and commitment of First Nation and Crown representatives to deal with the number of complex issues involved in treaty negotiations.[31] Although these and other issues in the negotiation process continue to challenge all parties, the success of several court decisions has revived the question – with considerable validity – of whether litigation is more effectual than negotiation. This is particularly pertinent in light of the judgment of the Supreme Court of British Columbia in *Tsilhqot'in Nation,* in which the court, while not making a declaration, stated in *obiter* that there was sufficient evidence for a finding of Aboriginal title over approximately two thousand square kilometres of land traditionally used by the Tsilqot'in Nation.[32]

The reasons for the lack of progress in negotiations, documented in 2001 in *Looking Back, Looking Forward: A Review of the BC Treaty Process,* are still valid today.[33] Chronic understaffing of the provincial negotiation teams continues to be a problem. In addition, the province has an unclear mandate in some matters, and provincial chief negotiators are sometimes left to depend on the personal whims of senior officials in responsible ministries as to what can be on the table. The situation is worse in the federal government. Federal government chief negotiators are frequently replaced by new and untested negotiators, and this has an impact on the issue of trust – a key ingredient required to achieve lasting agreements. And some federal mandates, particularly those related to tax and fiscal issues, are simply unworkable for most First Nations. On other matters, the federal government in its policies seems uncaring and inflexible. Often, after the parties have reached agreement on language, new positions are tabled and the federal chief negotiators have no mandate to deviate from the new language that has been predetermined by Ottawa mandarins. And more often than not, the new positions or language brought forward seem irreconcilable with the language of court decisions and some of the key objectives Aboriginal peoples are trying to achieve. This would appear to be because the treaty or land claims policies of the governments, and in particular the federal government, are simply out of step with the times and the law. They are status quo policies that reflect a failure of imagination for how treaties should look in the twenty-first century. All of this provides fodder for the cynics who believe the Crown has no intention of negotiating honourably.

Tough Issues

Certainty

Of all the issues that negotiators have to deal with, the topic of certainty may be the most gut-wrenching and enmity-provoking. It touches the core of what treaty making is all about and seems to deeply affect the rights of Aboriginal and non-Aboriginal peoples alike. In 1995, RCAP published a report on certainty entitled *Treaty Making in the Spirit of Co-Existence,* which provides a detailed analysis of the problems associated with the certainty paradigm preferred by governments and third parties.[34] The report also proposes solutions that try to balance the need to recognize and affirm the rights of Aboriginal peoples in a manner consistent with the language of section 35 of the Constitution Act, 1982, and at the same time address the needs of governments and industry. Shortly afterward, the Hon. A.C. Hamilton released his report on the same subject.[35] Both these reports agree that the fundamental problem is that there are two diametrically opposed views of what certainty is and how it should be achieved.

Certainty, from the point of view of the Crown and third parties, is to be achieved by the exchange of undefined Aboriginal rights for defined treaty rights. In the past, treaties have commonly required that the Aboriginal claimant cede, release, and surrender all rights to past and future claims with respect to the undefined rights not included in the treaty.[36] This type of certainty is referred to as extinguishment. Both reports seek to eliminate the concept of extinguishment from the certainty paradigm. The technique in modern land claims agreements has generally been to include a full release from all past and future Aboriginal rights and title claims. In addition, there are indemnity provisions that provide that if the Crown does not get it right in the certainty provisions, the First Nation will have to indemnify the Crown for any costs to the Crown associated with the exercise of rights that are not included in the final agreement but were not caught by the certainty provisions. Adding insult to injury, there has also been a requirement for a "backup release" in the event that the key certainty provisions and the indemnity provisions do not cast a broad enough net.

Aboriginal people view certainty differently from the Crown and third parties. Generally, Aboriginal people see the treaty as breathing life into existing rights and defining a new relationship. Aboriginal people also want to ensure that the treaty promises are kept in order to provide certainty for their future and for future generations. And, while the language in modern treaties attempts to provide certainty for governments and third parties, particularly those in the resource sector, there has been very little certainty of the kind envisioned by First Nations. The certainty provisions, with some variations, appear to amount to a release forever of claims for past and future infringements of Aboriginal rights and title in exchange for an untested treaty.

If treaties are to succeed in the twenty-first century, the approach to certainty will have to work for all parties. All aspects of the extinguishment model must be rooted out. Any certainty model must respect the fundamental values and beliefs of the particular Aboriginal community and reflect the recognition and affirmation language of section 35. As a result of the Nisga'a Final Agreement, the current certainty model in British Columbia uses the language of "modification" and "continuation" so that Aboriginal rights and title are modified and continue as reflected in the final agreement. The model attempts to balance the First Nation's aversion to extinguishment with the Crown's need for a release from future claims related to past infringements. Even so, if the certainty model results in the cessation of rights that are considered sacred and immutable, the treaty likely will not work. In the development of a certainty model, all parties need to keep in mind the clear directions from the court, which requires courts to consider, in those situations where infringements have occurred, "whether there has been as little infringement as possible in order to affect the desired result."[37]

RCAP and others have advocated a nonassertion model that would have all parties agree in clear and plain language that the concept of extinguishment is not a part of the treaty. At the same time, the only rights to be exercised would be those agreed to by the parties in the final agreement, and the First Nation would agree not to assert any other rights. There would also have to be a process in place to address rights that may arise in the future but had not been contemplated by the parties, and consequences for a fundamental breach of treaty obligations. This nonassertion model is more in keeping with the parties' stated common intention to root out all aspects of extinguishment.

In the context of discussions on certainty it is important to note that Canada has stated repeatedly that it rejects extinguishment as a method for achieving certainty. Yet, ironically, Canada insists that federal Crown lands be excluded from areas where First Nations can harvest pursuant to a section 35 right. In other words, although Canada is quite eager to agree to First Nation citizens exercising section 35 rights on provincial Crown lands, Canada's position is that, with minor exceptions, section 35 rights should not be exercised on federal Crown lands and that federal Crown lands should be, for the most part, excluded from the harvest areas where rights are to be exercised. For all practical purposes, this amounts to extinguishment on federal Crown lands. This raises the question of whether Canada is negotiating in good faith.

Section 91(24) of the Constitution Act, 1867, and the Tenure of Treaty Settlement Lands

As noted earlier, the federal government has the exclusive law-making authority over "Indians, and Lands reserved for the Indians." One of the issues linked with this authority is the question of the nature or tenure of treaty settlement lands – those lands the First Nation will own after the effective date of a final agreement.

The matter of the tenure of treaty settlement lands has become a problem area for some First Nations because of the dogmatic insistence by both the federal and provincial governments that the lands set aside as treaty settlement lands not retain their status as section 91(24) lands, or "Lands reserved for the Indians." Canada holds Indian reserve lands essentially in trust for and on behalf of the "Indians," and as such, Indian reserve lands can only be alienated or sold following a "surrender" under the Indian Act. The relationship between Indians and the federal government is said to be fiduciary. This broad fiduciary or trust-like relationship between Canada and Aboriginal people flows, at least in part, from the status of Indian lands and the discretion that Canada exercises over those lands.[38] By analogy, the same principles apply to unceded Aboriginal title lands since both Indian reserve lands and Aboriginal title lands fall within Canada's jurisdiction over "Indians, and Lands reserved for the Indians." Although it is difficult to engage federal negotiators in discussions about the reasons behind their mandates, it seems that Canada's approach to treaties may largely be driven by the desire to remove itself as a fiduciary.

The provincial position is largely driven by a desire to see provincial land use laws apply to treaty settlement lands acquired through treaty negotiations. Under the existing Indian Act regime, the provinces cannot regulate the use of "Lands reserved for the Indians."[39] From a provincial perspective, this creates a legislative void that the province would like to occupy in order to have its own land use laws apply. The province wishes to have its land use laws apply to treaty settlement lands, including the designation of Agricultural Land Reserve (ALR), so there is a consistent legislative regime in British Columbia with respect to land use planning and regulation. This may be a reasonable objective from a provincial point of view. However, it is anathema to some First Nations, particularly in regard to the application of the ALR. It is also an objective that can be achieved in other ways. In fact, the provincial objective has been achieved as a result of the "section 88 analogue" clause in the General Provisions chapter of final agreements, which allows provincial laws to apply on treaty settlement lands.[40] So both the province and Canada have insisted that treaty settlement lands not be "Lands reserved for the Indians." However, the constitutionality of the positions put forward by both British Columbia and Canada is questionable.

Structuring arrangements that would change the nature of Indian lands to something other than "Lands reserved for the Indians" raises constitutional difficulties. Over the years, the courts have looked at the meaning of the words "Lands reserved for the Indians" in section 91(24), and some questions about this now appear settled. It would appear that the words used are "sufficient to include all lands reserved, on any terms or conditions, for Indian occupation." This passage by Lord Watson speaking for the Privy Council in *St. Catherine's Milling and Lumber* was quoted with approval by former Chief Justice Lamer in *Delgamuukw*.[41] More recently, the BC Supreme Court adopted this position in *Tsilhqot'in Nation*.[42] Given the fairly clear and consistent direction from the courts, it is difficult to understand why treaty settlement lands held for a First Nation pursuant to a treaty should not fall within section 91(24). In some cases this will cause more of a problem of legal theory than a problem on the ground. However, many First Nations wish to maintain the special link with the federal government that section 91(24) offers, including Canada's exclusive law-making authority. The law seems to be in agreement.

Ironically, the federal claims policy is also in agreement. The current federal claims policy is not incorporated in a single document, but rather it is a series of policies beginning with the policy first announced in August 1973 following the decision in *Calder*.[43] Canada's Comprehensive Claims Policy was mandated by Cabinet document no. 570-73 entitled "Indian and Inuit Claims Policy" (June 5, 1973).[44] That policy specifically states, "Federal jurisdiction and responsibility would continue under section 91(24) of the BNA Act [now Constitution Act, 1982] – 'Indians, and Lands reserved for the Indians.'"

Notwithstanding the above, each of the final agreements and agreements in principle being negotiated in British Columbia includes a clause identical to this:

> There are no "Lands reserved for the Indians" within the meaning of the *Constitution Act, 1867* for the [First] Nation, and there are no "reserves" as defined in the *Indian Act* for the use and benefit of [the First Nation], and, for greater certainty, [First Nation] Lands ... are not "Lands reserved for the Indians" within the meaning of the *Constitution Act, 1867,* and are not "reserves" as defined in the *Indian Act*.[45]

Although it is written in sloppy language and somewhat redundant, the clause has been agreed to at a number of treaty tables. Constitutional theory does not support the language, but the practice of negotiation invokes the art of the possible. This often involves engaging in interest-based negotiations to find a solution that will work for all parties.[46]

Treaty settlement lands in final agreements are said to be fee simple lands. In Canada, the most common form of fee simple ownership is that in which such lands are held by individuals and registered in the provincial land title system. The underlying or "allodial" title of such lands is usually vested in the province. The provincial fee simple title system has several limitations that make the traditional concept of fee simple lands unattractive to First Nations. Fee simple lands are normally subject to escheat (reversion); that is, when there are no known heirs and the owner dies intestate, the lands normally escheat (revert) to the Crown in right of the province. Lands owned or held in fee simple are usually subject to forfeiture to the Crown in certain circumstances, such as if the landowner fails to pay the property taxes. In addition, under the common law, certain crimes against the state, such as treason, result in the forfeiture of the individual's estate to the Crown. Normally, fee simple lands are subject to the property tax regime of the Crown.

For these and other reasons, final agreement negotiators did not want to have treaty settlement lands held as provincial fee simple lands. Absent a final agreement, First Nations lands will continue to be Indian reserve lands under the Indian Act. But ownership as Indian reserve land also has limitations. Indian reserve lands are under the exclusive jurisdiction of the federal government and cannot be readily registered in the provincial land registry system. Normally, traditional mortgage financing depends on being able to register a mortgage against the title that is registered in the provincial system. "Indians" living on reserve often find it difficult to obtain personal loans because the lender will want to be able to seize the goods purchased if the loan payments are not made. Typically, personal property on Indian reserves cannot be seized by a lender because specific provisions of the Indian Act prevent assets located on Indian reserve lands from being seized. Under the Indian Act, almost all transactions conducted on Indian reserve lands, including the securing of permits and licences, require the permission of the minister of Indian affairs. This can make transactions more costly and time-consuming.

These limitations make it difficult for "Indians" living on reserve to participate in the commercial mainstream. However, some of these problems have been addressed by both the First Nations Land Management Act and the First Nations Fiscal and Statistical Management Act.[47] The latter enactment allows for First Nations to access the bond markets and makes and facilitates systems of financing that would not normally be available to First Nations. The former gives greater autonomy to First Nations by removing the role of the minister in the permit-approval process and makes traditional financing more readily available.

With respect to the creation or use of a land registration system for registering title, the final agreements currently in negotiation provide two options: the authority to make laws for the establishment and operation of a land title or land registry system, or a mechanism to allow for lands to be registered in the provincial land title system.[48] Even after such lands have been registered in the provincial land title system, they remain treaty settlement lands. There are some advantages to having lands registered in the province's title system. The provincial land title system facilitates commerce because when lands are registered in the land title system it is easy to get traditional financing such as mortgages. The provincial land title system also creates a system of priorities for creditors and guarantees indefeasible title through an insurance scheme.

Under the current treaty model – the model agreed to by the Tsawwassen, the Maa-nulth, and others – treaty settlement lands would be owned as First Nation fee simple lands. Treaty settlement lands are not provincial fee simple lands and they are not lands under the exclusive jurisdiction of the federal government pursuant to section 91(24). They are unique *(sui generis)* First Nation fee simple lands protected by section 35 and subject to First Nation law-making authority.

The current model for treaty settlement lands seems to work well for those First Nations with final agreements. However, because it is preferred by some First Nations should not be a pretext for imposing it on all others. Many First Nations are far more comfortable with the notion that treaty settlement lands are "Lands reserved for the Indians." From a constitutional point of view, these First Nations are on safer ground. However, both Canada and British Columbia seem to be locked in to a model that has somehow crept into their comfort zone even though it poses constitutional risk. But treaty settlement lands are what they are, and Canada and British Columbia should not impose a model that may be flawed simply because it suits them. If First Nations prefer that treaty settlement lands continue as "Lands reserved for the Indians," then the only logic to resisting this approach from the federal point of view is linked to Canada's role as a fiduciary. But the fiduciary relationship does not end with treaties; it is renewed by treaties, and Canada must realize this.

The province's interest is linked to the application of provincial land use laws on "Lands reserved for the Indians," but this can be resolved in other ways. In fact, both Canada and British Columbia are likely able to meet First Nations' preferences that treaty settlement lands remain "Lands reserved for the Indians" without risk to the underlying interests of the Crown.

Self-Government and the Concurrent Law Model

Self-government has been a politically volatile issue in the context of treaty negotiations in British Columbia. The political debate had turned on whether or not treaties should recognize a right to self-government, thereby providing such right with constitutional protection. The provincial government and then premier Gordon Campbell made this a political issue by refusing to include self-government as a treaty right. Indeed, the clumsy manner in which the early Campbell regime addressed Aboriginal issues was surprising. The provincial Liberals even managed to box themselves into a corner by committing to hold a referendum on Aboriginal matters, including the treaty right to self-government, and tying their political agenda to the results. Not surprisingly, as the result of a skewed referendum question, a low voter turnout, and the politics of fear, the citizens of British Columbia voted against allowing self-government to be a treaty right. And this was allowed to happen despite a decision by the BC Supreme Court finding that self-government is an inherent right and that the right has not been extinguished. Here, several passages from the court are relevant.

At paragraphs 95 and 96, Williamson J. said:

> In summary, these authorities mandate that any consideration of the continued existence, after the assertion of sovereignty by the Crown, of some right to aboriginal self-government must take into account that: (1) the indigenous nations of North America were recognized as political communities; (2) the assertion of sovereignty diminished but did not extinguish aboriginal powers and rights; (3) among the powers retained by aboriginal nations was the authority to make treaties binding on their people; and (4) any interference with the diminished rights which remained with aboriginal peoples was to be "minimal."
>
> A review of the cases in which Canadian courts, since Confederation, have considered enforcing laws which have their origins with aboriginal peoples rather than with Parliament or a Legislative Assembly, discloses that the above four points have been accepted.

And at paragraph 179, Williamson J. said:

> I have concluded that after the assertion of sovereignty by the British Crown, and continuing to and after the time of Confederation, although the right of aboriginal people to govern themselves was diminished, it was not extinguished.[49]

So, despite the law, the BC government held dogmatically and somewhat irrationally to the view that self-government is not a right and should not be protected as a treaty right. But politics makes strange bedfellows, and all of this is history. The right to self-government is now a part of treaties in British Columbia, and former premier Gordon Campbell is remembered by some as one of the strongest advocates of Aboriginal and treaty rights in the country.

However, treaties involve negotiations, and the exercise of the right to self-government in the context of Canadian federalism is complex. So although self-government is a treaty right, the concurrent law model provides that First Nation, provincial, and federal laws all apply on treaty settlement lands. The First Nations have no exclusive law-making authority. Federal laws are applicable because of Canada's jurisdiction over "Indians, and Lands reserved for the Indians." Provincial laws apply to "Indians" on treaty settlement lands because such laws are general in nature and by virtue of the section 88 analogue provision found in the final agreements.[50]

Section 88 of the Indian Act provides that provincial laws of general application apply to "Indians" either on their own or by virtue of section 88, which incorporates by reference valid provincial laws that touch on the "core of Indianness." The result is that such laws apply as federal laws. The section 88 analogue clause is found in the General Provisions chapter and is duplicated in all the agreements in principle and final agreements negotiated in British Columbia. The clause states:

> Canada will recommend to Parliament that Federal Settlement Legislation include a provision that, to the extent that a Provincial Law does not apply of its own force to [the First Nation, the First Nation Government, the First Nation Public Institutions, the First Nation Corporations, the First Nation Lands, or the First Nation Citizens], that Provincial Law, subject to the Federal Settlement Legislation, any other Act of Parliament and this Agreement, will apply to [the First Nation, the First Nation Government, the First Nation Public Institutions, the First Nation Corporations, the First Nation Lands, or the First Nation Citizens], as the case may be.[51]

This clause is analogous to section 88 of the Indian Act, with two important exceptions. Section 88 applies only to provincial laws "of general application" and not to all provincial laws, and section 88 does not apply to those laws that purport to regulate the use of land. The section 88 analogue clause advanced by the Crown in treaty negotiations goes far beyond section 88 of the Indian Act in incorporating the application of provincial laws.

Although some First Nations have justifiably expressed concern about the scope of the section 88 analogue clause, final agreement tables decided to focus on the issues linked to the scope of law-making authority and paramountcy rather than on the issue of concurrency. At this point, it is difficult to determine whether concurrency will be more than a theoretical problem. But not knowing creates a great deal of uncertainty – something that First Nations and the Crown have tried to eliminate from treaties. If treaties are truly intended to create certainty for all parties, then the concurrent law model needs revisiting.

As for the question of paramountcy, it is important that First Nations laws have paramountcy over key matters such as the management and administration of their own government and public institutions, education, children, language and culture, and lands, including land use planning and regulations. For the most part, the final agreements have been successful in this regard.[52] On the other hand, the Crown maintains paramountcy over some key matters, including health, social services, postsecondary education, and the administration of justice.

As for the scope of the law making, British Columbia has tried to insist that most First Nations' laws apply only to First Nations citizens. This race-based approach would have noncitizens subject to a different set of laws and different standards than First Nation citizens. At the same time, in the negotiations, the province fought hard to allow noncitizens to be elected to First Nations government. These seemingly irreconcilable policies are merely a residue of the politics of fear in which reason is overruled by political expediency. Fortunately, negotiators at the final agreement tables were able to find solutions that call for the application of First Nations' laws to First Nations' institutions. For example, if an individual is attending a First Nation's health or education institution, that individual will be subject to the First Nation's laws, regardless of race. Although the negotiated solutions will work, the positions tabled by the Crown are nevertheless bizarre, and the undercurrent of racism that drives some policies is frightening.

It would make much more sense if First Nations' laws on treaty settlement lands were to apply to "persons" and not just First Nations citizens, so that the model would be based on geography rather than on race. In addition, both Canada and British Columbia need to reconsider the section 88 analogue language so that the application of provincial laws becomes much less intrusive within treaty settlement lands. The current section 88 analogue language being proposed in treaty negotiations in British Columbia gives the province more constitutional authority over "Indians, and Lands reserved for the Indians" than exists under the Indian Act, in the nontreaty world.

Fisheries Negotiations

For many First Nations, particularly those that are on the coast, fisheries negotiations are the most critical aspect of reaching agreement. Fisheries issues are extremely complex, and stakeholders on all sides are seeking mandate changes to Canada's allocation model and fisheries management. In addition, many First Nations are passionate about ensuring that a section 35 commercial fishery is included as part of their treaty package. To date, both the federal and provincial governments have reacted by adamantly refusing to include a constitutionally protected commercial fishery in treaties. From the Crown's perspective, arrangements that allow the sale of fish under section 35 is simply not on the table. The reason for this has nothing to do with fisheries management. Rather, it is a question of political expediency catering to the radical right.

The government mandates on this issue seem out of touch with the Supreme Court of Canada's decision in *Gladstone,* though more recent court pronouncements have had mixed results.[53] More importantly, it is out of touch with Aboriginal culture and current practices. In many Aboriginal communities, there is a strong belief in the right to sell fish. In some cases, First Nations individuals will sell fish regardless of any treaty language. However, this does not seem to matter to policy makers. So, at least until now, the Crown has been willing to agree to treaty protection to fish for food, social, and ceremonial (FSC) purposes, and even to provide a nontreaty-protected volume of fish that would be available for commercial purposes. The Crown steadfastly refuses to allow for the constitutional protection of an Aboriginal commercial fishery. Ironically, from the point of view of the stocks, whether they are consumed pursuant to a constitutionally protected right or by an industrial licence is irrelevant. Once again, this has become a matter of the politics of fear driving policy.

In the few final agreements negotiated under the BC Treaty Commission process, the commercial volume for fish is mostly housed in a document called a harvest agreement. The harvest agreement is not a part of the treaty, nor is it a typical commercial licence. It is like a licence to sell fish for commercial purposes. But the harvest agreement is evergreen (perpetually renewable), and if not renewed, it is compensable. At the same time, as with other licences, the minister must authorize the fishery under the harvest agreement each year. If the minister does not, there will be no allowable commercial harvest. In fact, each year, the minister must authorize both the section 35 FSC fishery and the commercial fishery under the harvest agreement. So, under the fisheries component of the treaty negotiations, negotiations involve giving up an undefined Aboriginal right to fish in exchange for the privilege to fish at the discretion of the minister, subject to some negotiated checks and balances.

A unique feature of the language negotiated in the now stalled Lheidli T'enneh Final Agreement was the ability to move up to 50 percent of the FSC allocation in the harvest agreement or the commercial allocation. But as the mandates of governments narrow, particularly that of the federal government, this option will likely no longer be available. It was not included in either the Tsawwassen, the Maa-nulth, or the Tla'amin Final Agreements,[54] and it will likely not be included in others. In fact, fisheries negotiations are stalled to such an extent that mutterings of "bad faith" negotiations on the part of the federal government are never far from the surface. Continued delay of fisheries negotiations because of the Cohen Commission does not help matters.

Tax Powers and Exemptions

One of the more difficult and highly public issues in modern treaty negotiations is the question of First Nations tax powers and removal of the section 87 Indian Act tax exemptions. Status Indians – Indians registered under the Indian Act – have historically not paid taxes on personal and real property located on Indian reserve lands. The section 87 tax exemption flows from the special relationship between Aboriginal peoples and the Crown and from the unique nature of Indian reserve lands. In 1983, the Supreme Court of Canada reaffirmed the tax exemption and determined that income is personal property and that the section 87 exemption includes taxes on income earned on reserve.[55] The section 87 tax exemption is one of the few economic tools Aboriginal peoples have that can provide an edge in a market economy. For political reasons, rather than reasons based on valid Indian policy objectives, Canada has insisted that modern treaties must eliminate the section 87 exemption.

The final agreements under negotiation provide First Nations with authority to tax within treaty settlement lands, create tax exemptions for the First Nations government, and remove the section 87 tax exemptions. Collectively, these provisions provide First Nations governments with the authority to levy direct taxes, including property tax, sales tax, and income tax for persons living on treaty settlement lands. This includes provision for the negotiation of tax agreements to include tax authority over noncitizens and the harmonization of tax laws. The tax agreements also ensure that although the First Nation is taxing, federal and provincial tax laws over the same subject matter are inapplicable. The tax revenues raised are retained by the First Nation government, subject to some important exceptions, to be discussed later.

The mechanics of the tax chapter requires the First Nation to enact a tax law and enter into a tax agreement for the corresponding federal or provincial tax. The tax levy will have to be harmonized with any corresponding federal or provincial tax. For income tax, the revenues will actually be collected by Canada Revenue Agency and then transferred to the First Nation government. This will all be done through a tax agreement. The mechanism works the same way for the provincial share of income tax, except the province will transfer back only 50 percent of the provincial share of income tax revenues to First Nations. This seems to be because both Canada and British Columbia are of the view that once a modern treaty is in effect, and subject to any transition periods negotiated, First Nations ought to be sharing a greater burden in the costs of services provided to their citizens. British

Columbia will accomplish this by way of tax agreements that will ensure 50 percent of what is considered to be the provincial share of income tax and sales tax revenues is retained by the province. Canada achieves the same purpose by way of its Own-Source Revenue policy, which is also discussed later.

The mechanism for sales tax is slightly different. After the First Nation has enacted a sales tax law, and provided such laws are harmonized with the government sales tax, the Crown will withdraw the application of its corresponding sales tax as it applies to the First Nation citizens (or a broader category of taxpayers, depending on the scope of the tax agreement) on treaty settlement lands. Unlike income tax, in which the actual amount collected (or a percentage of the actual amount collected) is transferred back to the First Nations, a formula for sales tax is negotiated so that an estimated amount of the sales tax is returned, rather than the actual amount. This is because revenues from sales tax are harder to track than revenues from income tax. As in the case of income tax, the province is willing to share only 50 percent of the sales tax revenues collected, and the federal portion is subject to Canada's Own-Source Revenue policy.

Property tax is normally divided into three areas: local government tax, regional tax, and provincial or school tax. After the effective date of the final agreement, the First Nation will be the taxing authority for all three taxes. The local tax is to pay for municipal services. The regional tax is to pay for services provided by the regional government and would normally be transferred to the regional government if that is the service provider. The provincial share, or the school tax, typically goes to the provincial consolidated revenue fund. In the case of property tax, the province has backed out of the field entirely, and the First Nation keeps the full tax amount, but the actual amount collected will be offset by the cost of municipal services and services provided by the regional government.

In addition to the tax authorities negotiated, there are provisions that exempt the First Nation government from being taxed. These are contained in a tax treatment agreement. The tax treatment agreement essentially provides for the application of the exemptions under section 149(1)(c) of the Income Tax Act to the First Nation government because it is a public body performing governmental functions.[56] The taxation chapter in final agreements also provides for the removal of the section 87 tax exemptions after the expiry of a transition period – normally eight years for a consumption tax and twelve years for income tax. Although this tax regime has been reluctantly agreed to at the final agreement tables, it is not acceptable to most First Nations. In fact, the way the tax issue is addressed in final agreements was one of the reasons for the failure of the Lheidli T'enneh ratification process.

The most controversial issue of the tax model proposed by the Crown is the unilateral insistence that the exemption be removed within a fixed time period, regardless of the desires of the First Nation government or those being taxed. A model that would likely be more acceptable would involve the continuation of the equivalent of the section 87 tax exemption until the First Nation chooses on its own to exercise its tax authority. Once the First Nation becomes a taxing First Nation, the exemption would be removed for the particular tax levied by the First Nation. The incentive for the First Nation government to tax is the much-needed revenues. The province would also have to agree to allow the First Nation to retain 100 percent of the provincial share of sales and income tax. However, there is absolutely no need or policy rationale to force the elimination of the tax exemption within a fixed time period – be it eight years, twelve years, or twenty years – as a treaty precondition. The exemption can be removed when the First Nation begins to tax, pursuant to its own time frame. All parties' interests could be achieved through this mechanism.

Federal Own-Source Revenue Policy

Despite the difficult issues described above, the final agreement tables have made the political decision to move forward in treaty negotiations in order to bring to their communities the immediate and tangible benefits that come with a treaty. The First Nations with final agreements are focused on creating an environment within their communities that will foster economic development and greater independence. These communities seem ready to make the necessary compromises they hope will eventually lead to greater economic prosperity and freedom from the Indian Act regime. But Canada insists on drawing more blood from stone. And this insistence was another reason for the negative outcome of the Lheidli T'enneh ratification process.

In addition to the loan repayment requirements, one of the most difficult challenges in the negotiations is the federal Own-Source Revenue (OSR) policy. Subject to some exclusions and the transition period, the federal OSR policy essentially requires that 50 percent of any revenues generated by the First Nation in the post-treaty environment be clawed back through a corresponding reduction in the annual fiscal finance agreement.

In its most basic form and once fully phased in, the federal OSR policy reduces the transfer payments that a First Nation would receive to provide services to its members by a percentage of the revenues that the First Nation earns from its own sources. For example, if a First Nation has a revenue stream of $10,000 in one year from the equivalent of the federal share of income tax, the annual transfer payments would be reduced by 50 percent of the $10,000 (the inclusion rate), or $5,000. The same applies to income from tax-paying First Nations corporations that is transferred to a First Nation government. In other words, wherever a First Nation demonstrates initiatives to generate revenue streams, the transfer payments are reduced by 50 percent of the revenues generated, or by whatever the inclusion rate is on the particular revenue stream. The OSR policy is to be applied to all sources of revenues generated by the First Nation unless those revenues are specifically excluded. In fact, the OSR policy is to be applied to all potential revenue sources, though these cannot be calculated. The policy also applies to interest earned on the capital amount received in the settlement and to income earned on the revenue-sharing portion of the treaty assets.

It is important to understand that the policy will have different implications for different First Nations, depending on factors such as the level and nature of funding from Canada, the amount of own-source revenue generated, the sources of the revenue, and the manner in which the First Nation can structure itself to avoid own-source revenue offsets. However, a number of potential significant effects of this policy can be foreseen: the OSR policy reduces the real value of the settlement funds, reduces the potential economic advantages of treaty settlement lands, discourages economic growth, and drives the manner in which First Nations will choose to develop their resources. In addition, combined with the loss of the section 87 tax exemption, the OSR policy will likely reduce the incentive of First Nations governments to earn revenues. The consequent reduction in transfer payments to First Nations may undermine their ability to provide core social programs. In particular, this may compromise the ability of First Nations governments to fund the universal social programs that all Canadians are entitled to. In effect, the citizens of First Nations with modern treaties would then be disadvantaged in comparison with other residents of British Columbia. The OSR policy seems intent on keeping First Nations poor. Unless it is changed, the policy could lead to lost economic opportunities, slower economic growth rates, and greater reliance on federal transfers.

At the same time, there is no disagreement about the need for all governments to offset the costs of services through own-source revenues. That each government needs to pay its fare share is a cornerstone of self-determination. But where there is a severe fiscal imbalance, it is unhelpful to slavishly apply an untested model. An approach needs to be developed collaboratively by the parties to the negotiations and not imposed unilaterally. During the negotiations, some First Nations have suggested the use of comparability indicators as the trigger for the OSR offsets. This could include comparable wage levels, employment levels, child mortality rates, education levels, health statistics, and so forth. When comparability is achieved, there would be a policy rationale on the part of the federal government for the reduction of transfer payments.

There are, of course, ways of circumventing the federal OSR policy. First, it is important to note that the policy applies only when the income or income potential is generated by the First Nation government. In other words, as long as income is retained by the First Nation's corporations or in a partnership, the OSR offsets will not take effect. This can be achieved by OSR planning and through the proper structuring of business arrangements. Second, a number of important exclusions to the OSR offsets have been negotiated. For example, revenue streams that would normally accrue to the province but now go to the First Nation are excluded. The exclusion includes revenues derived from property tax or the provincial share of income tax and provincial sales tax. This would also include revenues from natural resource negotiations in which the province provides funds linked with consultation and accommodation. In addition, there are exclusions for the assets negotiated in the treaty, including the sale of treaty settlement lands, the initial capital transfer and resource revenue sharing, and the special purpose funds, as well as the income on portions of these funds. Funds received pursuant to the settlement of specific claims are also excluded from the policy. So although the policy itself is draconian, the final agreement tables have managed to negotiate around or are positioning themselves to avoid some of the negative economic aspects of the policy's more harmful aspects. However, this does not detract from the undercurrent of pettiness and the lack of sensitivity to First Nations' concerns about the policy by those in Ottawa responsible for developing the fiscal mandate. It is perhaps with these mandarins in mind that the Supreme Court of Canada made the following comments in the decision of *Mikisew Cree:*

> The fundamental objective of the modern law of aboriginal and treaty rights is the reconciliation of aboriginal peoples and non-aboriginal peoples and their respective claims, interests and ambitions. The management of these relationships takes place in the shadow of a long history of grievances and misunderstanding. The multitude of smaller grievances created by the indifference of some government officials to aboriginal people's concerns, and the lack of respect inherent in that indifference has been as destructive of the process of reconciliation as some of the larger and more explosive controversies. And so it is in this case.[57]

Shared Decision Making

Although the Supreme Court of Canada has made numerous rulings on Aboriginal rights and title and the duty to consult, with few exceptions none of these decisions on its own have had a significant impact on the conduct of the Crown. Undoubtedly, *Delgamuukw* was the case that should have had the most impact because the plaintiffs sought a declaration for Aboriginal title, and had they been successful, the jurisdiction of the province over lands and resource decision making would have been severely restricted. But *Delgamuukw* was sent back to trial for technical reasons. Since that time, the

Gitxsan and the Wet'suwet'en have been labouring away in negotiations without much progress.

Meanwhile, the Crown continues to have a predetermined treaty model in which it is offering a small amount of treaty settlement lands, particularly in less remote areas, with self-governing powers primarily over members but very little control over lands off treaty settlement lands. This is because the Crown continues to see itself as the only owner of the lands, and it views Aboriginal title as merely a burden on the title of the Crown that needs to be dealt with. All of this should have changed as a result of the decision in *Tsilhqot'in Nation*.[58] For the first time in British Columbia, a court has stated that the test for Aboriginal title has been proved, though the court did not provide the declaration sought. In addition, on the basis of earlier decisions of the Supreme Court of Canada, the BC Supreme Court made a nonbinding commentary that the province must reckon with when making lands and resource decisions over Tsilhqot'in traditional lands. Among other things, the court stated that the Tsilhqot'in people were able to prove Aboriginal title over an area of approximately two thousand square kilometres and that the provincial Forest Act cannot constitutionally apply to Aboriginal title lands, as these are "Lands reserved for the Indians" under section 91(24). The court also noted that where the legislative scheme under the provincial Forest Act does apply (where title has not been proved), the application of that regime does infringe Aboriginal rights, as British Columbia was unable to justify that infringement.

Although the opinion of Mr. Justice Vickers was nonbinding, the decision has serious implications for the Province of British Columbia's ability to manage lands and resources on Aboriginal title lands. The decision should have been a shot in the arm for the much-beleaguered treaty process and should force the province to get creative about lands and resource management issues, particularly co-management arrangements off treaty settlement lands.

The province has generally been more flexible in negotiations and has attempted to engage in creative interest-based problem solving, yet it has been reticent around sharing decision-making power over lands and resources that are not a part of the treaty settlement lands. This has not been helpful, particularly because the amount of lands the province is prepared to put on the table is relatively small when compared with negotiations that have taken place in northern Canada, Quebec, and Labrador.

The province is generally prepared to negotiate joint management arrangements in certain parks or protected areas. In addition, in the Lheidli T'enneh negotiations, the province agreed to negotiate and attempt to reach agreement on "shared decision making" in two specified watersheds. Unfortunately, the Lheidli T'enneh negotiations have not been able to bear fruit and enable a more detailed discussion of what the shared decision making would look like. As much of the concern among BC First Nations is generally linked with oil and gas development, forestry, and mining, it seems that genuine shared decision making should involve a decision-making role by First Nations in determining the annual allowable cut in a particular forest district, the approval of land use plans, and a decision-making role in the environmental assessment process. These areas are critical to any notion of shared decision making and joint stewardship of the land. In addition, it seems that where there is a proposed project that requires an environmental assessment or has a significant economic impact, a part of the project approval process should be a requirement for the proponent to enter into an impact-benefit agreement with the affected First Nation(s). This does not necessarily mean that First Nations would have a veto over all development, but arrangements could be made for specific watersheds that are important to a particular First Nation, and joint decision making would force serious negotiations on proposed projects. Much of this has been accomplished by the Haida outside of treaty,[59] but very little joint management or shared decision making is offered at the treaty table.

Another area in which the province should be able to show more creativity is the establishment of wildlife management boards and water boards. Currently, the province is refusing to engage in discussions that would require the establishment of management boards because of a fear that the minister's discretion will be fettered or that the costs might be prohibitive. This is old thinking, since the minister's discretion is already fettered by the constitutional protection of Aboriginal and treaty rights. As to the threat of increased costs, this is little more than a red herring. The fact is that the management boards established under the Nisga'a Final Agreement have proved cost-efficient, and the costs of not having treaties will be much more challenging to the provincial treasury. Again, this is an area in which creative solutions can be found, but this would require the requisite political will to honestly engage in interest-based problem solving, and this will is often lacking.

Interest-Based Negotiations

Cooperative interest-based negotiations play an important part in the negotiation process. Treaties cannot be successfully negotiated unless the parties are willing to come out from behind their positions and join together in dialogue. There are many instances in which the parties engage in a genuine structured problem-solving process in order to resolve issues in a manner that meets the interests of all. This happens in an atmosphere of trust and cooperation.

Yet, there are some tough issues, including the ones discussed above, in which the parties bring positions to the table that are illogical and that interest-based negotiations have not resolved. This is because there are structural impediments that do not easily lend themselves to interest-based problem solving. These structural impediments arrive because the negotiators, particularly the federal negotiators, do not have the decision-making authority required to allow the interest-based approach to work. The real decision makers are the mandarins in Ottawa who never appear at the table. They set the mandates, and once these mandates are set, there is very little that federal chief negotiators can do to change things. This is particularly true in relation to tax, fiscal issues, and fisheries. These structural impediments are the cause of much of the frustration at the negotiation table on the part of all parties, including the federal chief negotiators themselves. Although final agreement tables have concluded that the positions brought by the Crown are not necessarily irreconcilable with their own interests, many First Nations have concluded that not reaching an agreement or litigation would be better alternatives. The number of First Nations that have come to this conclusion is rapidly growing. This trend could prove fatal to the BC treaty process unless the structural impediments are removed.

Conclusion

Aboriginal title remains unextinguished in British Columbia, and the rights of First Nations and the settler populations remain to be reconciled. The BC Treaty Commission process is a brave undertaking in that it has made a sincere attempt at the process of reconciliation. But despite some successes by the final agreement tables, including the recent initialling of the Tla'amin Final Agreement, the process has stalled.[60]

The success achieved at the final agreement tables should not be underestimated. Their efforts successfully encouraged British Columbia to change its policies in some important areas,[61] but at the current pace, it will be another century or two before the majority of land claims in British Columbia are resolved. Resistance to negotiations under the status quo mandates is growing each day. By far, the great majority of First Nations are seeking fundamental changes to the mandates currently brought to the table by the Crown. This does not mean that First Nations are necessarily against the treaty process, but it does mean the status quo is unacceptable to most. It also means that it is time for change, and the following policy changes need to be considered:

- The parties need to revisit the certainty model to ensure that all aspects of extinguishment have been rooted out. A preference should be given to the nonassertion model as suggested by RCAP and Justice Hamilton.
- Canada and British Columbia need to be more flexible around the question of the status of treaty settlement lands by giving due consideration to First Nations' preferences, including the continuation of treaty settlement lands as section 91(24) lands.

- With respect to self-government, the concurrent law model and the section 88 analogue provisions need revision to ensure workability and to allow for greater autonomy by First Nations from the mechanical application of federal and provincial laws.
- Canada should adopt an approach to taxation that would allow the section 87 Indian Act tax exemptions to apply until First Nations governments are ready to draw down their own tax powers.
- Canada's fiscal policy needs revision. The loan repayment requirement is draconian, as is the OSR policy. In particular, the OSR exclusions need to be broadened and the OSR trigger should be linked to First Nations achieving socioeconomic parity.
- Canada's conduct and mandate at fisheries negotiations must change, and treaties will need to include a constitutional priority of the Aboriginal fishery, including an Aboriginal commercial fishery.
- British Columbia needs to be more flexible in its approach to shared decision making of treaty settlement lands, including giving First Nations a decision-making role in determining the annual allowable cut, approving land use plans and the creation of parks, and approving projects that require an environmental assessment.

Clearly, it is now time to do some constructive damage to the status quo when it comes to treaty making. British Columbia has shown some leadership in making modest changes to its mandates, but it must do better. Canada needs to be prepared to follow British Columbia's lead in risk taking and policy change. The federal land claims policy and the policies applied to the BC Treaty Commission process cry out for change. They are tired and outdated twentieth-century policies that seem out of step with the law and out of step with what is actually happening on the ground. Regardless of whether several more final agreements are concluded, it is time for an honest assessment of the policies and the process governing treaty making in British Columbia and elsewhere. But the honest assessment must be backed with the audacity of imagination and the political commitment required for making changes to the status quo.

Notes

1 Canada, *Report of the Royal Commission on Aboriginal Peoples: Looking Forward, Looking Back,* vol. 1 (Ottawa: Supply and Services Canada 1996), 137-91.

2 Ratification vote, March 30, 2007, BC government, http://www.gov.bc.ca/arr/firstnation/lheidli/down/final/lheidli_final_agreement.pdf.

3 Judith Lavoie, "Speed Up Treaty Process or Shut It Down Commission Chief Says," *Times Colonist,* October 13, 2011.

4 Constitution Act, 1867 (UK), 30 and 31 Vict., c. 3, reprinted in R.S.C. 1985, App. II, no, 5. "Indian" was the term commonly in use at that time. Its meaning is a complex legal question.

5 See, for example, the Nisga'a Final Agreement (August 4, 1998) or Lheidli T'enneh Agreement-in-Principle (October 29, 2006). See also the Comprehensive Land Claims Policy (Department of Indian and Northern Development, December 16, 1987), discussion online, Canadian Arctic Resources Committee, http://www.carc.org/pubs/v15no1/4.htm.
6 See generally Canada, *Report of the Royal Commission on Aboriginal Peoples: Restructuring the Relationship,* vol. 2 (Ottawa: Supply and Services Canada, 1996) [hereafter cited as *Restructuring the Relationship*].
7 Ibid., 538-39. However, as to interim and treaty-related measures, both British Columbia and Canada have policies that attempt to grapple with these issues. But these policies do not address in a genuine way the substantive issues that were at the basis of Recommendation 16 of the *British Columbia Claims Task Force Report* (June 28, 1991), BC Treaty Commission http://www.bctreaty.net.
8 *Restructuring the Relationship,* ibid., 538.
9 Traditionally, treaties have been characterized as an exchange of undefined Aboriginal rights for more defined treaty rights. Historically, on entering into a treaty, the Aboriginal group would have to "cede, release and surrender" all rights not included in the treaty. This surrender of rights results in extinguishment. The policy behind the "cede, release and surrender" language or the "certainty" provisions (terms common to the old numbered treaties) has been a cornerstone of modern treaties. The language in the Nisga'a Final Agreement and other final agreements departs from the objectives of the "cede, release and surrender" language.
10 These changes have been more of process than of substance.
11 *R. v. Sparrow,* [1990] 1 S.C.R. 1075, 70 D.L.R. (4th) 385 *[Sparrow].*
12 *R. v. Gladstone,* [1996] 2 S.C.R. 723, 137 D.L.R. (4th) 648 *[Gladstone].*
13 *Guerin v. R.,* [1984] 2 S.C.R. 335, [1984] 13 D.L.R. (4th) 321, *(sub nom. Guerin v. Canada)* 55 N.R. 161 (S.C.C.) *[Guerin].*
14 *Delgamuukw v. British Columbia,* [1997] 3 S.C.R. 1010, S.C.J. No. 108 *[Delgamuukw].*
15 *Haida Nation v. British Columbia (Minister of Forests),* [2005] 1 C.N.L.R. 72.
16 *R. v. Sappier,* [2006] 2 S.C.R. 686, S.C.J. No. 54.
17 *Mikisew Cree First Nation v. Canada (Minister of Canadian Heritage),* 2005 S.C.C. 69, [2005] 3 S.C.R. 388, S.C.J. No. 71 *[Mikisew Cree].*
18 *R. v. Powley,* [2003] 2 S.C.R. 207, 4 C.N.L.R. 321.
19 See note 5, above.
20 For details, see provincial legislative authorities listed under s. 92 of the Constitution Act, 1867.
21 See note 7, above.
22 See generally BC Treaty Commission Agreement (September 21, 1992), BC Treaty Commission, http://www.bctreaty.net/files/bctcagreement.php.
23 Ibid.
24 See Recommendations 2 and 16 of the task force's report, *British Columbia Claims Task Force Report* (see note 7, above). Recommendation 16 of the report provides that interim measures are to be negotiated when an interest is being affected that may undermine treaty negotiations.
25 *Where Are We?* BC Treaty Commission 2003 annual report, BC Treaty Commission, http://www.bctreaty.net, 37. See also *Looking Back, Looking Forward: A Review of the BC Treaty Process* (Special Report of the BC Treaty Commission, September 2001), BC Treaty Commission, http://www.bctreaty.net, 4.
26 See generally Maa-nulth First Nations Final Agreement, http://www.gov.bc.ca/arr/firstnation/Maa-Nulth/down/final/mna_fa.pdf; Tsawwassen First Nation Final Agreement; http://www.gov.bc.ca/arr/firstnation/tsawwassen/tfn_fa.pdf; Lheidli T'enneh Final Agreement, see note 2, above. Also, on October 21, 2011, the Sliammon First Nation, British Columbia, and Canada initialled the Tla'amin Final Agreement.
27 See generally *Looking Back, Looking Forward.*
28 Ibid., 3-5.
29 *BC Treaty Commission Annual Report* (1998), BC Treaty Commission, http://www.bctreaty.net/files/annuals.php, 5.

30 In this chapter, the term "final agreement table" refers to those treaty negotiations where a final agreement has been negotiated, or where a final agreement is near completion.
31 *Looking Back, Looking Forward,* 6-8.
32 *Tsilhqot'in Nation v. British Columbia,* 2007 B.C.S.C. 1700, [2008] 1 C.N.L.R. 112.
33 *Looking Back, Looking Forward,* 6-11.
34 *Treaty Making in the Spirit of Co-Existence: An Alternative to Extinguishment* (Ottawa: Supply and Services Canada, 1995).
35 *A New Partnership: Report of Hon. A.C. Hamilton, Fact-Finder for Minister of Indian Affairs and Northern Development* (Ottawa: Minister of Indian Affairs and Northern Development, 1995).
36 See generally BC Treaty Commission, "Aboriginal Rights: The Issues," http://www.bctreaty.net/files/issues_rights.php.
37 *Sparrow* at para. 82; see note 11, above.
38 See generally *Guerin;* see note 13, above.
39 *R. v. Dick,* [1985] 2 S.C.R. 309, 23 D.L.R. (4th) 33 at 49.
40 For example, clause 29 of the Nisga'a Final Agreement, Provincial Law, is analogous to s. 88, General Provincial Laws Applicable to Indians, of the Indian Act. See discussion in text referenced by notes 52 and 53 below.
41 *St. Catherine's Milling and Lumber v. R.* (1888), 14 App. Cas. 46 at 59; *Delgamuukw* at para. 174; see note 14, above.
42 See note 32, above.
43 *Calder v. British Columbia (Attorney General),* [1973] S.C.R. 313, 34 D.L.R. (3d) 145.
44 Obtained pursuant to an application under the federal Access to Information Act.
45 See, for example, Nisga'a Final Agreement at c. 2, s. 10, General Provisions.
46 Despite that the language is agreed to, it is difficult to reconcile the position that Canada and British Columbia bring to the table. That is, that extinguishment is no longer a part of the certainty model, but there will no longer be "Lands reserved for the Indians."
47 First Nations Land Management Act, R.S.C. 1999, c. 24; First Nations Fiscal and Statistical Management Act, R.S.C. 2005, c. 9.
48 See, for example the Land Title chapter in *Maa-nulth* and *Tsawwassen;* see note 26, above.
49 *Campbell v. British Columbia (Attorney General),* [2000] B.C.J. No. 1524, 2000 B.C.S.C. 1123.
50 See note 40, above.
51 See, for example, Lheidli T'enneh Final Agreement, c. 2, s. 50, General Provisions; see note 2, above.
52 None of the final agreement tables have been able to secure paramountcy with respect to the BC Agricultural Land Reserve (ALR) legislation, or paramountcy over child care. The ALR issue was addressed by ensuring that the ALR is removed on critical lands.
53 See *Lax Kw'alaams Indian Band v. Canada (Attorney General),* 2011 S.C.C. 56, which ruled against the Lax Kw'alaams claim of a commercial fishing right. But also see *Ahousaht Indian Band and Nation v. Canada (Attorney General),* 2011 B.C.C.A. 237, where the Court of Appeal upheld a finding of an Aboriginal right to harvest and sell fish but excluded the geoduck fishery.
54 See note 26, above.
55 See generally *Nowegijick v. R.,* [1983] 1 S.C.R. 29, 144 D.L.R. (3d) 193.
56 Income Tax Act, R.S.C. 1985, c. 1 (5th Supp.).
57 *Mikisew Cree* at para. 1, Binnie J.
58 See note 32.
59 Haida Gwaii Reconciliation Act, R.S.B.C. 2010, c. 17.
60 Tla'amin, British Columbia, and Canada initialled the Tla'amin Final Agreement on October 21, 2011.
61 Self-government, resource revenue sharing, tax sharing, and shared decision making.

5
Timber
Direct Action over Forests and Beyond
Rima Wilkes and Tamara Ibrahim

An earlier chapter by James S. Frideres explored the consequences of colonialism and the alienation of First Nations from ownership and decision making over forest lands. This theme has also been explored in this volume by M.A. (Peggy) Smith among others. In this chapter, we explore some of the strategies that First Nations have used to address land use and ownership conflicts over the past fifty years. We provide a brief overview and analysis of incidents that have involved "direct action," though other examples will also be considered. Direct action can be understood as a political tactic that legally or illegally disrupts the public interest in order to attract awareness and action to an issue or cause. In British Columbia, conflicts between First Nations rights and the forestry industry have often escalated to involve direct actions such as road blockades, rail blockades, and land occupations. In many cases, such disruptive and dramatic tactics have succeeded where legal challenges have failed (Notzke 1994).

Activists use direct action as a means of taking matters into their own hands when legal and political roadblocks prevent a satisfactory resolution. Direct action also allows activists to generate media coverage and thus public support for their cause. Some notable examples of direct action are the Nuu-chah-nulth, Haida, and Lillooet Nations' efforts to halt logging activities on their territory during the 1980s.[1] These groups have used road blockades and demonstrations as a means of drawing attention to Aboriginal ownership of disputed lands in British Columbia. Few treaties exist in British Columbia and, as of yet, very little land has been legally ceded.

Although each dispute arises within its own historical, geographical, and economic context, each must also be situated as part of a broader phenomenon of First Nation resistance in Canada. On the whole, direct action has been a successful tactic for many First Nations communities. A full list of the events covered in this chapter and their chronological order is provided in Appendix 5.1.

Political Mobilization by Indigenous Peoples in Canada

Resistance by indigenous peoples is not unique to the late twentieth and early twenty-first century. In 1864, members of the Tsilhqot'in nation resisted the encroachment of road builders and settlers attempting to use their land on the central west coast of British Columbia (Muckle 1998). Several decades later, five hundred First Nations peoples from northwestern British Columbia organized to prevent miners and the North West Mounted Police heading up to the Yukon to search for gold from entering their territory (Coates 2001). In the early 1920s, several rallies attended by more than a thousand people were organized by the League of Indians of Canada, a pan-Indian organization (York 1989, 246).

The roots of contemporary collective action by indigenous peoples in Canada can be traced back to 1969.[2] That year, the federal government drafted a White Paper entitled *The Statement of the Government of Canada on Indian Policy*. The purpose of the legislation was to do away with the Indian Act (in place since 1876), along with the Department of Indian Affairs. The White Paper aimed to eliminate "Indian status" and all of the special rights it entailed (Comeau and Santin 1990, 6; Sanders 1985). Purportedly, the goal of the White Paper was to improve the economic position of Native people by turning them into "ordinary" Canadians. Indigenous peoples responded with a statement to then Indian affairs minister and former prime minister Jean Chrétien highlighting the redundancy of the plan: "Take [our rights] away and you leave us nothing, you, and your just society ... We are Canadians by birth. We don't need a law passed to change us into Canadians" (cited in Comeau and Santin 1990, 8). Aboriginal people across Canada responded en masse, but not in the manner expected by the federal government. Rather than acquiescing to the proposed policy reform or attempting to combat it from within the system, indigenous peoples engaged in an unprecedented series of protests and demonstrations (Sanders 1985; York 1989). Eventually, the bill was pulled from the political agenda. Its demise marked the beginning of a new era of political mobilization by indigenous peoples.

Throughout the 1970s, direct action would become an increasingly common tool for First Nations groups to articulate their claims, especially when efforts to address problems through conventional channels failed. For example, in 1976, residents of the Whitedog and Grassy Narrows Reserves in Ontario resorted to direct action when their calls to investigate an environmental health risk were repeatedly ignored. They alleged that Reed Paper Ltd. had illegally disposed of toxic waste into the English-Wabigoon River off Lake Superior. The fish were found to contain exceptionally high levels of mercury. Mercury poisoning became a serious public health problem in these communities (York 1989). When their efforts to resolve the issue by other means failed, activists from Whitedog and Grassy Narrows took matters into their own hands by blockading a road running though their reserves (*Akwesasne Notes,* Early winter 1976, 13; York 1989).

In the summer of 1972, eighty Anishnawbek warriors from neighbouring reserves collaboratively occupied a Kenora park in protest of police brutality, poor conditions on reserves, and the unresolved ownership of the parkland. The Department of Indian Affairs, they claimed, had illegally sold the land to the city of Kenora in 1959 (*Akwesasne Notes,* Summer 1974; *New York Times,* August 20, 1974, 13). The 1970s also saw mobilization in western Canada. For example, parents of almost one thousand children from the Cold Lake, Saddle Lake, and Kehewin Reserves (Alberta) protested federal plans to close reserve schools and integrate their children into the provincial education system (*Akwesasne Notes,* October 1971). In addition to boycotting the schools, they occupied the Edmonton Department of Indian Affairs office.

On many occasions, disputes escalating into direct actions would become extended gridlock situations. On April 28, 1975, members of the Tl'azt'en First Nation began what would ultimately become a three-month rail blockade in an effort to secure adequate compensation for a proposed extension of the rail line onto Tl'azt'en reserves, as well as other infringements on Tl'azt'en territory by the BC government (Booth and Skelton 2006). Other key events in the 1970s include a cross-country British Columbia-to-Ottawa caravan and demonstration on Parliament Hill. The purpose of the caravan was to protest the poor conditions of reserves in Canada (Tait 2007). Members of the Piikani Nation would also become embroiled in conflicts about water rights. In 1978, to gain compensation for the use of their natural resources, they engaged in a three-and-a-half-week blockade of an irrigation weir that provided water for the Lethbridge Northern District that was located on their reserve (*Calgary Herald,* March 15, 1990, A3). This conflict would stretch into the next two decades.[3]

Indeed, by the 1980s and 1990s, the use of collective action had become even more frequent. Like the 1969 White Paper, which galvanized Aboriginal people across the country into action, a plan to patriate the Canadian constitution resulted in cross-Canada mobilization. The Canadian government, seeking to resolve the ongoing tension surrounding calls for Québécois sovereignty, argued that a patriated and amended constitution would provide a much-needed solution. Initial drafts, which began in 1978, almost entirely ignored the rights of Aboriginal peoples. Aboriginal peoples were not consulted in the process of drafting the constitution, and the draft contained little acknowledgment of their rights and roles as First Peoples. By 1981, when lobbying and public relations trips to Europe failed to garner the transnational advocacy they sought, Aboriginal organizations organized a series of protests to draw attention to what could become a systematic dismissal of their rights. Protests included an occupation of Vancouver's Museum of Anthropology, and an Edmonton demonstration attended by five thousand people (Sanders 1985, 175). Although they failed to halt the patriation in 1982, the final draft of the constitution did include several sections dealing specifically with Aboriginal rights. Activism surrounding this issue continued when Aboriginal people across Canada engaged in local protests when the Queen of England visited Canada to take part in ceremonies and celebrations that year (*Winnipeg Free Press,* April 19, 1982). These protests took place again in 1984, 1987, and 1997 during the first ministers' conferences that were required by the new constitution (*Windspeaker,* March 16, 1984, 2; *Akwesasne Notes,* Spring 1987, 7; *Vancouver Sun,* February 25, 1997, A3; *Globe and Mail,* February 27, 1997, A4; also see Sanders [1985] for a more in-depth account of the Indian lobby and the patriation of the Canadian constitution).

The mid-1980s would also be marked by several major conflicts over forestry in British Columbia. In 1985, the Clayoquot and Ahousaht First Nations of Vancouver Island engaged in a campaign of actions (including road blockades) to stop the logging of Meares Island (*Calgary Herald,* February 10, 1985). People from several Gitxsan villages in northern British Columbia also engaged in numerous blockades. Their associated legal challenge eventually led to the seminal 1997 *Delgamuukw* decision recognizing Aboriginal title (*Vancouver Sun,* October 6, 1988, A13; August 22, 1995, B3). Haida youth and Elders also worked together to block a logging road used by Western Forest Products at Lyell Island in the Queen Charlotte Islands in British Columbia (*Globe and Mail,* October 31, 1985, A5; January 3, 1986, A3; *Vancouver Sun,* November 1, 1985, A23; *Toronto Star,* November 15, 1985, A15). The Haida objected to the logging not only because it was destroying the land but also because their title to these unsettled lands had never been formally relinquished. Similar conflicts ensued between forestry companies and the Lytton First Nation and the Lillooet Tribal Council in the Stein Valley. Among the most predominant events they organized was the Stein Valley Voices for the Wilderness Festival (Notzke 1994).

That same year, the Innu of Labrador and Quebec began to adopt more militant tactics in their campaign to stop low-level military training flights on lands they claimed as their own (Armitage and Kennedy 1989).[4] They argued that the noise disturbed both them personally as well as the animals on which they depended for their livelihoods (ibid.; Wadden 1996). Because the Innu continue to engage actively in hunting, the flights posed a direct threat to their way of life and culture (Wadden 1996). That year, they conducted peace demonstrations. In subsequent years they would also pitch tents and occupy the runway of the air base (*Montreal Gazette,* April 22, 1987; *Calgary Herald,* October 5, 1988). Unfortunately, despite having successfully drawn international attention to their cause, the Innu have been unable to stop the military training activities. Nonetheless, Innu leaders have remarked that the campaign increased the community's sense of pride and agency (Wadden 1996).

In 1987, one First Nation used an especially innovative means of drawing attention to its concerns. Under the leadership of Chief Louis Stevenson, the Peguis First Nation invited South African ambassador Glenn Babb to visit its reserve (*Akwesasne Notes,* Spring 1987, 23). The purpose of this controversial visit was to draw attention to the deplorable living conditions faced by Peguis residents. By inviting the representative of a country that upheld an apartheid system, Stevenson ventured to draw a parallel between the lives of blacks in South Africa and Native peoples in Canada. Such a comparison apparently touched a nerve in the Canadian psyche and, as a result, garnered unprecedented international media attention – over three hundred journalists and camera crews came to the Peguis Reserve to cover Babb's visit (*Akwesasne Notes,* Spring 1987). The publicity also garnered attention from the Canadian government. After Babb's visit, the government marshalled significant funding for the reserve. The band built a $5 million shopping mall, a move that vastly helped the economy on the reserve (York 1989).

Other notable instances of protest during the late 1980s include a mobilization by the Lubicon people of northern Alberta, and the Temagami First Nation of Ontario. The Lubicon Lake Cree were able to mount an international campaign to draw attention to their cause (see Ferreira 1992; Churchill 1993). They were concerned about oil drilling on land that they had never ceded by treaty. They organized a boycott of an Aboriginal art display (sponsored by Shell) planned for the 1988 Calgary Olympics. When it was later anounced that a Japanese paper company was awarded logging rights and allowed to establish a plant on the claimed land, they set up road blockades and organized a second boycott of this company. They garnered support from Mohawks of Kahnawake and Six Nations in Ontario, who blocked traffic on their behalf (Goddard 1991). In Ontario, the Temagami First Nation was also forced to act when construction of a logging road began in territory that was part of a land claim under review by the Supreme Court of Canada (*Toronto Star,* November 16, 1989).

Despite the increasing size and frequency of protests during the 1980s, Aboriginal peoples and their concerns were largely ignored by both provincial and federal governments. By 1988, Assembly of First Nations leader Georges Erasmus forewarned in his re-election speech that the current cohort of Indian leaders "may be the last generation of leaders that are prepared to sit down and peacefully negotiate [their] concerns with you" (cited in York 1989, 259). His message failed to generate much action at the time. However, by the spring of 1990, it was evident that his speech was more prescient than previously estimated. In March 1990, the town of Oka in Quebec set in motion plans to expand a local golf course onto land known as the Pines. The Kanesatake First Nation had been trying to regain that same land for several hundred years. When their legal challenge failed, Kanesatake Mohawks erected a barricade (York and Pindera 1991). In response, on July 11, the town mayor ordered the Sûreté du Québec (the Quebec provincial police force) to storm the blockade. The Mohawks responded, resulting in the death of one police officer, Corporal Marcel Lemay. Lemay's death precipitated a seventy-eight-day standoff between the Mohawks and the police, who were later replaced by Canadian armed forces.

The conflict ended on September 25 when the Mohawks left their fortification. Some returned to their homes and others were arrested (York and Pindera 1991). Although many disputes involved the police, the Oka Crisis was of greater magnitude than any other in the late twentieth century.

Oka served as a rallying point for indigenous peoples across the country, especially in British Columbia, where few treaties have been signed. Several other First Nations, anxious to show their support for the Mohawks and to draw attention to their own grievances, set up blockades during this time. These included the Mount Currie First Nation of British Columbia (*Vancouver Sun,* October 20, 1990; *Globe and Mail,* November 7, 1990), the Big Cove First Nation in New Brunswick (*Globe and Mail,* September 7, 1990, A4), and the Laich-Kwil-Tach Indians of British Columbia (*Windspeaker,* August 3, 1990, 7). The summer of 1990 – which has since been referred to as "the summer of Native discontent" – would see dozens of Native protests around Canada.

Following this summer, indigenous peoples continued to use high-profile protest into the early 1990s. In 1992, Cree from James Bay (whose leader Matthew Coon Come was later to become the grand chief of the Assembly of First Nations) travelled by canoe from Ottawa to Montreal and then to New York City to garner publicity and support for their fight against a dam project having the potential to flood their lands. On their arrival in Manhattan they addressed thousands partaking in the 1992 Earth Day gathering (Niezen 1998). Their strategy of garnering international support worked. The hydro project was eventually cancelled (ibid.). Furthermore, the Cree people were able to make their cause and their right to sovereignty a central issue within debates about the role of Quebec in Canadian society (Ramos 2000).[5]

When the federal government drafted new legislation proposing to tax income earned off-reserve, Aboriginal people in urban centres across the country rallied (*Globe and Mail,* December 17, 1994, A3). In Toronto, the protest took the form of an occupation of a Revenue Canada office (*Globe and Mail,* December 17, 1994, A3). At that time, a group of Native youth also seized a mill on an island in the Ottawa River, declaring the land Aboriginal land and the mill an Aboriginal embassy (*Toronto Star,* March 19, 1996, A8). The Assembly of First Nations also encouraged and organized a mass protest to take place on April 17, 1997, to earmark the fifteenth anniversary of the signing of the Canadian constitution and associated failure by the government to live up to promises made at that time (*Vancouver Sun*, February 25, 1997, A3; *Globe and Mail*, February 27, 1997, A4).

In 1995, the Chippewas of Kettle and Stony Point set up a camp to reclaim land in Ipperwash Provincial Park. During the occupation, Anthony Dudley George, an unarmed protester, was shot and killed by provincial police (*Globe and Mail*, September 7, 1995, A1). That same year, protesters from the Nuxalk Nation in British Columbia set up a roadblock to stop International Forest Products from building a logging road on their territory. Nineteen people were arrested (*Vancouver Sun*, September 27, 1995, A3; *Calgary Herald*, September 28, 1995, C9).

Toward the end of the 1990s, several First Nations groups in New Brunswick became embroiled in a protracted logging dispute (*Toronto Star*, June 11, 1998). They continued to log and set up blockades in defiance of a 1998 New Brunswick Court of Appeal decision to revoke First Nations access to timber on Crown lands (*Halifax Chronicle Herald*, July 29, 1998; January 22, 1999; February 19, 1999; *Toronto Star*, May 6, 1998; June 11, 1998). As a result of their mobilization, several groups were able to gain some concessions from the government of New Brunswick.

Discussion and Conclusion

A number of scholars have analyzed these events both as a whole (Blomley 1996; Long 1997; Wilkes 2004, 2006; Alfred and Lowe 2005; Borrows 2005; Ramos 2006) and as individual cases. The individual cases given detailed scholarly attention include Oka (Skea 1993; Stuart 1993; Grenier 1994; Smith 2000; Kalant 2004) and Ipperwash (Edwards 2001; Miller 2005) and Gustafsen Lake (Lambertus 2004). One of the clearest messages to emerge from events such as those discussed here is the public articulation of indigenous people's right to self-determination and sovereignty. For example, the members of the Haida Nation were especially clever at making sovereignty understood by non-Natives when they distributed travel permits to tourists visiting the Queen Charlotte Islands (York 1989). This is just one of many tactics in the wide and varied "repertoire of contention" used by indigenous peoples to achieve their goals. A repertoire of contention is a time-varying collection of tactics groups used to achieve political ends (Tilly 1978). This contemporary repertoire has included marches and demonstrations, as well as road, train, and boat blockades. Although indigenous peoples are not the only groups in Canadian society to engage in collective action, they have certainly used direct action most frequently and, arguably, most legitimately, on both legal and moral grounds.

The most apparent cause of these escalated conflicts is the lack of willingness on the parts of provincial and federal governments to honour the indigenous rights enshrined in the Royal Proclamation of 1763, the Indian Act, and scores of nation-to-nation treaties signed over the course of centuries.

This disregard has been manifest in ongoing encroachments onto First Nations land. Governments have allowed the construction of access roads and rail lines, as well as logging by non-Native companies, on First Nations land. These practices have taken place throughout the country, though they have been especially pronounced in British Columbia, where the majority of "Crown" land is unceded First Nations land. These territorial encroachments involved forestlands as well as the annexation of reserve land for public roads, housing developments, and military bases. The severity of the problem of territorial loss by First Nations in Canada is well illustrated by the fact that, since the establishment of the federal claims commission in 1973, over eight hundred submissions remain unresolved as of 2008. And it has been demonstrated empirically that when land claims are settled, the use of collective action goes down (Ramos 2006).

Nevertheless, while injustices such as those discussed above are certainly one cause of collective action, injustice alone does not explain why only some First Nations members have participated in contentious political actions. Between 1981 and 2000, out of the over 600 First Nations in Canada, less than 25 percent have engaged in some form of direct action (Wilkes 2003). In general, larger First Nations, those with a higher proportion English/French as a mother tongue, those with high unemployment, and those with a past history of mobilization engaged in more collective action (Wilkes 2004). This suggests that the use of mobilization is far from random. However, the fact that there are hundreds of First Nations, each with its own unique history, does make systematic comparisons such as these a daunting task.

Overall, the use of collective mobilization has had a range of outcomes. In some cases, direct action has led to government action. Haida activism in the 1980s ultimately led to the creation of Gwaii Haanas National Park in 1993. Following the conflict at Oka, the federal government began buying parcels of the disputed land and eventually turned it over to the Mohawks of Kanesatake in 1997. These examples provide some evidence that direct action can work.

Nevertheless, the relative dearth of studies that have systematically compared the outcomes of various forms of political action remains an issue. In some cases, direct action has led at least to attention where there was none. Specifically, ongoing events have generated significant media attention to First Nations and their issues. For example, Oka and Ipperwash alone have generated hundreds of newspaper articles and television reports (Grenier 1994; Miller 2005). However, the general conclusion is that much of this attention is negative. For example, Henry and Tator (2002) show that media coverage of the Burnt Church/Esgenoopetitj fishing conflict placed greater emphasis on the impact of the disputes on non-Native people than on Native fishers and communities. Nevertheless, in the long run, it may be a case of "any publicity is good publicity."

In sum, the purpose of this chapter was to consider collective action both in reference to forestry and more generally within the larger Canadian context. Two final points are worth making. First, the fact that particular protests are about forestry is somewhat arbitrary, based on a group's geographic location. The conflicts could have just as easily been over other resources, such as water or fish. All of these conflicts have in common rights to control over territory and to self-determination. Second, part of the continued efficacy of the use of noninstitutional actions lies in their unpredictability. Those who engage in direct action as a means of drawing attention to their claims must use innovative and varied tactics. The most effective tactics are those that are either directly threatening to non-Native economic interests or that call into question Canada's international image as a tolerant nation. When and what form these events take in the future are questions worth pursuing.

Appendix 5.1

Chronology of key events

1968	St. Regis Mohawks block Seaway International Bridge, Canada-US border
1969	White Paper, cross-Canada protest
1970	St. Regis and Kahnawake people occupy Loon and Stanley Islands near Cornwall, Ontario
1971	Cold Lake, Saddle Lake, and Kehewin (Alberta) school boycott
1974	Native People's Caravan
	Bonaparte Reserve blockade
1976	Whitedog Reserve in Ontario – blockade over mercury poisoning
1981	Restigouche/Listuguj Mi'gmaq First Nation (Quebec) blockade over fishing
1982	Repatriation of Canadian constitution – protest across country
1984	Red People's Walk to Parliament Hill in Ottawa
1985	*Haida blockades over logging on Lyell Island*
	Conflict over Stein Valley – Lillooet/Lil'wat people
	Blockade of uranium mine by Lac La Hache First Nation in Saskatchewan
	Meares Island protest – Clayoquot and Ahousaht First Nations
	Innu begin protests and occupations of runway at Goose Bay military base (low-level flights)
1987	Peguis First Nation invites South African ambassador Glenn Babb to tour reserve
1988	*Lubicon boycott of Olympic art display, Daishowa paper company road blockade*
	Bear Island First Nation block logging road in Ontario
1990	Standoff at Oka, Quebec, between Mohawk warriors and Canadian army – death of Marcel Lemay (Sûreté du Québec officer)
	Numerous support blockades set up (especially in British Columbia)
	Peigan Lonefighters (Alberta) standoff with RCMP over construction of dam
1992	Cree participation in Earth Day to protest James Bay hydroelectric plans
1994	National protests against tax legislation

▶

◄ *Appendix 5.1*

Chronology of key events

1995	Kettle and Stony Point First Nation occupies Ipperwash Provincial Park – Anthony Dudley George (unarmed) is killed by Ontario Provincial Police
	Nuxalk Nation in British Columbia sets up roadblock to stop logging by International Forest Products
	Gustafsen Lake standoff
1997	*First Nations in New Brunswick log and blockade Crown land*
2005	*Haida people stop shipments of logs from Haida Gwaii*
2006	Six Nations people occupy Douglas Creek Estates in Caledonia, Ontario

Note: Forestry events in italic.

Notes

1 The term "Lillooet" is alternatively used to indicate a people (the Lillooet/St'at'imc people), a tribal council (the Lillooet/St'at'imc Tribal Council), and an individual First Nation (Lillooet First Nation/T'ít'q'et First Nation) that is a constituent member of these larger groups. The particular use of the term is specified throughout the chapter.

2 In early 1969, the Mohawks of St. Regis/Akwesasne had blocked the Seaway International Bridge at the Canada–US border to protest actions of Canadian customs agents who were charging duty on goods that they brought from the US side (*Akwesasne Notes,* February 10 and 11, 1969).

3 In 1990, Piikani Lonefighters engaged in a standoff with the RCMP over the construction of the Oldman Dam, which they said would lead to environmental damage and flood sacred land (*Windspeaker,* September 14, 1990, 1-2, 8, 11). The following year, the Peigan Nation founded the Indian Water Rights Coalition – an international (Canada–US) group whose purpose is to support Native water rights (Notzke 1994, 25).

4 The Innu include people from the communities of Davis Inlet, Sheshatshit, St. Augustin, La Romaine, Natashquan, Mingan, Matimekush-Lac John, and Kawawachikamach. Innu are non-status Aboriginal people.

5 The Cree have a long history of action dating back to the early 1970s, when they opposed Hydro-Québec's plan to construct an enormous dam project on several rivers (Eastmain, Caniapiscau, and La Grande) located in northern Quebec (see Churchill 1993; McCutcheon 1991; Niezen 1998; Saganesh 1997).

References

Alfred, Taiaike, and Lana Lowe. 2005. "Warrior Societies in Contemporary Indigenous Communities." Report to the Ipperwash Inquiry. 65 pages.

Armitage, Peter, and John C. Kennedy. 1989. "Redbaiting and Racism on Our Frontier: Military Expansion in Labrador and Quebec." *Canadian Review of Sociology and Anthropology* 26:798-817.

Blomley, Nicholas. 1996. "'Shut the Province Down': First Nations Blockades in British Columbia, 1984-1995." *BC Studies* 111:5-35.

Booth, Annie, and Norm Skelton. 2006. "A Case Study of Community Forestry in the Tl'azt'en Nation." Unpublished paper.

Borrows, John. 2005. *Crown and Aboriginal Occupations of Land: A History and Comparison.* Report to the Ipperwash Inquiry.

Churchill, Ward. 1993. *Struggle for the Land: Indigenous Resistance to Genocide, Ecocide and Expropriation in Contemporary North America*. Monroe, ME: Common Courage Press.

Coates, Ken. 2001. "Divided Past, Common Future." In *Prospering Together: The Economic Impact of the Aboriginal Title Settlements in BC,* edited by Roslyn Cunin, 1-35. Vancouver: Laurier Institution.

Comeau, P, and Santin, A. 1990. *The First Canadians: A Profile of Canada's Native People Today.* Toronto, ON: Lorimer.
Edwards, Peter. 2001. *One Dead Indian: The Premier, the Police and the Ipperwash Crisis.* Toronto: McClelland and Stewart.
Ferreira, Darlene. 1992. "Oil and Lubicons Don't Mix: A Land Claim in Northern Alberta in Historical Perspective." *Canadian Journal of Native Studies* 12:1-35.
Goddard, John. 1991. *Last Stand of the Lubicon Cree.* Vancouver: Douglas and McIntyre.
Grenier, M. 1994. "Native Indians in the English-Canadian Press: The Case of the 'Oka Crisis.'" *Media, Culture and Society* 16:313-36.
Henry, Frances, and Carol Tator. 2002. *Discourses of Domination: Racial Bias in the Canadian English-Language Press.* Toronto: University of Toronto Press.
Kalant, Amelia. 2004. *National Identity and the Conflict at Oka: Native Belonging and Myths of Postcolonial Nationhood in Canada.* New York: Routledge.
Lambertus, Sandra. 2004. *Wartime Images, Peacetime Wounds: The Media and the Gustafsen Lake Standoff.* Toronto: University of Toronto Press.
Long, David. 1997. "Culture, Ideology, and Militancy: The Movement of Native Indians in Canada, 1969-91." In *Organizing Dissent: Contemporary Social Movements in Theory and Practice,* edited by W. K. Carroll, 118. Toronto: Garamond Press.
McCutcheon, Sean. 1991. *Electric Rivers: The Story of the James Bay Project.* Montreal: Black Rose Books.
Miller, John. 2005. *Ipperwash and the Media: A Critical Analysis of How the Story Was Covered.* Report submitted to Aboriginal Legal Services of Toronto.
Muckle, Robert J. 1998. *The First Nations of British Columbia: An Anthropological Survey.* Vancouver: UBC Press.
Niezen, Ronald. 1998. *Defending the Land: Sovereignty and Forest Life in James Bay Cree Society.* Boston: Allyn and Bacon.
Notzke, Claudia. 1994. *Aboriginal Peoples and Natural Resources in Canada.* Toronto: Captus University Press.
Ramos, Howard. 2000. "National Recognition without a State: Cree Nationalism 'within' Canada." *Nationalism and Ethnic Politics* 6:95-115.
–. 2006. "What Causes Canadian Aboriginal Protest? Examining Resources, Opportunities and Identity, 1951-2000." *Canadian Journal of Sociology* 31:211-34.
Saganash, Diom Romeo. 1997. "James Bay II: A Call to Action." In *Justice for Natives: Searching for Common Ground,* edited by Andrea Morrison, 204-7. Montreal and Kingston: McGill-Queen's University Press.
Sanders, Douglas E. 1985. "The Indian Lobby and the Canadian Constitution, 1978-82." In *Indigenous Peoples and the Nation-States: Fourth World Politics in Canada, Australia and Norway,* edited by Noel Dyck, 151-89. St. John's: Institute of Social and Economic Research, Memorial University of Newfoundland.
Skea, W.H. 1993. "The Canadian Newspaper Industry's Portrayal of the Oka Crisis. *Native Studies Review* 9:15-31.
Smith, Heather. 2000. "The Mohawk Warrior: Reappropriating the Colonial Stereotype." *Topia: A Canadian Journal of Cultural Studies* 3:58-83.
Stuart, C. 1993. "The Mohawk Crisis: A Crisis of Hegemony: An Analysis of Media Discourse." MA thesis, University of Ottawa.
Tait, Patricia. 2007. "Systems of Conflict Resolution within First Nations Communities: Honouring The Elders, Honouring the Knowledge." National Centre for First Nations Governance, Vancouver, British Columbia.
Tilly, Charles. 1978. *From Mobilization to Revolution.* Reading, MA: Addison-Wesley.
Wadden, Marie. 1996. *Nitassinan: The Innu Struggle to Reclaim Their Homeland.* Vancouver: Douglas and McIntyre.
Wilkes, Rima. 2003. "First Nation Politics in Canada: A Study of Group Mobilization." American Sociological Association Conference. Atlanta, GA.
–. 2004. "First Nation Politics: Deprivation, Resources and Participation in Collective Action." *Sociological Inquiry* 74:570-89.

–. 2006. "The Protest Actions of Indigenous Peoples: A Canada-US Comparison of Social Movement Emergence." *American Behavioral Scientist* 50:510-25.

York, Geoffrey. 1989. *The Dispossessed: Life and Death in Native Canada*. Toronto: Lester and Orpen Dennys.

York, Geoffrey, and Loreen Pindera. 1991. *People of the Pines: The Warriors and the Legacy of Oka*. Toronto: Little, Brown.

Part 3
Differing Visions

6

Natural Resource Co-Management with Aboriginal Peoples in Canada

Coexistence or Assimilation?

M.A. (Peggy) Smith

In Canada, co-management has evolved as one of the most promising institutional arrangements to address the unresolved issue of Aboriginal peoples' involvement in natural resource development and management.[1] Given that resource management systems remain tightly controlled by the state, that Canadian court decisions increasingly direct the state to accommodate Aboriginal rights in natural resource development, and that Aboriginal peoples insist more vehemently that those rights be recognized, the problem is becoming more acute. Co-management has the potential to bring Aboriginal peoples and the state together to negotiate a common understanding of Aboriginal rights in relation to resource management and from there to work cooperatively toward sustainable resource management.

The negotiation of effective co-management regimes will require the state to recognize Aboriginal rights to lands and resources, including the right of self-determination equal to that of the state. Without this recognition, co-management becomes one more tool in the continued colonization by the state of Aboriginal peoples. This chapter explores the potential for co-management to become a tool for decolonization, offering a model of coexistence, rather than assimilation. This is premised on the theory that successful Aboriginal–state co-management regimes for sustainable resource management are more likely when the principle of the recognition of Aboriginal rights forms the basis for negotiated agreements. Integral to this approach is that such agreements are based on a shared understanding and incorporation of the aspirations both of the state and Aboriginal peoples related to the meaning and direction of natural resource management. Current co-management models are evaluated using the assimilation-coexistence distinction.

The Royal Commission on Aboriginal Peoples (RCAP 1996a, Volume 2, 27) referred to the renewal of the relationship between Aboriginal peoples and the Canadian state as "reconciliation" or "rapprochement," with a focus on reviving, revitalizing, and reintroducing Aboriginal values and concepts around natural resource stewardship. RCAP (1996a, 130) recommended a contemporary proclamation setting out the principles of a new relationship and outlining "the laws and institutions necessary to turn those principles into reality." The Royal Proclamation of 1763, issued by the British Crown, recognized Aboriginal peoples in Canada as sovereign nations and instructed agents of the Crown to enter into agreements seeking Aboriginal peoples' consent before any land was taken up for settlement or development (Borrows 1997). The proclamation formed the basis of historic treaties with Aboriginal peoples and continues to guide the state's direction in modern-day treaty making. RCAP (1996a, Highlights, 133) specified what should be in a new proclamation: "Principles to guide redistribution of lands and resources, the general scope of Aboriginal governments' core jurisdiction, principles of intergovernmental fiscal arrangements, *principles of co-management on public lands* and the character of interim relief agreements" (emphasis added).[2] Although the federal government has largely ignored RCAP recommendations, RCAP's examination of Aboriginal issues in Canada is the most comprehensive to date, with its recommendations still being explored and debated within Canada.

From among the variety of and often confusing definitions of co-management (Carlsson and Berkes 2005), this chapter adopts a definition that focuses on Aboriginal–state relationships incorporating the recognition of rights. McCay and Acheson (1987, 31, 32) defined co-management as Aboriginal communities' "political claim to the right to share management power and responsibility with the state ... an attempt to formalize a *de facto* situation of mutual dependence and interaction in resource management," recognizing co-management as "a challenge to the legitimacy of the state, especially in the context of widespread loss of confidence in it as the steward of common property resources." This definition lends itself to the Canadian situation in which Aboriginal peoples hold a unique historical, legal, and cultural position. Aboriginal peoples cannot be considered as a "user group"; they are "not just another stakeholder" (Smith and Bombay 1995; Stevenson and Webb 2003).

Not all Aboriginal communities or academics studying co-management, and certainly not the state, accept or pursue a rights-based approach to co-management. Institutional analysts now focus on co-management as a "learning by doing" approach or "adaptive co-management," where partnerships of stakeholders address situations of complexity and uncertainty (Berkes 2009). However, it has been noted that co-management in general is "weak in poverty reduction and empowerment of the marginalized" (Béné and Neiland 2004). Some Aboriginal communities have pointed out that a rights-based approach, especially when pursued through the legal system, can undermine Aboriginal peoples' traditional governance systems and culture (Daly and Napoleon 2003), steering Aboriginal communities away from self-determination. Napoleon responds to Daly that "the law can be a useful tool for Aboriginal communities, but only if its role is subsidiary to other processes, such as political and social visions, goals, and activism" (Daly and Napolean 2003, 124). Some Aboriginal communities continue negotiations with the state outside the rights framework, preferring to focus on more pragmatic approaches based on economic development or "economic renewal," as Pikangikum describes it (Pikangikum First Nation 2008).

So why take a rights-based approach to co-management? There are two reasons. The first is that we all share the land and together must be responsible for and benefit from its stewardship and development – land and resources are the ties that bind. Second, the rights-based approach offers an acknowledgement of the unique place of Aboriginal peoples within Canada enshrined in the Canadian constitution. Even if not embraced by all Aboriginal communities, the upholding of the Canadian constitution, and its recognition of the unique place and rights of Aboriginal peoples within Canada, should be the trigger that brings the state to the negotiating table, thus creating the space for Aboriginal peoples to assert their aspirations for self-determination within the Canadian confederation (Fleras and Elliott 1992, 12-38).

Notzke (1995a) provided one of the first overviews of co-management with Aboriginal peoples in Canada, addressing the issue of rights and exploring a range of catalysts and functions for co-management regimes, categorized as follows: a result of comprehensive claims settlements (regional agreements); a means of crisis resolution; a result of legal decisions on rights; applying to co-management of national parks; strategic co-management (attributed to Elias [1995] and referring to use of environmental and socioeconomic impact assessment); and a constitutional right. As Howitt (2001, 374) points out, based on Notzke's approach to co-management, it is time to "move discussion about co-management from dealing with indigenous participation in co-management regimes as a *concession to government,* towards dealing with it as a *constitutional (or indigenous sovereign) right*" (emphasis added). It is the rights aspect of co-management that requires further exploration and attention.

Assimilation versus Coexistence

Central to Aboriginal peoples' self-definition is their connection to the land. The alienation of Aboriginal peoples from the land as a result of colonization began in the sixteenth century. Initially, relationships between Aboriginal peoples and the colonizers were cooperative, based on a mutual need for the success of both the fur trade and military alliances. However, from the mid-1700s to present day, government policy has tended toward assimilation, one dimension of what Tully (2000, 39) describes as a process of "internal colonization":

> This form of colonization is "internal" as opposed to "external" because the colonizing society is built on the territories of the formerly free, and now colonized, peoples ... The essence of internal colonization, therefore, is ... the appropriation of land, resources and jurisdiction of the indigenous peoples, not only for the sake of resettlement and exploitation (which is also true in external colonization), but for the territorial foundation of the dominant society itself.

A new economy emerged that was incompatible with Aboriginal peoples' ways of life and connection with the land. The capitalist economy required state and private ownership of property, and the state became an enforcer of those property rights, engaging in "the process of alienating land from Aboriginal peoples as a collective group, transferring land to state control" (Mann 2003, 22). As part of the process of assimilation, Aboriginal peoples were removed to small land bases, or reserves, where it was the intention of the state to protect and civilize them in preparation for their eventual assimilation into the broader society "through education, religion and agriculture" (Richardson 1993, 53). McMillan (1988, 291) describes this dual state policy:

> Assimilation and paternalism have been the foundations of Canadian Indian policy. From the beginning, the goal was to protect Indians while attempting to "civilize" them and prepare them to enter mainstream society. Native populations were declining throughout the late nineteenth and early twentieth centuries, and the government plan was to encourage the gradual disappearance of Indians as Indians. The twin pillars of Indian policy conflicted, however, since the paternalistic protection of the Indian Act served to isolate Indians from the rest of the society they were expected to join.

In spite of assimilationist policies, Aboriginal peoples have not disappeared and have continually resisted state control of lands and resources, maintaining their assertion that they have collective rights to their traditional territories, namely those lands deemed by the state to be Crown or public land. The "Indian problem" in natural resources is at its core about finding a way to reconcile the conflicting goals of the state and Aboriginal peoples. One path is for the state to recognize and implement Aboriginal rights in a model of coexistence. "Coexistence" was espoused by RCAP (1996a, Highlights, 1):

> After some 500 years of a relationship that has swung from partnership to domination, from mutual respect and co-operation to paternalism and attempted assimilation, Canada must now work out fair and lasting terms of coexistence with Aboriginal people.

Coexistence implies diversity in cultural, social, economic, and/or political systems, with a conscious choice made to share and peacefully live side by side. In relation to nations, the verb "coexist" means "exist in mutual tolerance though professing different ideologies."[3] Mockus (2002, 21) elaborates that "coexistence means keeping common rules, having culturally rooted mechanisms of social self-regulation, respecting differences and complying with rules to process them; it is also learning to reach, comply with and amend agreements." Coexistence stands in stark contrast with "assimilation." "Assimilate" is defined as absorbing people "into a larger group, esp. by causing a minority culture to acquire the characteristics of the majority culture."[4]

In discussing the origins of co-management, RCAP (1996b, 27.3) observed that co-management as it exists today "represents a compromise between the Aboriginal objective of self-determination and governments' objective of retaining management authority," even while devolving that authority to third-party interests, such as private companies. Further, RCAP points out that "this compromise is not one between parties of equal power, however, and Aboriginal Peoples certainly regard co-management as an evolving institution." It is the contention of many Aboriginal groups that enter into co-management arrangements that they are not agreements for co-jurisdiction or coexistence based on recognition of Aboriginal rights and title, but simply a form of cooperation as an interim measure to more equal sharing of power (Treseder 2000).

Figure 6.1 illustrates the difference between the assimilation and coexistence approaches in relation to lands and resources, showing that under coexistence there is a nation-to-nation relationship with co-management resulting from a negotiation between equals, whereas under assimilation Aboriginal peoples are merely consulted on state-determined institutions, invited to participate by a state that does not recognize their sovereignty.

"Coexistence" implies agency or active influence involving choices to share responsibilities and benefits for a mutual purpose, whereas "assimilation" implies a giving up of influence and the domination of one party over another. The overarching concepts of coexistence or assimilation as a framework within which to assess co-management is an important one for Aboriginal communities that bring to the resource management arena what is often seen by representatives of the dominant society – be they politicians, bureaucrats, managers, planners, or scientists – as inferior knowledge and a lack of power and capacity. There is little recognition of the value of Aboriginal knowledge and that the lack of capacity and power is a result of colonial efforts to control Aboriginal lands and resources. Aboriginal peoples thus begin contemporary co-management negotiations and participation from a disadvantaged position. If mechanisms are not found to balance the power in the relationship, then assimilation is the inevitable result. "You cooperate,

Figure 6.1

Co-management as coexistence and as assimilation

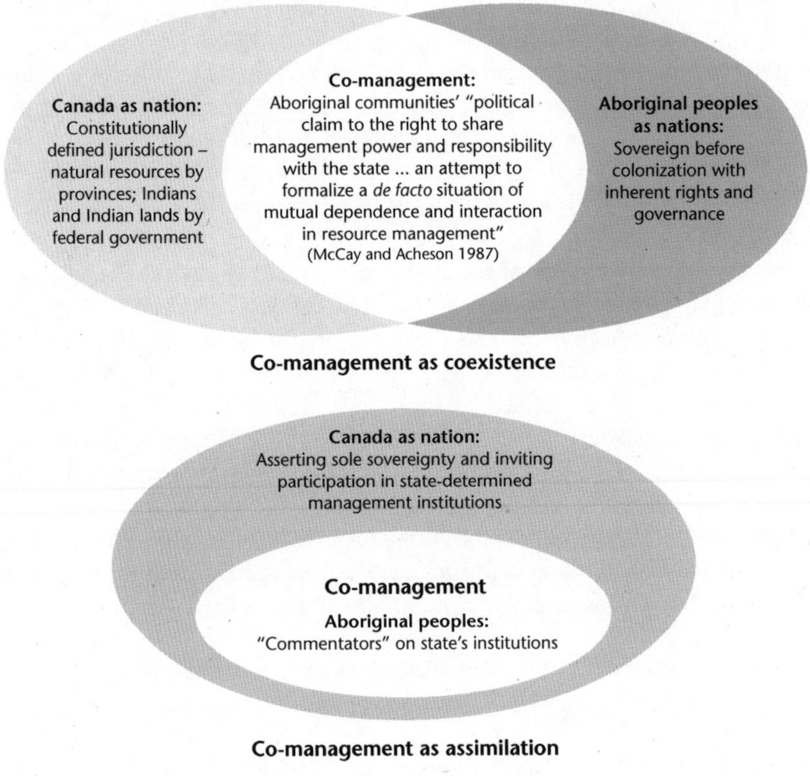

Canada as nation:
Constitutionally defined jurisdiction – natural resources by provinces; Indians and Indian lands by federal government

Co-management:
Aboriginal communities' "political claim to the right to share management power and responsibility with the state ... an attempt to formalize a *de facto* situation of mutual dependence and interaction in resource management"
(McCay and Acheson 1987)

Aboriginal peoples as nations:
Sovereign before colonization with inherent rights and governance

Co-management as coexistence

Canada as nation:
Asserting sole sovereignty and inviting participation in state-determined management institutions

Co-management

Aboriginal peoples:
"Commentators" on state's institutions

Co-management as assimilation

we'll manage" is one view of existing and proposed state–Aboriginal co-management schemes in Canada (Diabo, pers. comm. 2005).

Co-management as assimilation has its roots in the federal government's policy of assimilation proposed in the 1969 White Paper. Aboriginal people understood the policy as assimilationist because it denied Aboriginal and treaty rights (Rynard 2001). RCAP describes the doctrine of assimilation as being based on

> four dehumanizing (and incorrect) ideas about Aboriginal peoples and their cultures: that they were inferior peoples; that they were unable to govern themselves and that colonial and Canadian authorities knew best how to protect their interests and well-being; that the special relationship of respect and sharing enshrined in the treaties was an historical anomaly with no

> more force or meaning; and that European ideas about progress and development were self-evidently correct and could be imposed on Aboriginal people without reference to any other values and opinions – *let alone rights* – they might possess. (RCAP 1996a, 13; emphasis added)

The question of coexistence versus assimilation goes to the heart of why it is important to define co-management with Aboriginal peoples in Canada more narrowly and clearly. Although a difficult task, it is important to be clear about the criteria for assessing the effectiveness of co-management arrangements from both the state's and Aboriginal peoples' points of view or aspirations. Using coexistence versus assimilation as a point of analysis leads to an examination of whether there are mutually agreed-to terms based on a negotiated agreement at the highest level – the constitutional level – that recognize and affirm the rights of both parties – that is, as RCAP recommends, at the treaty level.

The coexistence concept is in keeping with other arguments for parallel development of Aboriginal and state management systems, with co-management being negotiated based on the exclusivity, distinctiveness, and mutual recognition of both systems, often termed the two-row wampum model (McGregor 2000; Ransom and Ettenger 2001; Stevenson and Webb 2003; Marc G. Stevenson, this volume). This concept of recognition is in keeping with international law as "the formal decision to establish a relationship with another government" (Barsh 2002, 232). Although there is a debate about the legal definition of "recognition" and its practice in Canada (Giokas 2002), the recognition by one government of the legitimacy of another reflects Aboriginal peoples' aspirations to be viewed as nations (Fleras and Elliott 1992, 24). As Tully (1995, 22) points out, "Recognition involves acknowledging in its own terms and traditions, as it wants to be and as it speaks to us." He sees this type of recognition as "the most important and difficult first step in contemporary constitutionalism" (ibid., 23).

State and Aboriginal Peoples' Aspirations

> We are the original people of this land and have the right to survive as distinct Peoples into the future;
> Each First Nation collectively maintains Title to the lands in its respective Traditional Territory;
> We have the rights to choose and determine the authority we wish to exercise through our Indian Governments;
> We have the right to exercise jurisdiction within our traditional territories to maintain our sacred connection to Mother Earth through prudent management and conservation of the resources for the economic survival and well-being of our citizens;

Figure 6.2

State's and Aboriginal peoples' aspirations in relation to natural resources

State's (provinces') aspirations	Aboriginal peoples' aspirations
Provide certainty for resource development	Seek recognition of and exercise Aboriginal and treaty rights
Carry out constitutional responsibility to sustainably manage resources in public interest through regulation, enforcement, and monitoring	Exercise responsibility for stewardship of natural environment
Implement regulations through management planning or devolve/decentralize responsibility	Assert inherent right to self-government, including jurisdiction over lands and resources
Ensure economic sustainability from resource development	Ensure community well-being by increasing share of economic benefits from resource development
Comply with court decisions to uphold Aboriginal rights based on consultation and conservation	Ensure community well-being by also preserving way of life, including values, knowledge, and connection to land
Develop shared understanding as basis for negotiating co-management agreements	

> Only through a process of informed consent may our governing powers of our land be shared. (UBCIC 2004, 1)

There may be many triggers for the state and Aboriginal peoples to consider co-management. A perceived ecological crisis, such as a shortage in stocks or degradation of environmental quality, is often cited (Pinkerton 1989), as are conflict (Castro and Nielsen 2001), court decisions, economic crises, or even policy initiatives (Notzke 1995a). However, these elements are only triggers or catalysts and do not always reveal the underlying aspirations of the parties. In a rights-based approach, the aspirations of Aboriginal peoples must be fully understood in order to assess ongoing satisfaction with co-management arrangements. At the same time, as shown in Figure 6.2, the state may have very different underlying goals and objectives, and these too must be understood, both to develop a shared understanding and to provide a basis for ongoing evaluation of the arrangements.

If the aspirations of only one party are used to design and evaluate co-management arrangements – and more often than not it is the state's aspirations that dominate – then Aboriginal peoples' aspirations will continue to be marginalized, ignored, or diminished. As can be seen in Figure 6.2, the state's aspirations focus on resource development and capture a frontier mentality.[5] In the bigger picture, this mentality has not included recognizing Aboriginal rights. Harman (1981 in Howitt 2001, 63) expands on the ideology behind the state's aspirations:

- Expanding the number of jobs and the level of income for citizens;
- Expanding settlement and extending civilized control (of both uncivilized Aboriginal peoples and undisciplined workers) to facilitate development;
- Incorporating the settler population as part of the destiny of building a new state, thus protecting and extending the state's rights;
- Contributing to nation building;
- Contributing to development in the underdeveloped world through the provision of resource commodities;
- Protecting capitalism and democracy; and
- Advancing the cause of civilization.

Even when the recognition of Aboriginal and treaty rights becomes part of the state's mandate, as increasingly directed by the courts in Canada, the state's tendency is to limit those rights within the context of its own aspirations. Therefore, a state interpretation of Aboriginal rights becomes a limited view of those rights – "a rigid, narrow, and literal interpretation," according to Fleras and Elliott (1992), rather than the full expression of Aboriginal peoples' views. As Elias (1995, viii) states,

> Since the late 1960s, aboriginal people have advocated a comprehensive approach to development that encourages simultaneous progress toward political, cultural, and economic goals ... Shared concerns for rights and territory, tradition, and well-being permeate comprehensive approaches to development, and make the entire construct uniquely aboriginal and, perhaps, Canadian.

Allowing both state and Aboriginal aspirations a full hearing in designing and evaluating co-management becomes essential before a shared understanding can be developed. Reconciling aspirations will help to avoid assimilation or "the co-optation or further marginalization of local interests" (Castro and Nielsen 2001, 236).

An Assessment of Some Current Co-Management Models Using a Rights-Based Approach

Contrary to what the Government of British Columbia contends, long-term economic certainty can be achieved only through the full recognition and accommodation of Aboriginal title interests, not by hawking cheap accommodation agreements designed to buy temporary peace in the woods, on the waters, and out in the oil patch (UBCIC 2004).

Across Canada there is a range of Aboriginal–state arrangements providing different contexts for the negotiation of co-management regimes. Table 6.1 summarizes these arrangements into two broad categories, depending on whether the Aboriginal right is based on Aboriginal title or treaty, though some First Nations argue that Aboriginal title cannot be extinguished through agreements with the Crown (UBCIC, n.d.). Aboriginal title applies to areas where no formal Aboriginal–state agreements have been negotiated. Title areas are subdivided into two categories: areas where claims have been filed and negotiations are underway under the federal Comprehensive Claims Policy (INAC 2003a) and areas where no claims have been filed. The treaty rights category is subdivided into historic and modern treaties, with modern treaties sometimes referred to as "regional agreements" (Richardson, Craig, and Boer 1994). These agreements were settled under the federal Comprehensive Claims Policy, beginning with the James Bay and Northern Quebec Agreement in 1975 and ending with the most recently settled – eight Yukon First Nations that signed individual agreements between 1995 and 2002 under the 1993 Council for Yukon Indians Umbrella Final Agreement and the 1999 Nisga'a Final Agreement. Each of these categories are briefly examined below to assess how rights are addressed within the various arrangements and whether the rights recognition is strong enough to support co-management as coexistence.

Although these categories are somewhat similar to those set out by RCAP (1996b, 7.3) – claims-based, crisis-based, and community-based – and modified by Notzke (1995a) and Shuter and Kant (2003) in their typologies, the approach taken here is distinguished by its focus on the recognition of Aboriginal rights. Arrangements examined exclude agreements in which Aboriginal peoples have entered into arrangements with the state strictly under state-determined institutions, such as those where Aboriginal groups have been granted state licences to harvest and manage resources and by virtue of those agreements are required to follow federal or provincial regulations (e.g., Meadow Lake Tribal Council's Forest Management License Agreement in Saskatchewan, Tl'azt'en First Nation's Tree Farm Licence in British Columbia, fisheries agreements in the Maritimes signed with the federal Department of Fisheries and Oceans, timber allocations to First Nations in New Brunswick, community forest pilot agreements in British Columbia).

Aboriginal Title

The Supreme Court of Canada's 1997 decision in *Delgamuukw* affirmed that the provincial Crown could not unilaterally extinguish Aboriginal title, stating that this power rests solely with the federal Crown.[6] In Canada, there are now only a few areas where agreements or treaties with the Crown have not been negotiated. In the case of resource development carried out under provincial authority, the *Delgamuukw* decision requires that where Aboriginal title exists the provincial Crown must justify any infringement to that title. Aboriginal title implies exclusive use and occupation of the land and the right to choose how to use that land based on proven historical practices. The Crown's justification carries with it a duty to consult the affected Aboriginal nation (Sharvit, Robinson, and Ross 1999), both to understand Aboriginal land use and, based on that knowledge, the extent of any potential infringement.

Aboriginal title is recognized in those provinces in which Aboriginal groups have not entered into formal agreements with the Crown. Aboriginal title is outstanding in the greater part of British Columbia, with the Innu in Quebec and Labrador, with the Algonquin of Barriere Lake in Quebec, and in isolated cases across Canada, including the Algonquin of Golden Lake and the Teme-Augama Anishnabai in Ontario: in most of these cases, the Aboriginal groups have filed claims under the federal Comprehensive Claims Policy (INAC 2003a). While these claims are being negotiated, many of these groups enter into interim measures agreements in the face of resource development to ensure that their claims to lands and resources are protected during the negotiation process. In these situations, there is more uncertainty, creativity, tension, and complexity, which put the Aboriginal party in a stronger negotiating position. In these cases, interim measures agreements may provide stronger protection for the Aboriginal party than a final agreement.

The Nuu-chah-nulth in Clayoquot Sound, British Columbia, offer a good illustration of how an interim agreement between the state and an Aboriginal people can provide the foundation for innovation. The Nuu-chah-nulth entered into a treaty-making process with the federal government in the 1970s. In the 1980s and 1990s, conflicts arose in the Clayoquot Sound region, between the BC government and private logging groups and the Nuu-chah-nulth and environmental groups opposed to logging practices, resulting in logging blockades and arrests in 1993 of over eight hundred people. Eventually, the provincial government established the Scientific Panel for Sustainable Forest Practices in Clayoquot Sound, which made recommendations to change to ecosystem-based timber harvesting. The Nuu-chah-nulth participated in this panel as "experts," based on their traditional ecological knowledge. In 1994, with the support of environmental groups, the Nuu-chah-nulth negotiated an interim measures agreement with the Province of British Columbia, a co-management agreement in essence, giving them a decision-making voice in land and resource planning in the Clayoquot Sound region. This co-management agreement facilitated the joint venture between

Table 6.1

Current Aboriginal rights framework and state policies in Canada

Aboriginal rights	Negotiations with the state	Nations and geographic area	State policy
Aboriginal title (no formal agreements with the state)	Land claim filed under federal Comprehensive Claims Policy	Innu of Labrador; Teme-Augama Anishnabai and Algonquin of Golden Lake, Ontario; Naskapi, Quebec; greater part of British Columbia (Nuu-chah-nulth, Haida)	The Federal Comprehensive Claims Policy is controversial because of title extinguishment provisions. In British Columbia, interim measures agreements negotiated while modern treaty-making process continues.[1]
	No claims filed under federal Comprehensive Claims Policy; trilateral agreements	Union of BC Indian Chiefs Algonquins of Barriere Lake	Some First Nations refuse to enter into agreements with the Crown because of the extinguishment provisions of the claims policy, preferring to exercise their inherent rights (UBCIC 2004), or, in the case of the Algonquins of Barriere Lake, negotiate a trilateral agreement outside the federal claims process (Notzke 1995b).
Treaty rights	Modern treaties/ land claims	James Bay and Northern Quebec Agreement (1975); Inuvialuit Final Agreement (1984); Gwich'in Comprehensive Land Claim Agreement (1992); Nunavut Land Claims Agreement (1993); Yukon Umbrella Final Agreement (1993); Sahtu Dene and Métis Comprehensive Land Claim Agreement (1994); Nisga'a Final Agreement (2000)	All of these agreements were negotiated under the federal Comprehensive Claims Policy, in which Aboriginal title is extinguished and nation consents to the limitations of its rights as set out in the agreement.

Historic treaties	Peace and Friendship, Atlantic Canada, seventeenth-eighteenth century	No land surrender. Treaties ignored by provinces until recent court decisions *(Paul, Marshall)*. Province: 5 percent timber allocation to First Nations in New Brunswick; federal Fisheries and Oceans: fisheries quota allocations to First Nations.
	Robinson and numbered treaties 1-11 (1850s-1920s): Ontario, Manitoba, Saskatchewan, Alberta, northeastern British Columbia, Vancouver Island, Northwest Territories, southeastern Yukon	Treaty rights ignored by provinces; interpreted to have ceded all lands in exchange for small reserve allocations under federal ownership. In Prairie provinces, treaty land entitlement being negotiated. More recently, forest management addressing state's limited view of treaty rights: hunting, fishing, trapping, and gathering, enshrined in the CCFM's framework of criteria and indicators for sustainable forest management.

1 According to the BC Treaty Commission, as of 2009 there were sixty First Nations in British Columbia participating in the treaty process. See http://www.bctreaty.net/ for a full list.

the Nuu-chah-nulth and Weyerhaeuser in which they adopted modified forest management practices recommended by the Clayoquot Sound Science Panel (Ross and Smith 2002, 27-30; see also Mabee et al. and Pechlaner and Tindall, this volume). Their treaty with the Crown is still under negotiation (Iisaak 2000).

In a few cases, Aboriginal nations have refused to enter into claims negotiations with the Crown, preferring to exercise what are called "inherent rights," rights that existed prior to colonization and that do not require state recognition (UBCIC 1998-2004). In either situation, whether Aboriginal nations are negotiating with the provincial and federal governments or whether they are boycotting such negotiations, the recognition of their Aboriginal title by the Canadian courts has put them in a strong bargaining position. The exercise of inherent rights is strongest in British Columbia. Several First Nations have taken action described as "a determined assertion of self-regulation" (Notzke 1995a, 202). The Haida in 1990 asserted their control over the sports fishery, a step that two years later led to a joint stewardship agreement with the Province of British Columbia. Then the Haida asserted their authority in the establishment of a national park in their territory, which led to the Government of Canada passing legislation that acknowledged "a parallel statement of ownership" (ibid., 207).

The Algonquin of Barriere Lake (ABL) chose to negotiate a trilateral agreement with the federal government and the Province of Quebec outside the federal Comprehensive Claims Policy, based on their understanding that the claims policy would extinguish their Aboriginal title. Notzke (1995b, 6) states that the ABL trilateral agreement is

> not a co-management agreement in the sense that it immediately effects [sic] the establishment of co-management institutions and co-management procedures, concerned with the joint management of a particular species or area. Rather it is designed to lay the groundwork for the cooperative development of an integrated resource management plan for a region comprising 1 million hectares, the major portion of the traditional use area of the Algonquins of Barriere Lake.

Although the integrated resource management plan has yet to be completed, the agreement does lay the basis for sharing rights and responsibilities in resource management and blending Algonquin values and knowledge with the provincial management system (Notzke 1995b, 77).

Treaty Rights

In the treaty rights category, both historic treaties, triggered by the Royal Proclamation of 1763, and modern-day treaties, negotiated under the federal

Comprehensive Claims Policy, are included. Although each agreement negotiated is unique, treaties, whether historic or modern, have common elements. In relation to the recognition of rights, all treaties, from the state's point of view, limit and define narrowly Aboriginal rights, though this perspective may not be shared by Aboriginal signatories. Common to Aboriginal peoples' understanding is that the spirit and intent of treaties were to share lands and resources, not cede Aboriginal title to land or give up access to resources (Treaty 7 Elders and Tribal Council 1996). This difference in viewpoints leads to ongoing debate, and sometimes conflict and legal challenges in attempts to have the courts hear and recognize Aboriginal peoples' understanding of these agreements.

Comprehensive Claims

The federal Comprehensive Claims Policy, formulated in 1973, was used in the negotiation of the first modern-day treaty in Canada – the 1975 James Bay and Northern Quebec Agreement (JBNQA) – and has remained the preferred policy tool of the state for addressing Aboriginal rights and title issues. From the beginning, the policy was controversial because of the federal government's insistence that settlements must include the extinguishment of Aboriginal rights not specifically defined in the treaty. The Task Force to Review Comprehensive Claims Policy addressed the government's "refusal to include political rights, decision-making power on land and resource management boards, revenue sharing from surface and subsurface resources, and offshore rights in the negotiations" (INAC 1985, 13) and made recommendations to address some of these shortcomings. The Comprehensive Claims Policy was revised in 1986. However, the changes did not allay Aboriginal peoples' concerns, even though the federal government claimed that there was an improvement in the negotiation process, greater flexibility in land tenure, and a clearer definition of topics for negotiation (INAC 2003a).

The criticism that the Comprehensive Claims Policy extinguishes Aboriginal title and rights continues within the Aboriginal community. The criticism is focused squarely on the issue of Aboriginal rights, claiming that the policy denies the existence of Aboriginal title, requires "a complete surrender and extinguishment of any of the rights that 'may exist'" (Algonquin Nation Secretariat 1999, 4), and does not recognize Aboriginal ownership in resource-revenue-sharing components, arbitrarily capping the Aboriginal portion of revenues and denying that Aboriginal peoples own subsurface resources (AFN 1990a, 1990b; Venne 1997).

Much of the co-management literature in Canada has focused on regimes enabled by comprehensive land-claim settlements, with many authors contending that these agreements have led to sound co-management arrangements (Notzke 1995c; RCAP 1996b, 7.3; Natcher 2001; Campbell 2003).

However, other authors see co-management under these agreements as political compromises that lead to a "range of systemic institutional and economic barriers," in fact inhibiting Aboriginal self-determination and self-reliance (Mulvihill and Jacobs 1991, 36). Rynard (2000, 2001) critically examined the JBNQA and compared it with the most recently settled claim, the Nisga'a Final Agreement, in relation to land and resource rights and progress made on the recognition of these rights in Canada since 1975. He concludes:

> First Nations leaders have consistently said that the uniqueness of their peoples is bound up with their relationship to their lands. Yet it is precisely this relationship which is contained and redefined by the certainty or extinguishment provisions, combined with the land regimes, of the Nisga'a and James Bay agreements ... In ongoing and future negotiations federal and provincial governments must end this approach. If they do not, the reconciliation between First Nations and non-Aboriginal Canadians envisioned by the Royal Commission on Aboriginal Peoples will likely not be possible because agreements and treaties will either be unreachable or be tainted as coerced surrenders. (Rynard 2000, 240-41)

Given the Aboriginal rights extinguishment provisions of the Comprehensive Claims Policy, the co-management regimes enabled by the agreements settled under the policy are structurally flawed, with Aboriginal signatories in an advisory role only, subject to the provincial, territorial, and federal laws of general application (Usher, Tough, and Galois 1992) and therefore the final decision of government ministers (Smith 1991; Notzke 1995a). In a coexistence model, Aboriginal rights aspirations would be recognized and respected.

Historic Treaties

If modern treaties are controversial because of the state's extinguishment of Aboriginal title, historic treaties are even more so, revealing conflicting views about their spirit and intent. The Peace and Friendship treaties signed in Atlantic Canada in the seventeenth and eighteenth centuries did not involve "land transfers, annuities, trading rights or compensation for rights limited or taken away" (Frideres and Gadacz 2001, 169), but in some cases provided for rights to harvest natural resources for commercial purposes, rights that have largely been ignored until recent court cases like the *Marshall* decision (McDonald 2003). Land transfers became part of later treaty making, particularly the numbered treaties modelled after the Robinson treaties, signed in the mid-1800s in northern Ontario. These treaties contain language defining what Aboriginal peoples' rights to resources include, usually a continuation of traditional activities such as hunting, fishing, trapping, and gathering. In exchange, the Crown elicited what it contended were concessions to land:

> The said chiefs and principal men do freely, fully and voluntarily surrender, cede, grant and convey unto Her Majesty, Her heirs and successors forever, all their right, title and interest in the whole of the territory above described. (INAC 2003b)

These historic treaties cover most of Ontario, the Prairie provinces, a small portion of Vancouver Island, northeastern British Columbia, Yukon, and the Northwest Territories.

For the state, these historic treaties were based on a recognition of Aboriginal nations as sovereign at the time of contact, as outlined in the Royal Proclamation of 1763. The treaties were designed to clearly define Aboriginal peoples' rights, limit their access to lands and resources, and pave the way for settlement. However, for Aboriginal nations, the treaties were considered to be a means to preserve their way of life and to formalize a sharing of lands and resources with the colonists. Pratt (1995, 2) describes this clash of perspectives:

> To Indian Treaty Nations, the land is not something that can be sold or traded. The exchange of rights described in the written Treaty text is one that is literally inconceivable in Aboriginal cultures. The essential difference may be described as a clash of world-views. One view regards Aboriginal title as being comprised of property rights, and thus capable of sale, exchange or extinguishment assuming equivalent value is provided. The other view regards them as fundamental human rights which form the foundation of their very being, and are thus inherently inalienable.

Debate continues about the application of treaty rights in natural resource management. Several legal decisions have given credence to Aboriginal interpretations of these treaties, contributing legal uncertainty about the state's role in upholding these treaties in natural resource management. These decisions have found similarities between Aboriginal and treaty rights, concluding that in the same way that the infringement of Aboriginal title must be justified by the state, including provincial governments, so must any infringement of treaty rights. Treaty rights are also unique because they bestow on the Crown a fiduciary obligation to Aboriginal nations; involve the honour of the Crown in their fulfillment; must be interpreted broadly, taking into account the understanding of the Indians at the time of signing; and are constitutionally protected. Development decisions in historic treaty areas therefore impose on the state the same duty to consult that was stipulated by *Delgamuukw* (Pratt 1995; McDonald 2003). The Supreme Court of Canada decision in *Mikisew Cree* in 2005 confirmed this duty (Lawson Lundell 2005). In spite of these court decisions indicating the duty to consult applies in historic treaty areas, provincial governments in historic treaty areas insist that they have no obligations to uphold treaty rights (OMNR 2002, 3-120). In this climate, co-management is very difficult. Campbell (2003, 51) describes this web of constitutional confusion:

> The guarantee of hunting, trapping, fishing, and gathering rights protected by treaty can be seen as trapped within a vicious cycle of government policy. Provincial governments view treaty matters as a federal responsibility, while the federal government sees jurisdiction over natural resources, either as a result of the Natural Resources Transfer Agreement, 1930 or through the Terms of the Union, as a provincial responsibility. First Nation people have concerns about exercising their treaty rights on provincially regulated Crown land when treaty rights are a federal matter. Further, it is extremely difficult to exercise federal treaty rights on provincial Crown land when neither government level has any desire to step over the carefully delineated lines of constitutional jurisdiction.

Although the provinces resist consideration of co-management in historic treaty areas, and the federal government ignores its obligation to uphold Aboriginal interests in the provincial arena of natural resource management, Aboriginal nations that were signatories to the historic treaties continue to insist on an approach based on sharing lands and resources. In some cases, treaty obligations were not fulfilled, and so some First Nations are in the process of negotiating specific land claims or treaty land entitlements. In this context, given the uncertainty of the nature of their treaty rights, as with interim measures agreements negotiated under the Comprehensive Claims Policy, these groups have somewhat more bargaining power to establish a stronger voice in natural resource management. Anderson (1994, 10), in his testimony on behalf of Manitoba Keewatinowi Okimakanak, a political advocacy organization representing thirty northern Manitoba First Nations, to the House of Commons Standing Committee on Aboriginal Affairs and Northern Development studying co-management, gives further voice to these aspirations:

> First Nations are interested in cooperative management for some very specific and clear reasons. The first is to promote, protect and implement benefits from treaty fishing, hunting and trapping rights. That is, within this area [Manitoba], where so much development has occurred without attention to these interests and rights, the co-management agreements and co-management activities are hoped to acknowledge, recognize and implement these rights.

Anderson (1994, 20) went on to recommend to the committee that the ideal co-management process would be to develop, in the first place, a global co-management accord, "not as some means of offsetting First Nations' interests or having First Nations agree to set aside rights that have now been recognized by the courts and the constitution, but as a genuine effort to make the balance." This "ideal" process is, of course, the coexistence model.

In Alberta, the Little Red River Cree Nation (LRRCN), signatory to Treaty 8, negotiated a cooperative management agreement with the Province of Alberta and, as Treseder (2000, 80) points out, although Aboriginal rights issues are "not 'officially' part of the discourse in the forest co-management process in northern Alberta," they are "a prominent underlying motivating factor for the Little Red River and Tallcree First Nations and the Province of Alberta." LRRCN uses the term "co-operative management," defining it as

> an *interim measure* which leaves the larger Treaty/Constitutional issues undisturbed, and concentrates on use of existing consultation, planning, management and resource-tenure processes to implement pragmatic initiatives for self-reliance and self-determination. (Treseder 2000, 25)

Whitefish Lake First Nation in Alberta, also signatory to Treaty 8, shares this pragmatism:

> Aboriginal communities are recognizing the strategic value in establishing interdependent relationships with government and industry as a means of enacting fundamental change in the institutions most responsible for the management of their traditionally used land and resources. (Natcher 2003, 171)

Even in the absence of treaty land entitlement negotiations, other historic treaty nations continue to assert their world view and insist that provinces negotiate resource development with them.

Given the assimilation practices of provincial governments, illustrated by their reluctance to share authority over natural resources on publicly owned land, even though those lands are also subject to historic treaties, one has to wonder if there is any hope for the emergence of a treaty renewal process as recommended by RCAP. The US 1974 legal decision *United States v. Washington,* more commonly referred to as the Boldt decision, is one case that shows it is possible for radically different state–tribal relationships to develop. Judge Boldt, after the state's marginalization of the tribes of western Washington for over a hundred years, recognized their prior treaty rights (signed in the 1850s), gave those tribes the right to harvest up to 50 percent of the salmon and authority to manage on- and off-reservation tribal fisheries, and stipulated that the tribes develop expertise and departments of natural resources to manage the fisheries. Funding was provided to the tribes to help them meet those obligations (Ebbin 2002). Over time, the state and tribes learned to work together, each bringing their own institutions to the mix, but not by equal compromise. Singleton (1998, 76) observes that, from a rocky beginning,

> near-total reliance on formal rules laid down by the court ... has given way to the current system of cooperative management through jointly created management plans and projects ... It was not, by and large, a process in which both sides made significant compromises and gradually came to occupy a middle ground. The position of the tribes changed very little: throughout this period they demanded their full allocational share and the right to exercise management authority over their fisheries and over decisions that directly affected their fisheries. State managers, on the other hand, substantially altered their decision-making rules and procedures as the bureaucratic culture of the organization was changed from above.

This US example points to the possibilities of developing a coexistence model for a renewed relationship between Aboriginal peoples and the state. Court decisions in Canada, state advisory initiatives such as RCAP, and Aboriginal peoples themselves have called for a renewal of treaties. However, it will take political leadership and will in the federal Parliament and provincial legislatures, not solely legal decisions, to facilitate this renewal.

Conclusion

Co-management in Canada has continued a process in which Aboriginal rights and aspirations are made subservient to those of the state. In these arrangements the state has attempted, through the imposition of its own policies, to extinguish, ignore, or minimize Aboriginal rights. Historic treaties, modern treaty making, and land claims have become mechanisms of colonization. In their present form, co-management arrangements can be seen as neocolonial institutions. Is it possible for co-management arrangements in Canada to become institutions of decolonization?

Some proponents of institutional change argue that change is easier when it is built on past institutions (North 1981). The question must then be asked, do provincial natural resource management regimes contain any elements that lend themselves to co-management as coexistence based on recognition of Aboriginal rights? The historic treaties signed between the Crown and Aboriginal peoples in Canada constitute an institutional arrangement that at the time seemed designed to serve the interests of both parties: ensuring for the state certainty for settlement and resource development and, for Aboriginal peoples, continuation of a way of life, even if limited to some extent. These original compromises, based on the recognition of the authority and sovereignty of both the Crown and Aboriginal peoples, can provide the basis for a renewal of the relationship, precisely as recommended by the Royal Commission on Aboriginal Peoples.

At the constitutional level, the state – both federal and provincial levels – must ensure that Aboriginal peoples are equal partners in negotiating changes to rules to ensure mutually satisfactory outcomes. In this role, the state will be a facilitator rather than regulator or enforcer. In this role, the state facilitates coexistence rather than promoting assimilation. M'Gonigle (1999, 23) points to the need for the modern state to act as a supporter for the development of local institutions. He recommends that the modern state be reinvented with a new constitution, which would mean

> the communalization of productive territorial wealth, the equalization of access to the bases of social and economic power, and the strengthening of the territorial economy through enhanced self-reliance, the enhancement of a "use" (in contrast to exchange) economy, and the development of regionally controlled markets.

Ideal co-management regimes will "draw their legitimization not from necessity, common sense, legislation, policy or a sense of social justice, but from a constitutionally entrenched right" (Notzke 1995a, 207). This rights-based approach forms the foundation of the coexistence model. In this model, successful co-management arrangements will emerge, based on a shared understanding and shared aspirations of both the state and Aboriginal peoples, when constitutional-level agreements are negotiated between Aboriginal peoples and the state, both federal and provincial. These agreements will contain not the usual language of limitation or non-derogation clauses on Aboriginal rights common to so many existing agreements, but wording such as "enable," "give meaning to," and "uphold and strengthen" the Aboriginal and treaty rights enshrined in the Constitution Act. Co-management will be successful when state–Aboriginal arrangements are founded on the principle of coexistence rather than assimilation.

Notes

1 This chapter does not cover Métis and Inuit involvement in resource management, although some of its arguments may apply to these Aboriginal peoples. Therefore, most arguments apply to First Nations, defined as status Indians under the Indian Act. "Aboriginal" is used in the inclusive sense when discussing rights as defined under s. 35 of the Constitution Act.

2 RCAP's full recommendation (1996b, Appendix A) on co-management, 2.4.78, is: "The following action be taken with respect to co-management and co-jurisdiction: (1) the federal government work with provincial and territorial governments and Aboriginal governments in creating co-management or co-jurisdiction arrangements for the traditional territories of Aboriginal nations; (b) such co-management arrangements serve as interim measures until the conclusion of treaty negotiations with the Aboriginal party concerned; (c) co-management bodies be based on relative parity of membership between Aboriginal nations

and government representatives; (d) co-management bodies respect and incorporate the traditional knowledge of Aboriginal people; and (e) provincial and territorial governments provide secure long-term funding for co-management bodies to ensure stability and enable them to build the necessary management skills and expertise (which would involve cost sharing on the part of the federal government)."

3 *Canadian Oxford Dictionary,* 2nd ed. 2004., s.v. "coexist."

4 Ibid., s.v. "assimilate."

5 In this case, because of their constitutional jurisdiction over natural resources, the "state" refers to provincial and territorial governments.

6 The question about whether the federal Crown can unilaterally extinguish Aboriginal title without negotiation remains unanswered, with some Aboriginal groups arguing that, even when stipulated in agreements, it is unjust and impossible to unilaterally extinguish Aboriginal title (Borrows 2002, 135-37).

References

AFN (Assembly of First Nations). 1990a. *Critique of Federal Land Claims Policies.* Vancouver: Union of BC Indian Chiefs. http://www.ubcic.bc.ca/.

–. 1990b. *Doublespeak of the 90's: A Comparison of Federal Government and First Nation Perceptions of Land Claims Process.* Vancouver: Union of BC Indian Chiefs.

Algonquin Nation Secretariat. 1999. *Land Rights and Negotiations: Measuring Canada's 1986 Comprehensive Claims Policy against the Supreme Court of Canada's 1997 Delgamuukw Decision.* http://gsdl.ubcic.bc.ca/collect/ubcicupd/import/1999-12-newsletter/TXT/1999-12-newsletter%20-%200003.txt.

Anderson, Michael. 1994. Testimony to the House of Commons Standing Committee on Aboriginal Affairs and Northern Development respecting, pursuant to Standing Order 108(2), a study of the co-management of natural resources. Minute no. 20, October 25. First session of the thirty-fifth Parliament.

Barsh, Russel Lawrence. 2002. "Political Recognition: An Assessment of American Practice." In *Who Are Canada's Aboriginal Peoples: Recognition, Definition, and Jurisdiction,* edited by Paul L.A.H. Chartrand, 230-57. Saskatoon: Purich Publishing.

Béné, C., and A.E. Neiland. 2004. "Empowerment Reform, Yes ... but Empowerment of Whom? Fisheries Decentralization Reforms in Developing Countries: A Critical Assessment with Specific Reference to Poverty Reduction." *Aquatic Resources, Culture and Development* 1:35-49. (Cited in Berkes 2009.)

Berkes, Fikret. 2009. Evolution of Co-Management: Role of Knowledge Generation, Bridging Organizations and Social Learning. *Journal of Environmental Management* 90:1692-1702.

Borrows, John. 1997. "Wampum at Niagara: The Royal Proclamation, Canadian Legal History, and Self-Government." In *Aboriginal and Treaty Rights in Canada,* edited by Michael Asch, 155-72. Vancouver: UBC Press.

–. 2002. *Recovering Canada: The Resurgence of Indigenous Law.* Toronto: University of Toronto Press.

Campbell, Tracy. 2003. "Co-Management of Aboriginal Resources." In *Natural Resources and Aboriginal People in Canada: Readings, Cases and Commentary,* edited by Robert B. Anderson and Robert M. Bone, 47-53. Concord, ON: Captus Press.

Carlsson, Lars, and Fikret Berkes. 2005. Co-Management: Concepts and Methodological Implications. *Journal of Environmental Management* 75:65-76.

Castro, Alfonso Peter, and Erik Nielsen. 2001. "Indigenous People and Co-Management: Implications for Conflict Management. *Environmental Science and Policy* 4:229-39.

Daly, Richard, and Val Napoleon. 2003. "A Dialogue on the Effects of Aboriginal Rights Litigation and Activism on Aboriginal Communities in Northwestern British Columbia. *Social Analysis* 47(3):108-29.

Ebbin, S.A. 2002. "Enhanced Fit through Institutional Interplay in the Pacific Northwest Salmon Co-Management Regime." *Marine Policy* 26:253-59.

Elias, Peter Douglas. 1995. *Northern Aboriginal Communities: Economies and Development.* Concord, ON: Captus Press.

Fleras, Augie, and Jean Leonard Elliott. 1992. *The "Nations Within": Aboriginal-state Relations in Canada, the United States, and New Zealand.* Toronto: Oxford University Press.
Frideres, James S., and René R. Gadacz. 2001. *Aboriginal Peoples in Canada: Contemporary Conflicts.* 6th ed. Toronto: Pearson Education.
Giokas, John. 2002. Domestic Recognition in the United States and Canada. In *Who Are Canada's Aboriginal Peoples: Recognition, Definition, and Jurisdiction,* edited by Paul L.A.H. Chartrand, 126-90. Saskatoon: Purich Publishing.
Harman, Elizabeth J. 1981. "Ideology and Mineral Development in Western Australia 1960-1980." In *State Capital and Resources in the North and West of Australia,* edited by E.J. Harman and B. Head, 167-96. Perth: University of Western Australia Press. (Cited in Howitt 2001.)
Howitt, Richard. 2001. *Rethinking Resource Management: Justice, Sustainability and Indigenous Peoples.* London: Routledge.
Iisaak (Iisaak Forest Resources). 2000. "Our History." http://www.iisaak.com/.
INAC (Indian and Northern Affairs Canada). 1985. *Living Treaties, Lasting Agreements: Report of the Task Force to Review Comprehensive Claims Policy.* Ottawa: Minister of Supply and Services.
–. 2003a. *Comprehensive Claims Policy and Status of Claims.* http://www.ainc-inac.gc.ca/
–. 2003b. *Copy of the Robinson Treaty Made in the Year 1850 with the Ojibewa Indians of Lake Superior Conveying Certain Lands to the Crown.* http://www.aadnc-aandc.gc.ca/eng/1100100028984.
Lawson Lundell. 2005. *The Mikisew Cree Decision: Balancing Government's Power to Manage Lands and Resources with Consultation Obligations under Historic Treaties.* Vancouver: http://www.lawsonlundell.com/.
M'Gonigle, R. Michael. 1999. "Ecological Economics and Political Ecology: Towards a Necessary Synthesis." *Ecological Economics* 28:11-26.
Mann, Michelle. 2003. "Capitalism and the Dis-Empowerment of Canadian Aboriginal Peoples." In *Natural Resources and Aboriginal People in Canada: Readings, Cases and Commentary,* edited by Robert B. Anderson and Robert M. Bone, 18-29. Concord, ON: Captus Press.
McCay, Bonnie J., and James M. Acheson. 1987. "Human Ecology of the Commons." In *The Question of the Commons: The Culture and Ecology of Communal Resources,* edited by Bonnie J. McCay and James M. Acheson, 1-34. Tucson: University of Arizona Press.
McDonald, Michael J. 2003. "Aboriginal Forestry in Canada." In *Natural Resources and Aboriginal People in Canada: Readings, Cases and Commentary,* edited by Robert B. Anderson and Robert M. Bone, 230-56. Concord, ON: Captus Press.
McGregor, Deborah. 2000. "From Exclusion to Co-Existence: Aboriginal Participation in Ontario Forest Management Planning." PhD diss., University of Toronto.
McMillan, Alan D. 1988. *Native Peoples and Cultures of Canada.* Vancouver: Douglas and McIntyre.
Mockus, Antanas. 2002. "Co-Existence as Harmonization of Law, Morality and Culture." *Prospects* 32(1):19-37.
Mulvihill, Peter R., and Peter Jacobs. 1991. "Towards New South/North Development Strategies in Canada." *Alternatives* 18(2):34-39.
Natcher, David C. 2001. "Land Use Research and the Duty to Consult: A Misrepresentation of the Aboriginal Landscape." *Land Use Policy* 18:113-22.
–. 2003. "Institutionalized Adaptation: Aboriginal Involvement in Land and Resource Management." In *Natural Resources and Aboriginal People in Canada: Readings, Cases and Commentary,* edited by Robert B. Anderson and Robert M. Bone, 159-73. Concord, ON: Captus Press.
North, Douglass C. 1981. *Structure and Change in Economic History.* New York: W.W. Norton.
Notzke, Claudia. 1995a. "A New Perspective in Aboriginal Natural Resource Management: Co-Management." *Geoforum* 26(2):187-209.
–. 1995b. *The Barriere Lake Trilateral Agreement.* Research report prepared for the Royal Commission on Aboriginal Peoples. Libraxus CD-ROM for Seven Generations: An

Information Legacy of the Royal Commission on Aboriginal Peoples. Ottawa: Minister of Supply and Services Canada.
–. 1995c. "The Resource Co-Management Regime in the Inuvialuit Settlement Region." In *Northern Aboriginal Communities: Economies and Development,* edited by Peter Douglas Elias, 36-52. Concord, ON: Captus Press.
OMNR (Ontario Ministry of Natural Resources). 2002. *State of the Forest Report, 2001.* Toronto: Queen's Printer.
Pikangikum First Nation. 2008. "Welcome to the Whitefeather Forest Initiative." http://www.whitefeatherforest.com.
Pinkerton, Evelyn. 1989. "Introduction: Attaining Better Fisheries Management through Co-Management – Prospects, Problems, and Propositions." In *Co-operative Management of Local Fisheries: New Directions for Improved Management and Community Development,* edited by E. Pinkerton, 3-33. Vancouver: UBC Press.
Pratt, Alan. 1995. *The Numbered Treaties and Extinguishment: A Legal Analysis.* Research report prepared for the Royal Commission on Aboriginal Peoples. Libraxus CD-ROM for Seven Generations: An Information Legacy of the Royal Commission on Aboriginal Peoples. Ottawa: Minister of Supply and Services Canada.
Ransom, James W., and Kreg T. Ettenger. 2001. "'Polishing the Kaswentha': A Haudenosaunee View of Environmental Cooperation." *Environmental Science and Policy* 4:219-28.
RCAP (Royal Commission on Aboriginal Peoples). 1996a. *People to People, Nation to Nation: Highlights from the Report of the Royal Commission on Aboriginal Peoples.* Ottawa: Minister of Supply and Services Canada. http://www.aadnc-aandc.gc.ca/eng/1307458586498.
RCAP. 1996b. *Report of the Royal Commission on Aboriginal Peoples.* Volume 2, Restructuring the Relationship, Part 2, Lands and Resources. Ottawa: Minister of Supply and Services Canada. http://www.collectionscanada.gc.ca/webarchives/20071124125812/http://www.ainc-inac.gc.ca/ch/rcap/sg/shm4_e.html.
Richardson, Benjamin, Donna Craig, and Ben Boer. 1994. "Indigenous Peoples and Environmental Management: A Review of Canadian Regional Agreements and Their Potential Application to Australia – Part 1." *Environmental and Planning Law Journal* 11(4):320-43.
Richardson, Boyce. 1993. *People of Terra Nullius: Betrayal and Rebirth in Aboriginal Canada.* Vancouver: Douglas and McIntyre.
Ross, Monique, and Peggy Smith. 2002. *Accommodation of Aboriginal Rights: The Need for an Aboriginal Forest Tenure.* Synthesis report. Edmonton: Sustainable Forest Management Network.
Rynard, Paul. 2000. "'Welcome in, but Check Your Rights at the Door': The James Bay and Nisga'a Agreements in Canada." *Canadian Journal of Political Science* 33(2):211-43.
–. 2001. "Ally or Colonizer: The Federal State, the Cree Nation and the James Bay Agreement." *Journal of Canadian Studies* 36(2):8-48.
Sharvit, Cheryl, Michael Robinson, and Monique Ross. 1999. *Resource Developments on Traditional Lands: The Duty to Consult.* CIRL Occasional Paper no. 6. Calgary: Canadian Institute of Resources Law.
Shuter, J.L., and S. Kant. 2003. "A Multi-Dimensional Framework and Its Application to Aboriginal Co-Management Arrangements in the Forest Sector of Canada." Paper 0942-C1. Presented at the twelfth World Forestry Congress, Quebec City, September 21-28. Rome: Food and Agriculture Organization. http://www.fao.org/forestry/.
Singleton, Sara. 1998. *Constructing Cooperation: The Evolution of Institutions of Comanagement.* Ann Arbor: University of Michigan Press.
Smith, Peggy A. 1991. *A Survey and Evaluation of Natural Resource Agreements Signed with Aboriginal People in Canada: Do They Result in Autonomy or Dependence?* HBScF thesis, Lakehead University.
Smith, Peggy, and Harry Bombay. 1995. *Aboriginal Participation in Forest Management: Not Just Another Stakeholder.* Ottawa: National Aboriginal Forestry Association.
Stevenson, Marc, and Jim Webb. 2003. "Just Another Stakeholder? First Nations and Sustainable Forest Management in Canada's Boreal Forest." In *Towards Sustainable Management of the Boreal Forest,* edited by Philip J. Burton, Christian Messier, Daniel W. Smith, and Wiktor L. Adamowicz, 65-112. Ottawa: National Research Council of Canada.

Treaty 7 Elders and Tribal Council. 1996. *True Spirit and Original Intent of Treaty 7*. Montreal and Kingston: McGill-Queen's University Press.

Treseder, Leslie Caroline. 2000. *Forest Co-Management in Northern Alberta: Conflict, Sustainability and Power.* MSc thesis, University of Alberta.

Tully, James. 1995. *Strange Multiplicity: Constitutionalism in an Age of Diversity*. Cambridge, UK: Cambridge University Press.

–. 2000. "The Struggles of Indigenous Peoples for and of Freedom." In *Political Theory and the Rights of Indigenous Peoples,* edited by Duncan Ivison, Paul Patton, and Will Sanders, 36-59. Cambridge, UK: Cambridge University Press.

UBCIC (Union of BC Indian Chiefs). 1998-2004. *Aboriginal Title Implementation.* http://www.ubcic.bc.ca/

–. 2004. "Complete Failure of Aboriginal Policy." Press release, Chief Stewart Phillip, March 24. Vancouver: UBCIC.

–. n.d. *Certainty: Canada's Struggle to Extinguish Aboriginal Title.* http://www.ubcic.bc.ca/Resources/.

Usher, Peter J., Frank J. Tough, and Robert M. Galois. 1992. "Reclaiming the Land: Aboriginal Title, Treaty Rights and Land Claims in Canada." *Applied Geography* 12:109-32.

Venne, Sharon. 1997. "Understanding Treaty 6: An Indigenous Perspective." In *Aboriginal and Treaty Rights in Canada: Essays on Law, Equality, and Respect for Difference,* edited by Michael Asch, 173-207. Vancouver: UBC Press.

7

Aboriginal Peoples and Traditional Knowledge

A Course Correction for Sustainable Forest Management

Marc G. Stevenson

> The Whiteman said, *"What will happen if any of your people may someday want to have one foot in each of the boats we have placed parallel?"* The Onkwehonweh replied, *"If this so happens that my people wish to have their feet in each of the two boats, there will be a high wind and the boats will separate and the person that has his feet in each of the boats shall fall between the boats; and there is not a living soul who will be able to bring him back to the right way given by the Creator, but only one: The Creator Himself."*
>
> – From the *Record of the Two Row Wampum Belt*, translated by Huron Millar

Current social and economic conditions in most forest-dependent Aboriginal communities are neither desirable nor sustainable. Although workable solutions to the many problems faced by these communities may be hard to come by, three future scenarios can be envisioned: (1) things get worse, (2) things remain the same, or (3) things get better. In the first scenario, Aboriginal peoples and communities fail to thrive, and assimilation into the Canadian mainstream becomes the only viable alternative. In a country that values cultural and biological diversity, this loss to Canada would be incalculable. The cost of maintaining the status quo, currently characterized by ineffective institutional and policy responses to Aboriginal needs, rights, and interests, among other things, is also unacceptable and doubles with each generation. Canada simply cannot afford to keep its Aboriginal peoples dependent. The only acceptable outcome for Canada is one in which Aboriginal peoples and their representative governments negotiate arrangements based on a coexistence (see Smith, this volume) or a nation-to-nation relationship, thereby becoming true partners in confederation. Such a scenario will not happen over night, or maybe not even in our lifetimes. But we cannot ignore the challenge before us. This chapter directly takes up this challenge of the engagement of Aboriginal peoples and their knowledge in sustainable forest management, planning, and decision making. Specifically, it addresses the issue of incorporating traditional knowledge, and the holders of that knowledge, into sustainable forest management (SFM).

As forest companies move from sustained yield forestry toward SFM, they encounter a range and complexity of issues new to the industry. Foremost among these is increased engagement with the public, environmental organizations, Aboriginal communities, and First Nations governments. SFM may mean different things to different people, but the concept usually entails "managing the forest" to sustain a mix of economic, ecological, and social values (Burton et al. 2003).[1] Just whose values receive priority is an issue that currently dominates the discourse and political arena among these stakeholders. Forest companies, which hold timber cutting rights to Crown land allocated to them by the provinces, often find themselves between a rock and a hard place when trying to balance their tenure obligations with the needs, rights, and interests of other stakeholders, especially environmental groups and First Nations peoples. Conflict inevitably results. Subsequently, in some provincial jurisdictions, such as British Columbia, forest policy has been changed or amended to address issues of social equity and ecological sustainability. Arguably, these measures fall well short of meeting the needs and rights of British Columbia's First Nations.[2]

There are, however, a few forest companies in British Columbia that appear to be making a sincere effort to engage BC First Nations and to involve them in forestry operations and management. For those companies that are leading the way, capacity and economic development for First Nations to participate in forestry operations and the integration of traditional knowledge and values in forest management have become focal points around which dialogues with Aboriginal peoples are framed. Although the two can and should be related, this chapter focuses primarily on the latter.[3]

The current state of the art for integrating traditional knowledge into environmental resource management and SFM is reviewed. The systemic deficiencies identified in the status quo force consideration of an alternative, and arguably more robust, effective and collaborative framework to incorporate Aboriginal peoples and their knowledge into decisions taken regarding their traditional forested lands and resource use activities. Adoption of the approach advocated creates the space necessary to meaningfully and effectively engage Aboriginal peoples, along with their knowledge and values, in SFM and, as such, can lead only to more informed decision making.

Why Traditional Knowledge?

The inclusion of traditional knowledge (TK) into environmental resource management has received considerable attention in recent years, and has been recognized by the World Commission on Environment and Development:

> Lifestyles of tribal and Indigenous peoples ... can offer modern societies many lessons in the management of complex ... ecosystems. Their disappearance is a loss for the greater society, which could learn a great deal from their traditional skills in sustainably managing very complex ecological systems. (World Commission on Environment and Development 1987, 12, 114-15)

Canada is also signatory to numerous international agreements that promote the use of TK (e.g., the Convention on Biodiversity), and federal acts of Parliament are being amended to incorporate TK into environmental impact assessment (e.g., the Canadian Environmental Assessment Act). At the same time, government policies in some jurisdictions (e.g., Northwest Territories and Nunavut) mandate the use of TK in territorial-led processes and initiatives.

There are various reasons for the elevated status of TK among many non-Aboriginal Canadians, institutions, and advocates of SFM. Certainly, the current limitations of scientific knowledge and environmental resource management to deal effectively with environmental issues of increasing magnitude and complexity (e.g., global warming, multiple and cumulative impacts, biodiversity conservation) have opened the door to alternative sources of knowledge. In addition, as Aboriginal peoples and their governments become more assertive of their rights and seek greater equity from natural resource allocations and developments, political concessions are often made in which TK is considered, even mandated, in environmental decision making. Not surprisingly, Aboriginal peoples champion the use of their local knowledge in order to promote their active involvement in environmental resource management and SFM, usually through the creation of some form of cooperative management regime. In turn, TK has become a nexus around which industry and government frame their dialogues with Aboriginal peoples.

What Is Traditional Knowledge Anyway?

> Traditional environmental knowledge [TEK] is a body of knowledge and beliefs transmitted through oral tradition and first-hand observation. It includes a system of classification, a set of empirical observations about the local environment and a system of self-management that governs resource use. Ecological aspects are closely tied to social and spiritual aspects of the knowledge system ... With its roots firmly in the past, TEK is both cumulative and dynamic, building upon the experience of earlier generations and adapting to the new technological and socio-economic changes of the present. (Dene Cultural Institute 1995)

Many definitions have been advanced to capture the essence of what TK is or is not. None of these definitions, even the oft-cited Dene Cultural Institute's definition above, is particularly empowering to those Aboriginal peoples who have this knowledge or useful to those non-Aboriginal peoples who wish, for whatever reasons, to access and apply it. Definitions tend to limit or pigeonhole the contributions of Aboriginal peoples and their knowledge to decisions required to achieve ecological, social, cultural, and economic sustainability. This is especially so if we exclude from consideration knowledge *not directly* related to environmental or ecological phenomena. This realization has led some to use other, but still problematic, terms such as "traditional ecological knowledge," "traditional environmental knowledge," "indigenous knowledge," "indigenous science," "naturalized knowledge," "users' knowledge," and so on.[4] TK, unlike the supposedly objective, value-free raw data and information used to construct positivistic knowledge of Western science, relies heavily on its social and cultural context for meaning and value.

The Context of Traditional Knowledge

TK, whether its intended use by non-Aboriginal parties and institutions targets "ecological" or "traditional" knowledge held by Aboriginal peoples, rightfully resides in the broader, articulated system of knowledge, meanings, values, and understandings that most Aboriginal peoples possess, and from which it cannot be easily separated. In other words, the knowledge systems of most Aboriginal peoples include not just TK but also nontraditional and nonecological knowledge, all of which are intricately related and articulated into a system of understandings that are individually constructed and often held collectively. Moreover, all Aboriginal peoples' knowledge is contemporary. Although great value is attached to land- and resource-based knowledge handed down from previous generations, even this knowledge is recast and its utility reevaluated in the light of contemporary experience, needs, and values:

> Many Aboriginal people feel that requests to access their traditional knowledge represent another form of exploitation, because this knowledge can easily be taken out of context and misinterpreted. Moreover, viewing the knowledge that Aboriginal people possess as essentially traditional invites denial of the relevance and efficacy of applying their knowledge to present-day issues and problems. (Stevenson 1996, 280)

On the Wrong Course?

The practice to date, with rare exception, has been to cherry-pick certain elements of TK – most notably, some form of specific environmental

Figure 7.1

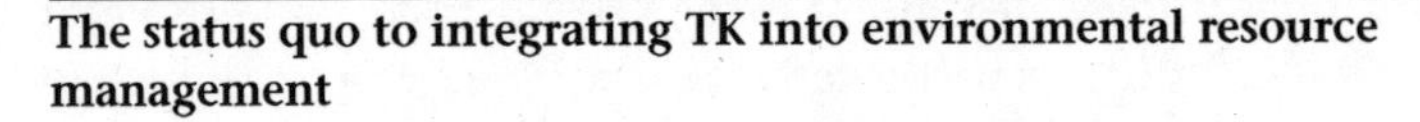

The status quo to integrating TK into environmental resource management

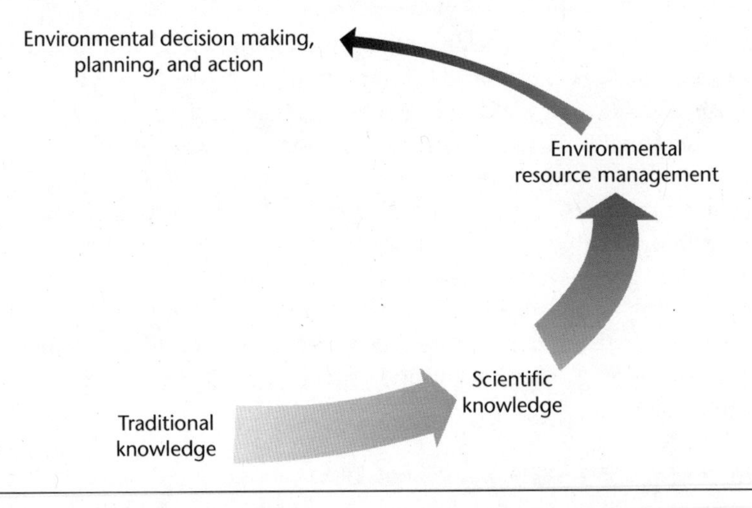

knowledge that can be easily accommodated within the perspectives of Western science and environmental resource management – from their broader sociocultural context and to merge them with existing data sets to inform well-established information requirements and management procedures (see Figure 7.1). The end result, despite all the recent attention and rhetoric, has not allowed TK nor Aboriginal peoples to make significant contributions to the way resources are managed, frustrating both those who possess this knowledge and those wishing or required to use it.

There are at least two systemic reasons why this is so. The first has to do with the fact that the knowledge of Aboriginal peoples did not evolve to inform Western positivistic science or environmental resource management. Certainly, many Aboriginal Elders and TK holders have developed extensive knowledge about the spatial and temporal distributions, composition, health, conditions, behaviours, and so on of many species, and the factors that influence these phenomena. This knowledge, being the product of both personal experience and knowledge passed down from previous generations, may reveal much about natural variation over time and space in valued ecosystem components. At the same time, many forest-dependent Aboriginal peoples have witnessed the specific and combined impacts of natural and human disturbances and their subsequent effects on forest resources. It is these types of information and knowledge that forest managers and planners require, but usually do not have, in order to make informed decisions over broad temporal and spatial scales.

However, such uses of TK in the absence of models such as that described below are intellectually stifling; they access only a fraction of the knowledge held by Aboriginal peoples required to achieve ecological, social, economic, and cultural sustainability of Canada's forests. Moreover, they focus on information and knowledge about the "resource" or "management unit" to the virtual exclusion of knowledge and understanding of ecological (including human use) "relationships" critical to the resource, while inviting a plethora of problems that inevitably follow for both Aboriginal owners and non-Aboriginal users of this knowledge.

A second and related reason why TK has made little impact on the way forest and other natural resources are managed is that TK is invariably taken out of context, or decontextualized. This process, which has profound management implications, is familiar to many researchers and Aboriginal research participants, and goes something like this:

1 The research issue or problem is usually, but not universally, identified by non-Aboriginals (e.g., government managers, company scientists, researchers).
2 The research questions almost always originate with those "cultured," or having a significant investment, in the positivistic science and environmental resource management traditions.
3 The knowledge sought to answer these questions requires that it be compatible with Western science; i.e., usually some understandable or usable form of environmental knowledge to which non-Aboriginals do not have ready access (see below).
4 Elders and other TK holders are then interviewed using Western information-gathering techniques that ignore the richness and complexity of Aboriginal narratives (e.g., Natcher and Hickey 2002).
5 Local interpreters are used to filter and translate complex concepts and issues originating in one culture into the language of another.
6 The interview is recorded or videotaped and then transcribed, in whole or in part, onto paper and/or maps.
7 These formats are subjected to analyses that single out specific elements that contribute new information to established scientific data sets and environmental resource management procedures.
8 This "information" then becomes the authoritative reference on which decisions are made (Stevenson 2006).

As many research projects that purportedly document TK have taught us, such processes are fraught with methodological and ethical pitfalls. In each of the later steps there is a progressive loss of knowledge, meaning, and context. Not only is knowledge increasingly divorced from the social/cultural context where it more properly resides but its original owners are increasingly

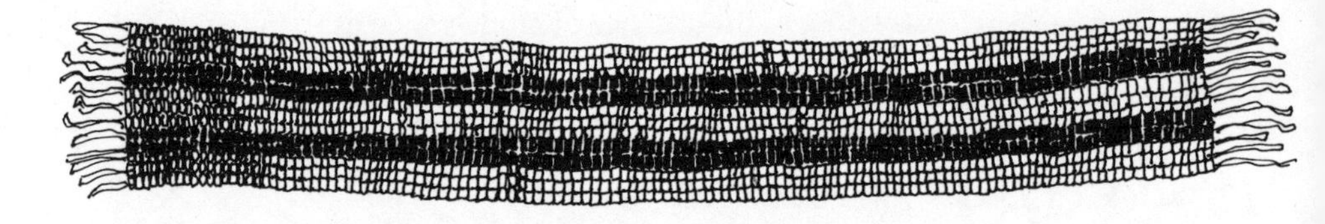

Figure 7.2 **On the right course: The two-row wampum approach**
Reproduced with authorization of the Six Nations/Hodenushonee Confederacy Council from Grace Woo, *Ghost Dancing with Colonialism* (Vancouver: UBC Press, 2011), 6-7.

separated from knowledge that they once owned and controlled, effectively excluding them from decision making. In this all too common scenario, TK is valued primarily for its information contribution to Western science and environmental resource management. However, through its progressive sanitization, or dumbing-down, TK assumes the role of handmaiden to Western science, and alternative ways of knowing, seeing, and relating to the natural world are devalued, diminished, and dismissed. This process not only reflects the predominant positions of Western scientific knowledge and environmental resource management in environmental decision making but also strengthens the existing institutional arrangements and power relationships that support them. In the end, everybody loses. This chapter proposes an alternative approach to effectively integrating Aboriginal peoples and their knowledge into SFM.[5]

The knowledge of Aboriginal peoples has evolved, and continues to evolve, to inform ways and philosophies of life different from those in which positivistic science and environmental resource management emerged and achieved their currency. Western scientific knowledge is heavily imbued with western European religious, social, cultural, and economic thought and tradition, whereby specific resources, or some other arbitrarily defined management unit, are managed using the best available scientific data and information. Western scientific knowledge is, in fact, despite its claims to neutrality and objectivity, culturally constructed, just like the knowledge of Aboriginal peoples. Most indigenous peoples worldwide traditionally did not manage resources per se but their relationships to and with resources (Stevenson 2006). Control or dominance over nature was a foreign concept, whereas maintaining reciprocity and a robust and resilient relationship between human beings and the natural world was humankind's responsibility. Not until well after contact with Europeans did most indigenous peoples become familiar with the concepts and practices of positivistic science and environmental resource management. It is relationships with the natural world, not resources, that Aboriginal peoples managed, and they did this by constructing knowledge necessary to sustain core social and cultural

Figure 7.3

Preferred path to incorporating Aboriginal peoples and their traditional knowledge in SFM (modelled after the two-row wampum)

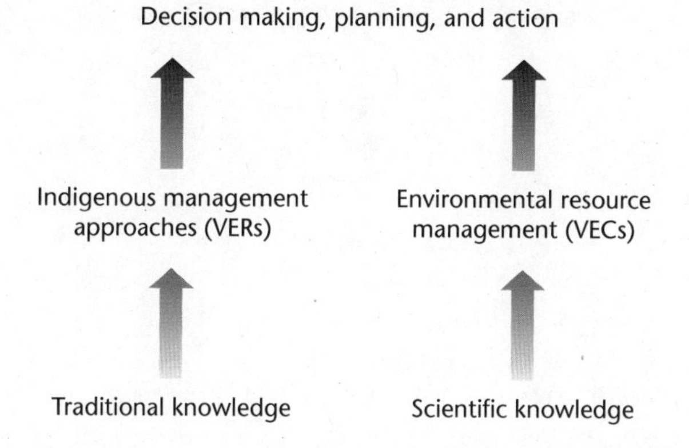

values and critical relationships; that is, elements rarely considered in environmental resource management (Lewis and Sheppard, this volume).

In this light, and contrary to the claims of many researchers, environmental resource managers, and even Aboriginal peoples who have been forced to co-opt the environmental resource management paradigm for marginal gains, TK may have little to offer conventional Western scientific thought and practice, and even SFM as it is currently practised. However, the knowledge of Aboriginal and indigenous peoples may have much to contribute to understanding and developing sustainable relationships with the natural world. The full contributions of Aboriginal peoples, and their knowledge and management systems, to managing for sustainability of Canada's forests will not be realized until environmental policy makers and managers consider them equally with Western scientific knowledge and environmental resource management systems in forest decision making (Figure 7.3).

In the two-row wampum approach, TK is not decontextualized and recast into a currency that can be used in the service of Western scientific thought or practice (see Figure 7.1). Rather, it is used to inform indigenous management approaches, which focus on managing and sustaining valued ecosystem relationships (VERs). In this model, the two management approaches are not mutually antagonistic but complementary – one contributing knowledge and wisdom relevant to managing appropriate and desirable VERs, the other information and knowledge relevant to managing valued ecosystem components (VECs). Theoretically, the latter, which is common to current environmental assessment policy and praxis, could include VERs,

Figure 7.4

Dialogue process built into Aboriginal and non-Aboriginal knowledge and management systems incorporation model

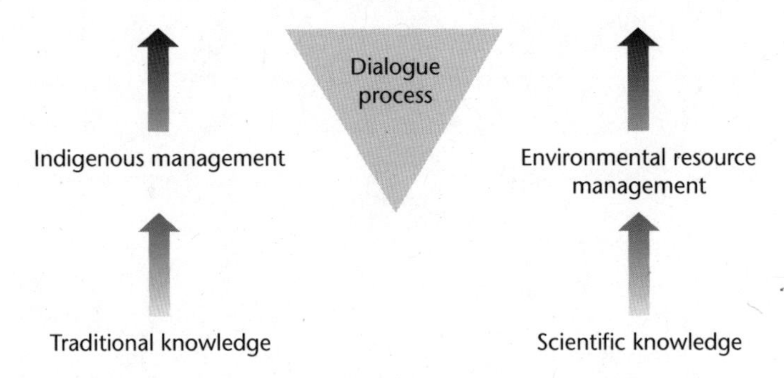

and in some environmental assessments Aboriginal values and relationships with key species have been identified as such. Nevertheless, management considerations almost always focus on information about the resource/species to the exclusion of knowledge about maintaining appropriate relationships with or to the resource. At a minimum, both should be considered equally in decision making. In reality, however, each should be accorded consideration commensurate with their respective contributions to achieving ecological, social, cultural, and economic sustainability of Canada's forests. This does not mean that knowledge holders of both traditions should not talk to, or share their knowledge with, each other. On the contrary, the process of learning from one another should start at the beginning of the relationship and build from there in order to facilitate discussions at the planning and decision-making stages (Figure 7.4).

Figure 7.5

Factors influencing decision making in SFM

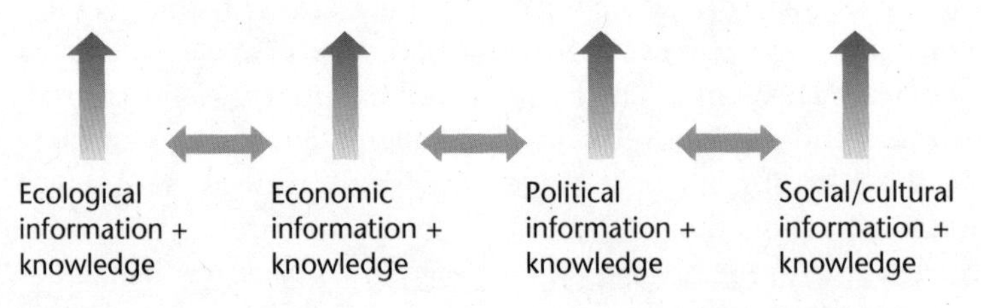

The fact of the matter is that, in order to achieve ecological integrity, political certainty, social stability, and economic viability of Canada's forests, serious consideration must be given not just to the environment or ecology but also to political, economic, social, and other factors, all of which are interrelated and mutually influencing (Figure 7.5).

At the same time, decision-making models for achieving sustainability of Canada's forests must allow for the contributions that Aboriginal peoples and their knowledge will make on all these fronts (Figure 7.6). In this model, the environmental and ecological knowledge of Aboriginal peoples is not appropriated in the interests of the other agendas but recontextualized to become part of a larger comprehensive strategy to achieve ecological, social, cultural, and economic sustainability of Canada's forests.

Steps Toward the Sustainability of Aboriginal Cultures and Forested Landscapes

Many forest companies and government agencies, because of existing regulatory requirements, operational momentum, or commitments made in forest management plans, may be unable, even if willing, to adopt the approach advocated above. However, there are steps they can take right now that may eventually lead to the implementation of models such as that illustrated in Figure 7.6.

First, forest companies can begin, if they haven't already, to consult with Aboriginal peoples about traditional resource uses and critical habitats (e.g.,

Figure 7.6

A model to realize the full contributions of Aboriginal peoples and their knowledge to decision making in SFM

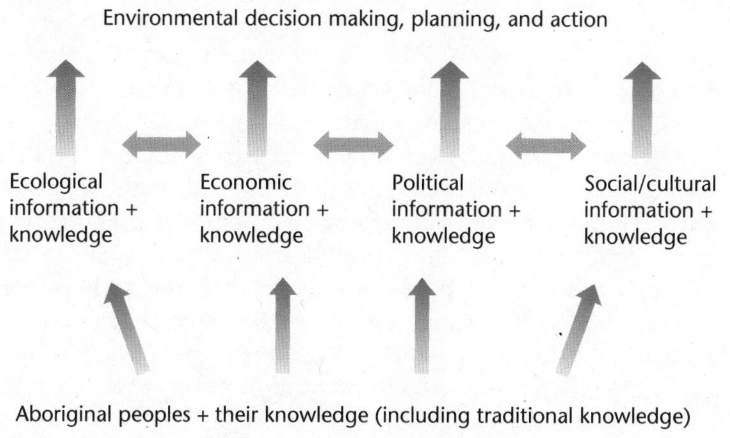

Figure 7.7

Predicted extent of problems and benefits associated with various approaches to incorporating Aboriginal peoples and their knowledge into SFM

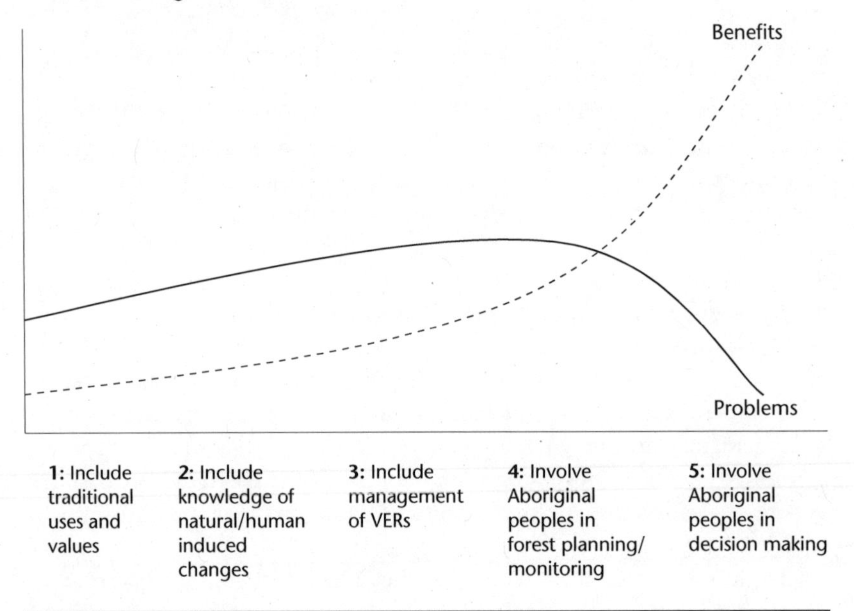

Note: Assumes that each step builds on the one before.

traplines, cabins, hunting grounds, trails, berry-picking sites, sacred sites) within their respective management areas in an effort to incorporate this knowledge into forest planning and regulatory requirements (Step 1 in Figure 7.7).[6] Traditional use areas are perhaps the first thing that forest companies consider when they think of TK. Incorporating Aboriginal peoples' knowledge about natural variation and anthropogenic changes in resource characteristics (e.g., distributions, numbers, behaviours) would constitute another meaningful step toward the development of sustainable forestry practices (Step 2). Forestry plans that include knowledge of Aboriginal "valued ecosystem relationships" would constitute perhaps the penultimate use of the knowledge of Aboriginal peoples in forest planning (Step 3). However, without involving the "knowers" into planning and decision-making processes, these steps may engender problems for both parties relating to the ownership and use of intellectual property. Including Aboriginal peoples into forest planning and the monitoring of both valued ecosystem relationships and components would go a long way toward mitigating problems associated with taking the knowledge without the knower (Step 4). Finally, incorporating Aboriginal peoples into the decision-making processes, commensurate with their needs, rights, and interests, would lead to win-win arrangements that few SFM supporters, and certainly no forest product certifier, would find contentious (Step 5).

Conclusions

The choices before us are obvious. The status quo is not working and will lead either to complete assimilation or the continuing disempowerment and marginalization of Aboriginal peoples. This chapter has outlined an approach that will hopefully reverse these trends for the benefit of all Canadians. Creating space for Aboriginal peoples and their knowledge in SFM makes sense from scientific, legal, moral, and arguably, other perspectives. However, in order to create the conditions of ecological, economic, social, and cultural sustainability, as well as certainty, to which we all aspire, there are a number of other things that we can do, both collectively and individually.

Be More Self-Critical and Reflective

Representatives from government, industry, and the Aboriginal, ENGO (environmental nongovernmental organization), and research communities can be more aware of their cultural biases and assumptions when they consider the roles of Aboriginal peoples and their knowledge in forest use planning and decision making. Resource managers and researchers who work with TK holders, in particular, must avoid being influenced by the positivistic paradigm to the extent that the traditional management and knowledge systems of Aboriginal peoples are muted, or not recognized at all by those "cultured" in the Western scientific and economic traditions. Without such critical self-reflection, these individuals may become unwitting agents for the authoritative knowledge systems and processes now in power. Aboriginal peoples too, who have been co-opted, or forced into accepting the environmental resource management paradigm, must weigh the advantages of doing so against rebuilding and reinstituting their own knowledge and management systems in the context of asserting their constitutionally protected rights. After all, it is these systems – with their focus on managing and sustaining appropriate relationships with and carrying out Aboriginal peoples' stewardship responsibilities to the natural world – that have stood the test of time; environmental resource management (and its various permutations, including SFM) is the new kid on the block. Finally, we can all be more reflective of our own involvement in maintaining the systemic barriers, and the power relationships supporting them, that keep Aboriginal peoples and their knowledge on the sidelines of forest decision making and planning.

"Get Up to Speed" on Aboriginal and Treaty Rights

In order to create the political certainty and conditions necessary for sustainability of Canada's forests, all stakeholders may wish to inform themselves of Aboriginal and treaty rights. The exercise of these rights is not only protected under law but is also critical to sustaining Aboriginal relationships, knowledge, and management contributions to forest decision making and, arguably, forest biodiversity.

Support Aboriginal and First Nations Efforts

There is much to be gained in the interests of sustainability by supporting Aboriginal efforts to find a greater voice in forest resource and land use planning. Elevating the status of Aboriginal knowledge and management systems in forest decision making makes both good political and ecological sense – elsewhere it has been argued that traditional hunting and burning practices of indigenous peoples played a significant role in maintaining the biodiversity of forests (Stevenson and Webb 2004). Also, in the interests of sustainability we need to support Aboriginal efforts to (1) document, assess, and prioritize their uses, values, and needs of forest resources, and (2) develop forest resource and land use plans for the future based on these. Only through such processes will the needs and aspirations of Aboriginal peoples find a balance with conventional forestry. Finally, for reasons that now should be obvious, we need to support activities and practices that promote Aboriginal values and their transmission from one generation to the next.

Support Policy Reform and the Creation of New Institutional Arrangements

In order to create the political certainty and ecological, social/cultural, and economic formulae for sustainability that we all seek, government, Aboriginal, industrial, public, and ENGO proponents of SFM must work together to reform existing policies, practices, regulations, and institutions. Only through the design of new tenure regimes and institutional arrangements will the needs, rights, interests, and knowledge contributions of Aboriginal peoples and other forest users be given a fair hearing and weighted accordingly in forest decision making. For true coexistence to occur, government must not only accept its fiduciary obligations to Aboriginal peoples but also endeavour to develop new institutional arrangements and administer policies that accommodate Aboriginal and treaty rights into SFM. Industry partners not wishing to sit on the sidelines may, in addition to lobbying governments for policy reform, enter into agreements with Aboriginal groups that share management authority over and the economic benefits from forestry and the development of forest resources, including nontimber and value-added products. In the design of new institutions, space must be found for Aboriginal values (Lewis and Sheppard, this volume), concepts, terms, language, and approaches related to managing valued ecosystem relationships.

Support Research

In addition to supporting Aboriginal efforts to develop forest and land use plans, forest companies, in their shift from sustained yield to SFM, may wish to entertain variable retention, innovative zoning, and other approaches that might meaningfully involve Aboriginal peoples and their knowledge in forest decision making, and in the management and monitoring of VERs and VECs. Once a forest company moves away from clear-cutting, the questions become, what values and resources do we retain, and why? Companies may also wish to work with Aboriginal communities to become certified through certification schemes that encourage Aboriginal involvement and to support Aboriginal research efforts to develop plans that sustain their values and sustainable uses of forest resources. Finally, employing the perspectives advanced in this chapter, researchers must continue to undertake research with and driven by Aboriginal peoples and their industry partners to inform their roles and contributions to SFM.

Notes

Epigraph: Record of the Two Row Wampum Belt, translated by Huron Millar, http://www.akwesasne.ca/tworowwampum.html.

1 For reasons that should become clear below, I do not believe that SFM or "managing forests" are appropriate concepts or descriptors of what we are attempting to do. About all we can do is manage forest uses and impacts, that is, our relationship to the forest, in order to sustain the broadest range of values possible based on the accommodation of every stakeholder's needs, rights, and interests. This perhaps not so subtle difference has profound implications for engaging Aboriginal peoples and their knowledge in managing forest uses and impacts.

2 Letter to Premier of British Columbia, the Hon. Gordon Campbell, and Minister of Forests, the Hon. Michael de Jong, 27 November 2003, from Justa Monk, committee coordinator, Title and Rights Alliance.

3 For a thorough examination of the former, consult Marc G. Stevenson and Pamela Perreault, *Capacity for What? Capacity for Whom? Aboriginal Capacity and Canada's Forest Sector* (Edmonton: Sustainable Forest Management Network, 2008).

4 I employ the term "traditional knowledge" (TK) with reservation and with the understanding that (1) most Aboriginal and non-Aboriginal people who use this term recognize its inherent deficiencies and limitations, (2) it has different meanings to different people, and (3) its utility is context dependent.

5 The two-row wampum, produced by First Nations peoples as a depiction of their negotiated relationship with early European settler society, is based on a nation-to-nation relationship that respects the autonomy, authority, and jurisdiction of each nation (see Figure 7.2). The two rows symbolize two paths or two vessels travelling down the same river of life together. One, a birch bark canoe, represents the Original peoples, their laws, their customs, and their ways. The other, a ship, is for the European peoples, their laws, their customs, and their ways. They travel down the river together, side by side, each in their own boat, neither trying to steer the other's vessel.

6 In the recent 2003 Supreme Court decision in *Haida,* the obligation to consult, which arises from the honour of the Crown, resides with the provincial government, not forest companies. Nevertheless, it would be expedient and in the best interests of the forest companies to initiate consultation with First Nations to accommodate their needs, rights, and interests prior to receiving harvesting rights and developing forest management plans.

References

Burton, P.J., C. Messier, D.W. Smith, and W.L. Adamowicz, eds. 2003. *Towards Sustainable Management of the Boreal Forest.* Ottawa: NRC Research Press.

Dene Cultural Institute. 1995. Written submission to the BHP Diamond Mine Environmental Assessment Panel, April 6. Available from BHP Diamond Mine Environmental Assessment Panel, Information Office, 2nd Floor, 5-5120 49th Street, Yellowknife, Northwest Territories, X1Z 1P8.

Natcher, D., and C. Hickey. 2002. "A Criteria and Indicators Approach to Community Sustainability." Working Paper 2002-2, Sustainable Forest Management Network.

Stevenson, M.G. 1996. "Indigenous Knowledge in Environmental Assessment." *Arctic* 49(3):276-91.

–. 2006. "The Possibility of Difference: Rethinking Co-Management." *Human Organization* 65(2):167-80.

Stevenson, M.G., and J. Webb. 2004. "First Nations: Measures and Monitors of Biodiversity." *Ecological Bulletins* 51:83-92.

World Commission on Environment and Development, 1987. Our Common Future. Oxford University Press, 400 pages.

8

Accommodation of Aboriginal Rights

The Need for an Aboriginal Forest Tenure

Monique Passelac-Ross and M.A. (Peggy) Smith

The Forest Tenure System

The forest tenure system in Canada encompasses legislation, regulations, tenure agreements, permits, and government policies and guidelines that together define the rights and obligations of third parties to harvest and grow trees on public lands. We suggest that major aspects of provincial tenure systems are a structural and systemic impediment to the recognition and protection of Aboriginal and treaty rights in forest management in Canada. In particular, current tenure systems impair Aboriginal peoples' (First Nations and Métis in the case of forest management) ability to pursue traditional land uses and to translate their underlying forest values into a contemporary expression essential to the exercise of their rights. As such, for the most part, forest tenures have been instruments of assimilation rather than instruments of coexistence between Aboriginal peoples and the Canadian government. Others in this volume (Mason; Lewis and Sheppard) have pointed to other aspects of forest management that fit the assimilation approach, such as cultural resource management.

Forest tenures, as instruments of public policy, are designed to serve particular public objectives. The overriding concern of forest policies across Canada has been "the utilization and management of timber resources to provide wealth ... and a continuing, stable source of employment on a regional basis" (Haley and Luckert 1998, 129). To further the objectives of industrial development and employment, provincial governments have until recently granted long-term forest tenures to companies on two key conditions: (1) that they build, operate, or supply wood-processing facilities, and (2) that they practise sustained yield forestry. Sustainability of wood supplies is to be achieved mainly "by regulating the annual rate of harvest to ensure a continuous supply of mature timber on a crop rotation-basis" (Dellert 1998, 255). An annual allowable cut (AAC) sets the annual rate of harvest. Although AAC calculations in the twenty-first century have been modified to address a broader range of social and ecological goals (Andison 2003, 438), the determination of the AAC has generally not been sensitive to Aboriginal and treaty rights.

These two basic tenets of forest management – the allocation of long-term forest tenures to provide a continuous, secure timber supply to mill owners, and the reliance on the concept of sustained yield to manage Crown forests – remain central to forest policies. Even though the sustained yield paradigm was replaced in the 1980s by the multiple use sustained yield (MUSY) paradigm, which sought to ensure that forests are managed for both timber and nontimber uses, the goal of maximizing timber production did not change. Objectives for managing nontimber values, such as freshwater, wildlife species, and wildlife habitat, were simply added on to timber production requirements and perceived by forest companies as constraints on production. By the 1990s, the objectives of forest management had evolved from MUSY to sustainable forest management (SFM), a concept that addresses the integration of economic, environmental, and social concerns. To date, neither the MUSY nor the SFM concept has translated into any significant change in the tenure system, with the few exceptions noted in this chapter. Tenure reform and the discarding of sustained yield as the core principle of forest policies are key to achieving the new objectives of sustainability (Dellert 1998, 273).

This chapter examines two key features of forest tenure systems that most impede the accommodation of Aboriginal values, rights, and uses: (1) the determination of the AAC and (2) the process of allocating large-scale, long-term tenures to industrial interests. First, the importance of addressing Aboriginal and treaty rights in forest management is explored. Second, we examine how the determination of the AAC and the tenure allocation process affect Aboriginal and treaty rights, values, and land uses. Case studies of the Nisga'a in British Columbia and the James Bay Cree in northern Quebec illustrate the impacts of these forest policies and how these communities negotiated changes to protect their rights and interests. We conclude by making a case for alternative tenure arrangements that accommodate Aboriginal rights.

The Overarching Problem: Lack of Recognition and Protection of Aboriginal and Treaty Rights in Forest Management

Following the recognition of Aboriginal and treaty rights in the Constitution Act, 1982, the Supreme Court of Canada and lower courts across the country have been at work defining the nature and scope of these rights and governments' obligations toward Aboriginal peoples. Over the past thirty years, court decisions have changed the legal landscape in Canada and have challenged governments to amend their laws and policies to acknowledge and protect Aboriginal and treaty rights (Marc Stevenson, this volume).

In the forest sector, government and industry have responded by involving Aboriginal peoples in forest development through avenues such as the allocation of forest tenures to First Nations (for the most part, short-term, nonrenewable licences); agreements involving some say in forest management decision making; joint ventures with industry; and the negotiation of social and economic benefits in the form of training, employment, and subcontracts (NAFA and IOG 2000; Wyatt, Fortier, and Hébert 2009). Forsyth, Hoberg, and Bird (this volume) describe interim measures in British Columbia that included a reallocation of tenures to First Nations and the limitations of this approach in the absence of a full recognition of Aboriginal rights.

In all of these models of participation – what the Royal Commission on Aboriginal Peoples (RCAP) termed the "integration approach" (RCAP 1996, 851-57) – Aboriginal peoples are expected to operate within the framework of the existing industrial tenure and forest management systems. With very few exceptions, the fundamental tenets of forest policies and the tenure system have not been modified to accommodate the particular rights, values, needs, and knowledge systems of Aboriginal peoples. By drawing Aboriginal peoples into the industrial tenure system and compelling them to operate according to industrial management practices that are incompatible with their values and cultures, governments contribute to creating internal tensions and crises in many Aboriginal communities.

In British Columbia, the Haida Nation's long-standing efforts through the courts to establish that the provincial government should both consult with it and seek its consent before the replacement and transfer of Tree Farm Licence (TFL) 39 from MacMillan Bloedel to Weyerhaeuser were finally met with success. The licence area allocated by TFL 39 constitutes almost one-quarter of the total land area of Haida Gwaii. The Haida had long objected to the rate at which old-growth forests, particularly the large red cedar trees that they highly value, were being logged, as well as the methods and environmental effects of logging. The lower-court judge found that "consultation at the operational level does not permit the Haida to influence the quantity of the annual allowable cut on Block 6" (*Haida Nation v. BC,* 2000, paragraph 29). In February 2002, the BC Court of Appeal unanimously found that both the provincial government and Weyerhaeuser had a legal obligation "to consult with the Haida people and to seek an accommodation with them at the time when the processes that were under way for a replacement of T.F.L. 39 and Block 6 and for a transfer of T.F.L. 39 from MacMillan Bloedel to Weyerhaeuser" (*Haida Nation v. BC,* 2002, paragraph 58).

The Province of British Columbia and Weyerhaeuser appealed the decision to the Supreme Court of Canada. In 2004, the court ruled that the province had a duty to consult as part of its responsibility to uphold the honour of the Crown in dealings with Aboriginal peoples. However, this same obligation did not extend to the third party, Weyerhaeuser: "The Crown alone remains legally responsible for the consequences of its actions and interactions with third parties, that affect Aboriginal interests ... The honour of the Crown cannot be delegated" (*Haida Nation v. BC*, 2004, paragraph 53). The Haida's legal battle eventually brought the Province of British Columbia to the negotiating table and a new agreement (Forsyth, Hoberg, and Bird, this volume) provided for the development of a new land use plan, the establishment of a new tenure, and the determination of a new AAC for Haida Gwaii.

The time is overdue for provincial governments – as the Crowns whose honour must be upheld and as forest resource managers – to revise their forest policies and legislation to ensure that the rights of Aboriginal peoples are recognized and protected, and to find accommodation between these constitutional rights and those of forest tenure holders. The courts and the RCAP have provided guidance to governments on how this might be done. In *Haida* (2004, II A14), the Supreme Court of Canada stated that "negotiation is a preferable way of reconciling state and Aboriginal interests." The RCAP (1996, Appendix A, 2.4.48) recommended, among other things, that the federal and provincial governments establish a national code to "recognize and affirm the continued exercise of traditional Aboriginal activities (hunting, fishing, trapping and gathering of medicinal and other plants)," with the provinces adopting supporting legislation. Over fifteen years later, the RCAP recommendations have not been implemented, even though they are still relevant (McGregor 2011).

Some national forest policies began to address the recognition of Aboriginal and treaty rights, in particular the Canadian Council of Forest Ministers' Criteria and Indicators (CCFM C and I) for framework of sustainable forest management (CCFM 2003) and National Forest Strategies beginning in 1992 until 2008 (Hessing, Howlett, and Summerville 2005, 75). It is worth noting that the current National Forest Strategy, *A Vision for Canada's Forests 2008 and Beyond* (CCFM 2008), makes no mention of Aboriginal and treaty rights, making a jarring break with the 2003-8 strategy that had a comprehensive approach to addressing Aboriginal and treaty rights (NFSC 2003).

Although past national forest strategies and the CCFM C and I established a framework and commitment, to date, governmental initiatives have focused mainly on supporting Aboriginal employment and business opportunities in the forest sector and on so-called capacity building. Although economic development and capacity building are crucial for Aboriginal communities (see Allen and Krogman, this volume, for a further discussion on capacity building), progress on the more contentious issues of the recognition and protection of Aboriginal and treaty rights, and the legislative reforms that such recognition calls for, has been slow. Economic development and capacity building in the absence of a recognition of Aboriginal and treaty rights fit the assimilation model. There have been limited moves by some provinces to address changes to forest tenure, and these efforts are increasing, especially in the face of the recent forest-sector crisis. However, two fundamental features of tenure – the determination of the AAC and the process of allocating tenures – still do not consider Aboriginal and treaty rights.

Two Fundamental Forest Tenure Features That Impede the Accommodation of Aboriginal Rights

Determination of the AAC

Provincial governments regulate the annual rate of timber harvest on Crown land by setting an allowable annual cut. The AAC is the mechanism to ensure sustained yields – simply put, that harvest volumes match growth volumes. The AAC is normally calculated for a forest management unit, and specific allocations are then made to various tenure holders, with the AAC being adjusted based on more detailed local information provided by tenure holders. Calculating the AAC is a complex task. It involves the identification of a productive forest land base, the collection of wood supply inventories, and the use of statistical estimates of growth and harvest rates over time, along with the predicted effect of proposed silvicultural treatments. Problems that have plagued AAC calculations include the inadequate and/or outdated inventories of timber supplies in several provinces, and estimates of second-growth yields are not very reliable due to lack of adequate growth and yield information and monitoring (Senate Subcommittee on the Boreal Forest 1999, chapter 3, 8). Further, the calculation of an AAC is as much a political as it is a scientific endeavour. In addition to purely biophysical inputs, the AAC is also based on economic and social considerations that may considerably affect the "scientific" nature of the determination. A former BC deputy chief forester has stated that "the selection of the AAC was a subjective choice within a range of uncertainty" (Dellert 1998, 256). This assessment appears to be shared by many in British Columbia and other provinces, where confidence in AAC determinations is low, especially among environmental nongovernmental organizations, and there is widespread concern that cut levels remain too high (May 1998, 32-38).

Long-term area-based tenure holders not only acquire the right to harvest the AAC or an agreed portion, they also assume an obligation to cut all of the approved volume of timber, even if the harvesting becomes uneconomical at times (the "cut it or lose it" rule). The setting of harvest volumes in either tenure agreements or government-approved forest management plans is designed to ensure that the wood supply requirements of the mills are met. Cut control provisions are inserted in forest legislation or in tenure agreements to monitor the flow of wood. Typically, over a cut-control period of five years, the volume harvested must not be greater than 110 percent and no less than 90 percent of the volume allocated. Penalties for underharvesting (e.g., loss of the allocated timber volumes or reduction in the size of the tenure area) or excess harvesting are imposed on tenure holders (e.g., British Columbia's Forest Act, Division 3.1 – Cut Control).

The Crown retains the right to either increase or decrease the volumes allocated to tenure holders under certain circumstances specified in legislation or in tenure agreements. For example, in British Columbia, the AAC was increased in areas struck by mountain pine beetle in order to salvage beetle-killed timber (Pousette and Hawkins 2006). In principle, cut levels can be readjusted downward should they prove to be too high to accommodate environmental or social concerns. In contrast to land withdrawals, decreases in allocated volumes typically do not entitle tenure holders to be compensated. However, in British Columbia, when a government-appointed commission inquired into issues of compensation for the taking of resource interests, the forest industry argued that reductions in AAC levels and allocated volumes constitute "takings" that should be compensated (BC Forest Industry Task Force on Resources Compensation 1992, 19). Recent amendments to the Forest Act that provide for a one-time reduction in volume from the tenures of major companies also acknowledge that licensees are entitled to compensation (Forest [Revitalization] Amendment Act, 2003). In Quebec, the government can revise the AAC and the allocated volumes every five years according to certain criteria set out in the Forest Act (sections 77-78). A tenure holder will be compensated for any damage suffered as a result of the revision only if an arbitrator finds that the minister's decision does not meet the criteria set out in the act. It is noteworthy that in 2005, as a result of the recommendations of the Coulombe Commission, the Government of Quebec reduced the overall AAC in Quebec by an average of 20 percent. Quebec's new Sustainable Forest Development Act, set to go into force in 2013, continues the practice of preventing agreement holders from seeking compensation if the minister deems a reduction in AAC is in the public interest (sections 88-114).

The AAC and Aboriginal Values and Uses

The calculation of the AAC has moved somewhat beyond the sustained yield approach. Jurisdictions across Canada now consider environmental and social objectives in the determination of the AAC. For example, the AAC may be reduced to protect a species at risk as required by federal and/or provincial legislation other than forestry acts. However, provincial commitments to forest industry licence holders still constrain the AAC. In British Columbia, limits were placed on the reduction of the AAC as a result of changes in the amount of protected areas, new forest practices regulations, and species-at-risk strategies (Cashore and McDermott 2004, 34-35).

Although other values are now being considered in the determination of the AAC, it is not common practice to incorporate Aboriginal land and resources uses, nor Aboriginal rights. In order to incorporate Aboriginal concerns and interests at this critical stage in the decision-making process, governments and industry have an obligation to consult with Aboriginal peoples early in the calculation process. Only Aboriginal communities can provide the necessary information about their current land and resource uses, as well as about important sites and areas. Their input is critical to the determination of the land base available for commercial timber production and the subsequent assessment of the timber supply in a given area. It is also critical to the determination of a sustainable rate of harvest.

The most commonly accepted ways to obtain this cultural information are through either cultural heritage inventories or land use and occupancy studies. Archaeologists and anthropologists catalogue significant sites. Land use and occupancy studies evolved from work Aboriginal organizations undertook in support of land and legal claims. In several provinces, cultural heritage inventories and land use studies are now being used in forest management, usually during the planning stage. However, both of these methods have limitations (Mason, this volume). Aboriginal communities are often left out of the design of these systems; funding may be inadequate; standards for the ownership, collection, and storage of the data are usually lacking; and, most importantly, direction on how to incorporate the results into forest management plans are woefully inadequate (Tobias 2000, 2010; Wyatt et al. 2010).

An example of provincial attempts to address Aboriginal values in forest management is the Province of Ontario. The Ministry of Natural Resources (MNR) is required through the regulated Forest Management Planning Manual (FMPM) to prepare Aboriginal background information reports, values maps, and reports on the protection of identified Aboriginal values. In addition, the authors of forest management plans must discuss how those Aboriginal values will be affected by forest operations (OMNR 2004, A129-38).[1] The MNR also states that the steps taken to address Aboriginal involvement in the FMPM may be considered in addressing the obligations the province has under section 35 of the Constitution Act (OMNR 2009, A4.1). First Nations do not agree that these forest management planning provisions meet the province's legal duty to consult on forest management planning, and some, such as the Ojibway of Pic River, have developed their own consultation and accommodation guidelines that they expect the Crown to respect (Michano 2009).

Cut levels that do not permit the accommodation of both timber production and a more traditional economy based on hunting, trapping, fishing, and gathering, as well as the cultural and spiritual activities based on these land uses, let alone Aboriginal claims to the land itself, are too high and unsustainable. The impact on affected Aboriginal communities has often been devastating. As Lewis and Sheppard indicate in this volume, "If public land managers are to be responsive to the needs and values of indigenous cultures, they must work toward a fuller understanding of those needs and values, in a spatial, landscape-specific context that resonates with local communities and can be translated into forest management planning." We would take this a step further by stating that the first stage of forest management planning, the determination of the AAC, must be sensitive to Aboriginal values and rights.

The Process of Tenure Allocation

As a rule, and in contrast with volume-based tenures that tend to be allocated by public competition, the allocation of long-term area-based tenures is the outcome of private negotiations between government and industry. The only two provinces where the allocation of new area-based tenures involves a public process are British Columbia and, to a limited extent, Ontario.

In British Columbia, applications for tree farm licences (TFLs) are advertised, and public hearings may be held on the applications (Forest Act, section 33). In Ontario, the Crown Forest Sustainability Act, 1994, requires that the minister give public notice of his intention to grant a forest licence and that a competitive process be used for the allocation (section 24). However, the act also provides that another process may be used, as required by an agreement or as authorized by Cabinet (section 24(3)), so there is no hard and fast rule in Ontario for public notice of licensing.

These provisions apply only to the issuance of new area-based tenures, not to the replacement or extension of existing tenures. As in most other provinces, long-term area-based tenures in British Columbia and Ontario are protected by evergreen clauses; that is, their term can be extended at regular intervals for another period of time. The extension or replacement of the tenure is not a right. It is subject to ministerial discretion and, in Ontario, Cabinet approval. In practice, governments rarely refuse to renew or replace these tenures. Since the majority of productive forest lands have already been allocated to forest companies under some form of evergreen agreement in both British Columbia and Ontario, it is unlikely that new area-based tenures will be allocated by means of a public process.[2] An exception in Ontario is the area north of the current forest management zone (called "the area of the undertaking"), which was opened to forest development under the umbrella of the Northern Boreal Initiative in 2001 based on a community-based land use planning approach. This approach was extended in 2010 with the passing of the Far North Act.

New tenures opening up is rare; it is more common for tenures to be transferred when forest companies close mills or consolidate. In Ontario, the transfer or assignment of a long-term area-based tenure to a new tenure holder requires the consent of government, whereas in British Columbia, this is no longer the case. The BC Forest Act was amended to remove the need for ministerial consent for tenure transfers and licensee change of control (Forest [Revitalization] Amendment Act, 2003). However, the licensee must inform the minister, and the minister can cancel an agreement under certain conditions (Forest Act, section 54). Further, these transactions occur without any public involvement. Clogg (2003, 1) contends that the changes in BC legislation were designed by the province to "limit its duties to consult and accommodate First Nations, through legislative changes which reduce or eliminate statutory decisions about tenure, planning and practices and by removing opportunities for future tenure redistribution."

The fact that the process of allocation and renewal of large area-based tenures is in most provinces discretionary and shielded from public review raises questions about procedural fairness. For Aboriginal peoples, whose traditional territories and resources are thus disposed of, and who lose access to and control over lands and resources that are vital to their existence, this allocation process raises even more serious issues.

Lack of Consultation with Aboriginal Peoples in the Allocation of Forest Tenures
At the critical point of licensing in forest resource development, consultation with potentially affected Aboriginal communities would require the political will to address Aboriginal concerns and rights. Governments would have to acknowledge that Aboriginal uses of lands and resources, as well as Aboriginal and treaty rights, may be negatively affected by proposed allocations and that Aboriginal peoples are entitled to be involved in the allocation process.

In all provinces, whether or not treaties were signed, governments have shied away from such acknowledgement. Provincial governments have assumed that public forests and timber resources, which ownership the provinces claim, could be freely allocated to the private sector for industrial development. The potential impacts of forestry development on Aboriginal communities as a result of forest tenure allocations have been routinely ignored, and issues of Aboriginal rights carefully avoided. If any "consultation" with Aboriginal peoples takes place, it is only as part of a general consultation process with the local population and affected "stakeholders" during the forest management planning process. Such consultation is not considered meaningful by Aboriginal communities and does not meet the conditions for consultation to protect Aboriginal and treaty rights laid out by recent court decisions (*Haida Nation v. BC,* 2004). Again, in the absence of acknowledgment of Aboriginal and treaty rights, such consultation falls into the assimilation model.

Case Studies on Forest Tenure Systems That Accommodate Aboriginal Rights, Values, and Forest Uses

In its final report, the Royal Commission on Aboriginal Peoples (RCAP 1996, 639) stated: "It will not be enough simply to incorporate aboriginal people into existing systems of forest tenure and management. It is important to give proper consideration to aboriginal values." The RCAP commissioners encouraged the provinces to show greater flexibility in their timber management policies and guidelines. The commission provided some indication of the way in which harmonization (or coexistence) of two very different value and management systems, as an alternative to assimilation, would translate in practice:

> Some jurisdictions are already reducing their annual allowable cut requirements and the size of clearcut areas. Continued experimentation with lower harvesting rates, smaller logging areas and longer maintenance of areas left unlogged would allow greater harmonization with generally less intensive Aboriginal forest management practices and traditional Aboriginal activities. (RCAP 1996, 640)

First Nations across Canada are developing their own responses to the problems of industrial forest development. These responses vary widely depending on political, social, and economic factors both within individual communities and at a provincial level. The following examples of the Nisga'a Nation in British Columbia and the James Bay Cree in northern Quebec illustrate how Aboriginal peoples, with cooperation from provincial governments willing to explore alternative arrangements, are attempting to move beyond the strictures of the current industrial tenure system to achieve this greater harmonization of values and practices contemplated by the RCAP. Other authors in this volume also discuss provincial policy changes, particularly in British Columbia.

The Forestry Provisions of the Nisga'a Final Agreement, British Columbia

The Nisga'a have occupied the Nass Valley on the northwest coast of British Columbia since "time immemorial" (Nisga'a Lisims Government, British Columbia, and Canada, n.d.). The Nisga'a homeland, the Nass River watershed, was described by Frank Calder in the Nisga'a appeal for recognition of their Aboriginal title to the Supreme Court of Canada (*Calder et al. v. Attorney-General of British Columbia*, 1973, 349):

> From time immemorial the Nishgas have used the Naas River and all its tributaries within the boundaries so submitted, the lands in Observatory Inlet, the lands in Portland Canal, and part of Portland Inlet. We still hunt

> within those lands and fish in the waters, streams and rivers, we still do, as in time past, have our campsites in these areas and we go there periodically, seasonally, according to the game and the fishing season, and we still maintain these sites and as far as we know, they have been there as far back as we can remember. We still roam these territories, we still pitch our homes there whenever it is required according to our livelihood and we use the land as in times past, we bury our dead within the territory so defined and we still exercise the privilege of free men within the territory so defined.

The Supreme Court of Canada acknowledged that the Nisga'a had never given up rights to their territory. This decision eventually led to negotiations between the federal and provincial governments and the Nisga'a. Almost thirty years after the court ruling, on May 11, 2000, the first modern-day treaty in British Columbia came into effect. The trilateral agreement, between the Nisga'a Nation and the Governments of Canada and British Columbia, gave the Nisga'a fee simple title to roughly two thousand square kilometres of land in the Nass River valley of west-central British Columbia. Prior to the agreement, the Nisga'a were wards of the federal government under the Indian Act and restricted to sixty-two square kilometres of reserve lands, including the villages of New Aiyansh, Canyon City (Gitwinksihlkw), Greenville (Laxgalts'ap), and Kincolith (Gingolx).

The Nisga'a Final Agreement (2000) changed the path the Nisga'a had walked since European colonization sought to control their destiny and lands. In the midst of controversy about treaty making in British Columbia – from those who think the Nisga'a gave up too much to those who contend that governments gave up too much – the Nisga'a now have to demonstrate to the world that control of their territory will benefit not only their own society but also that of the province and country. This is especially true in forest management, where onlookers wait to see if there is any difference in how the saws are wielded when First Nations hold them.

What kind of forest did the Nisga'a gain control over? As early as 1985, the Nisga'a Tribal Council had laid a formal complaint with the BC Ombudsman's Office about forestry practices, including "highgrading, overcutting and lack of reforestation," that began in the 1950s when large forest companies started decimating the forests in the Nass Valley. They also proposed a plan in 1982, "Forests for People: A Nishga Solution," that called for the Nisga'a to take over the Tree Farm Licence (Notzke 1994, 101). Much later, a timber supply analysis conducted for the Nass (Kalum North) Timber Supply Area by the BC Ministry of Forests (BC MoF 1993) supported the contention that cutting in the area was unsustainable. The analysis concluded that "under current management practices, the current harvest can be maintained for one decade" and "the rates of decline in harvests shown in this analysis are about 10 percent per decade."

The agreement the Nisga'a signed did not give them immediate control over their land base. Chapter 5 of the Nisga'a Final Agreement (NFA) covering "forest resources" outlined a five-year transition period during which time the Nisga'a Nation and the Province of British Columbia would jointly oversee forestry on Nisga'a lands. Before the Nisga'a could take control of their lands, the Nisga'a had to develop a forest legislation that would meet or exceed provincial guidelines. During the transition period, the preexisting allowable cut of 155,000 cubic metres per year to current licensees was maintained, with the Nisga'a receiving an additional allocation of 10,000 cubic metres. After year five, the AAC was revised to 135,000 cubic metres.

Critics have questioned whether the agreement gives the Nisga'a the flexibility to establish anything other than an industrial model of forest management, which would counter the system that led to severe overcutting in the region. Given this history of overcutting, the Nisga'a face a double challenge in implementing the forestry provisions of the treaty: to use the revenues from timber harvesting to support their communities and develop their government structures while ensuring that forestry practices reflect the broad range of Aboriginal uses and values.

The NFA prohibited the Nisga'a from establishing a primary processing facility for a period of ten years. Given the commitments to maintain licences and cut levels to the existing mill owners during the five-year transition period, particularly Skeena Cellulose, which controlled almost half of the cut, and the requirements imposed on existing licensees to award logging contracts to the Nisga'a, one could easily imagine that the Nisga'a might become dependent on those mill owners. However, the forest sector downturn that had already begun in 2000 did not result in such dependence. In fact, mill closures, including the Skeena mill, meant that the Nisga'a cut much less than the AAC. Actual harvesting went from about 100,000 cubic metres in 2000 to a low of 21,000 cubic metres in 2004, climbing to 87,000 cubic metres in 2007 with a value of $5.5 million (Nisga'a Final Agreement Implementation Report 2006-2007, 2007-2008). In spite of the industry downturn, the Nisga'a were able to find markets for cedar and sawlogs with the help of Lisims Forest Resources LLP (LFR), established by the Nisga'a Lisims Government in 2006. LFR looks after the sale of all forest resources, including timber and nontimber products, to new markets in China, Japan, and Korea. In the management area, in 2007-8, the Forestry Transition Committee was superseded by the Nisga'a-BC Forest Service Forestry Committee, which meets periodically to address forest issues on Nisga'a lands (Nisga'a Final Agreement Implementation Report 2006-2007, 2007-2008).

The Nisga'a have used the law-making powers given to them in the treaty to develop a better forest management system based on values and land use practices that have long held them in good stead. The Nisga'a Forest Act, 2000, was prepared prior to the ratification of the final agreement and establishes the basis for regulation of forest practices. The Nisga'a have chosen a results-based, rather than prescriptive, approach. For example, rather than prescribing reserves along waterways, management is based on maintaining the integrity of those waterways and avoiding downstream damage. This approach requires detailed knowledge of lands and waters and places a greater degree of responsibility on managers who are answerable to the Nisga'a Lisims Government.

The Nisga'a Forest Act addresses both timber and nontimber products. Pine mushrooms are a significant source of revenue, bringing in anywhere from $50,000 annually to $1.3 million (Nisga'a Lisims Government, British Columbia, and Canada, n.d.). The Nisga'a have put in place a regulatory system to ensure both sustainable harvests and a fair share of economic benefits. Permits are issued for the harvest of pine mushrooms, with at least 80 percent of mature timber being retained within pine mushroom harvesting areas. They have also set aside no-cut zones in the Nass bottomlands, within any archaeological site, within any ecological reserve, and around their village sites.

The combination of fee simple title; the requirement to manage to a standard equal to or higher than the provincial government's; the ability to make, monitor, and enforce their own laws; and a continuing source of revenue from timber harvesting and nontimber forest products are all conditions that support the ability of the Nisga'a to set a new standard for sustainable forest management. But these conditions do not guarantee that goal and any one of them, as evidenced by the forest sector economic crisis, can cause a setback. It will be some time before the Nisga'a results-based approach to forest management can be fully analyzed.

The James Bay Cree and the Province of Quebec's Bilateral Agreement

The James Bay Cree have occupied the homeland they call "Iiyiyuuschii" for thousands of years. The region covers about 400,000 sq km in northern Quebec in the eastern James Bay and southeastern Hudson Bay watersheds (GCC 1999, 14). Approximately fifteen thousand Cree live in nine communities in northern Quebec: five communities along the coast – Waskaganish, Eastmain, Wemindji, Chisasibi, and Whapmagoostui – and four inland – Waswanipi, Nemaska, Oujé-Bougoumou, and Mistissini. They claim their identity has been shaped by their relationship to the land and the greatest challenge they now face is "how to be Cree in this rapidly changing world (ACCI, n.d.). Collectively, the James Bay Cree are governed by two bodies, the Grand Council of the Crees (GCC) and the Cree Regional Authority, "managed and operated as one organization by the Cree Nation" (GCC, n.d.).

In the early 1970s, the James Bay Cree rose up in opposition to Quebec's plans for a massive hydroelectric project on James Bay (CBC 2009). Their opposition led to the signing of the first modern land claim settlement in Canada in 1975, the James Bay and Northern Quebec Agreement (JBNQA). Despite the guarantees that were provided in the JBNQA that their way of life and traditional economy would be maintained, for more than twenty-five years the Cree experienced the disastrous environmental and social consequences of forestry developments in their territory. Entire hunting territories were logged by forest companies in the southern part of Iiyiyuuschii, leading to displacement of Cree families from their traditional hunting territories and loss of their ability to subsist from the land. These developments threatened their way of life and resulted in serious social problems in the affected communities.

The Grand Council of the Crees conducted research to measure the impact of industrial logging on their traplines north of the fiftieth parallel, where logging had been occurring for over ten years. The Nemaska Trapline Project (Forestry Working Group, GCC, n.d.) assessed regeneration on one trapline in northern Quebec in 1999, the only trapline where sufficient time had passed since logging for the establishment of regeneration. The research team included the trapper, as well as forestry and environmental technicians. Although this type of study will need to be repeated across other traplines, this single study did provide what the researchers describe as "directional signposts." The study found that only two out of eleven cut blocks sampled were satisfactorily regenerated. For the Cree, the study raised serious concerns, including excessive alder growth, poor drainage leading to erosion, siltation of nearby watercourses, and rutting damage from machinery. The research also pointed to the lack of a "formal scientific vehicle" for trappers' observations on how forestry operations are changing the local environment.

After many years of negotiations and having launched several lawsuits against the Province of Quebec and forest companies operating in their territory, in 2001 the James Bay Cree negotiated an agreement with the Province of Quebec, La Paix des Braves.[3] This landmark "nation-to-nation" agreement instituted changes identified by the Cree to protect their traditional land uses while still allowing industrial forestry. The agreement in principle, signed in October 2001, was approved by 72 percent of the Cree in a referendum held across the nine Cree communities in northern Quebec, with 53 percent of eligible voters casting a ballot. In the communities most directly affected by the proposed diversion of the Rupert River, 80 percent of eligible voters supported the agreement (Scott 2005).

The agreement deals with issues as diverse as forestry, hydroelectricity, mining, economic and community development, financial provisions, and legal proceedings. The forestry provisions are found in chapter 3 of the agreement. This chapter provides for the application of a different forestry regime in the James Bay Territory, one that should enable the Cree to manage development "so that it respects the natural productivity of the territory" and allow them to decide, in cooperation with the provincial government, "what is appropriate development and what is not acceptable."[4] The agreement states that a new forestry regime will apply in Cree territory, respecting both the Forest Act (R.S.Q., chapter F-4.1 as amended by S.Q. 2001, chapter 6) and the JBNQA. This adapted regime will allow (1) adaptation to better take into account the Cree traditional way of life; (2) greater integration of concerns relating to sustainable development; and (3) participation, in the form of consultation, by the James Bay Cree in the various forest activities, operations planning, and management processes.

One of the most fundamental changes to the provincial forestry regime is the adoption of the Cree traplines as the basis for delimiting forest management units. The delimitation of the trapline boundaries was entrusted to the Cree Regional Authority. The agreement anticipated that the next generation of forest management plans, to be adopted in April 2005, would be configured on the basis of these new trapline management units. A provisional Cree-Quebec working group would determine the limits of the new management units, which would be approved by the minister of natural resources, who would then notify the agreement holders in conformity to the Quebec Forest Act. The AAC calculation would be determined on the basis of these traplines in keeping with the rules established in the agreement.

Other forestry provisions allow for the protection of sites and areas of special interest to the Cree. For example, for each trapline, the Cree will identify and map areas where no forestry activity can be undertaken unless authorized by the tallyman, the equivalent of the resource manager or game warden, who is a key player in the Cree traditional land management system. This system is based on family or group hunting territories, or traplines. It is the responsibility of the tallyman to ensure that the trapline has enough fish and game to support the group.

Other provisions of the agreement address wildlife habitat management and stipulate that, within designated areas, "specific management standards will be applied to maintain or improve the habitat of very important wildlife species (moose, marten, beaver, hare, fish, caribou, partridge)" (3.10.1). Forest management activities in these areas will be planned with the goal of maintaining and improving the diversity of forest stands, with specific measures applying and rates of harvest as defined in the agreement. Over the remaining area of each trapline, specific measures must be taken "to ensure the protection of a residual forest cover" (3.11.1) such as the use of mosaic cutting, limiting the size of cutblocks to a maximum of one hundred hectares, disallowing logging in heavily logged traplines, and defining annual rates of harvest in each trapline according to the level of previous disturbances. Additional forest protection measures (e.g., buffer strips of a certain width) apply along watercourses and lakes. Finally, the road access network is to be planned jointly by tenure holders and tallymen in order to limit the number of road connections between traplines.

The agreement provides for the creation of two types of co-management structures to facilitate Cree involvement in the forest planning and management decision making. The first is the Cree-Québec Forestry Board, comprising equally representatives of the Cree and the Quebec government. The chair is appointed by the Quebec government after consultation with the Cree Regional Authority. The board is responsible for monitoring, analyzing, and assessing the implementation of the agreement forestry provisions and for making recommendations to the parties as to required modifications. The board is also responsible for reviewing forest management plans prior to their approval by the minister of natural resources. A second community-based co-management structure is the joint working group established in each Cree community affected by commercial forestry. These working groups, with equal representation from the community and government, have input into the development and monitoring of forest management plans and, when required, elaborate harmonization measures flowing from the agreement forestry provisions.

The Quebec government also commits to allocating to the Cree, within the commercial forest located in the James Bay Territory, timber volumes increasing from 70,000 cubic metres per year in 2003 to 350,000 cubic metres in 2006 and thereafter. These allocations will be made primarily under long-term area-based tenures, and the Cree Regional Authority will distribute them among their communities.

The agreement forestry regime, which takes precedence over conflicting or incompatible provisions of the Forest Act and regulations, represents a marked departure from past practice in the James Bay Territory. It creates conditions for a co-management regime that is meant to enable the Cree to achieve both the protection of a traditional way of life and economic benefits for their communities. To date, the Cree Regional Authority, in consultation with the tallymen, has finalized the boundaries of the traplines, and the two types of co-management structures (the Cree-Québec Forestry Board and the local joint working groups) have been established. The Cree-Québec Forestry Board published a status report on the implementation of the forestry-related provisions of the agreement (2002-8), which stated that the large majority of these provisions have been implemented (CQFB 2009). In consultation with the Cree, the AAC across the James Bay Territory has been reduced by 25 percent, from some 5 million cubic metres to about 3.75 million cubic metres per year. This has relieved the development pressure on the land. Former grand chief Dr. Ted Moses (2005) describes the benefits of the new forestry regimes as follows:

> Implementation of the Paix des Braves and Cree involvement in the management of the forests [have] reduced the number and extent of roads; increased the number and size of protected areas (including Muskuuchii, our sacred mountain, that has sustained us with its bounty of resources since time immemorial); increased the size and efficacy of buffer strips along lakes and watercourses; engaged Cree communities and the Ouchimauch in day to day management of the forest regime; and helped to protect and restore vital habitat.

Quebec changed its forest legislation in 2010: the Sustainable Forest Development Act (R.S.Q., chapter A-18.1) will take full effect in March 2013. The act implements major transformations to the existing regime, including the abolition of the long-term area-based tenures, the resumption of planning responsibilities by the government, the auctioning of public timber, and the decentralization of decision making. The impact of this new regime on the adapted forestry regime instituted by the 2002 Paix des Braves is uncertain. The Cree and the provincial government are currently involved in negotiations to decide how to harmonize the regime applicable on the Cree territory and the new provincial forestry regime.

Toward an Alternative Forest Tenure That Accommodates Aboriginal and Treaty Rights

The Nisga'a and James Bay Cree case studies indicate the beginning of a departure from the systematic application by provincial governments of the industrial model of forestry to Aboriginal communities. These innovative models have been enabled by agreements between federal, provincial, and Aboriginal governments or by provincial governments willing to modify the industrial tenure system. The case studies also clearly demonstrate the determination of Aboriginal communities and organizations to exercise on their own, or to influence in partnership with others, different forms of forest stewardship incorporating Aboriginal forest values and land uses. Both the Nisga'a and the James Bay Cree are committed to implementing new forestry practices that reflect both their need for contemporary economic development and their desire to protect traditional land use values, even to the point of proposing reduced levels of harvesting.

But these case studies are the exception, not the rule. There are still too few examples of innovative approaches to forest use, and those that are leading the way face what seem to be insurmountable odds of continued structural impediments. Their gains have been the result of long, hard-fought battles, often following acts of civil disobedience, legal actions, long and tedious negotiations with governments, and appeals to international and national media, as was the case with Clayoquot Sound (see Smith, this volume). Once underway, those involved with these experiments face uncertainty and unforeseen setbacks because the pressure is still on to fit within the current industrial model of forest management. If these obstacles are overcome, and provincial governments fully embrace a recognition of Aboriginal and treaty rights, alternative forest tenures might help move Aboriginal peoples toward "shared power" or co-management as coexistence, as proposed by Smith in this volume. Provincial and federal governments might also find that such new Aboriginal forest tenures offer innovative options for achieving sustainable forest management.

Notes

1 See McGregor (2000) for recommendations on improving values mapping in Ontario.

2 Since this chapter was originally written, British Columbia in 2003 began awarding forest and range agreements with a revenue-sharing component to First Nations and in 2010 put in place an area-based, twenty-five-year First Nations woodland licence only for those First Nations with an interim measures agreement with the province. Because this is such a new form of tenure, the authors are not in a position to evaluate the effectiveness of these new tenure arrangements at this point. For commentary from BC First Nations on these agreements, visit the BC First Nations Forestry Council at http://www.fnforestrycouncil.ca/.

3 Agreement Concerning a New Relationship between Le Gouvernement du Québec and the Crees of Québec, February 7, 2002, http://www.gcc.ca/gcc/.

4 Ibid., 4.

References

ACCI (Aanischaaukamikw Cree Cultural Institute). n.d. "Iiyiyuuschii – Our Land." CreeCulture.ca, http://www.creeculture.ca/.

Andison, D.W. 2003. "Tactical Forestry Planning and Landscape Design." In *Towards Sustainable Management of the Boreal Forest,* edited by Philip J. Burton, Christian Messier, Daniel W. Smith, and Wiktor L. Adamovicz, 433-75. Ottawa: National Research Council of Canada.

BC Forest Industry Task Force on Resources Compensation. 1992. *A Submission in Response to the Report of the Commission of Inquiry into Compensation for the Taking of Resource Interests.* Vancouver: Council of Forest Industries of BC.

BC MoF (British Columbia Ministry of Forests, Integrated Resources Branch). 1993. *Nass (Kalum North) TSA Timber Supply Analysis Executive Summary*. Victoria: Ministry of Forests. http://www.for.gov.bc.ca/.

Calder et al. v. Attorney-General of British Columbia, [1973] S.C.R. 313.

Cashore, B., and C. McDermott. 2004. "Global Environmental Forest Policies: Canada as a Constant Case Comparison of Select Forest Practice Regulations; International Forest Resources." http://www.llbc.leg.bc.ca/.

CBC. 2009. "James Bay Project and the Cree." CBC Digital Archives, http://archives.cbc.ca/

CCFM (Canadian Council of Forest Ministers). 2003. *Defining Sustainable Forest Management in Canada: Criteria and Indicators 2003*. Ottawa: Canadian Forest Service, Natural Resources Canada. http://www.ccfm.org/.

–. 2008. *A Vision for Canada's Forests 2008 and Beyond.* Ottawa: Canadian Forest Service, Natural Resources Canada. http://www.ccfm.org/.

Clogg, J. 2003. *Provincial Forestry Revitalization Plan – Forest Act.* Vancouver: West Coast Environmental Law Research Foundation. http://wcel.org/resources/.

CQFB (Cree-Quebéc Forestry Board). 2009. *Agreement Concerning a New Relationship between the Gouvernement du Québec and the Crees of Québec: Status Report on the Implementation of Forestry-Related Provisions 2002-2008*. Abridged version. Québec City and Waswanipi: Cree-Quebéc Forestry Board. http://www.ccqf-cqfb.ca.

Crown Forest Sustainability Act, 1994. S.O., c. 25.
Dellert, L.H. 1998. "Sustained Yield: Why Has It Failed to Achieve Sustainability?" In *The Wealth of Forests: Markets, Regulation and Sustainable Forestry,* edited by C. Tollefson, 255-77. Vancouver: UBC Press.
Far North Act. S.O. 2010, c.18.
Forest Act. R.S.BC. 1996, c. 157.
Forest Act. R.S.Q. 1996, c. F-4.1.
Forest (Revitalization) Amendment Act, 2003, S.BC., c. 30.
Forestry Working Group, GCC (Forestry Working Group, Grand Council of the Crees). n.d. *The Nemaska Trapline Project: A Preliminary Review of Forest Regeneration on Traplines North of the 50th Parallel.* http://www.gcc.ca/cra/.
GCC (Grand Council of the Crees). 1999. Importance of Neebee ("Water") and Neebee Management in James Bay Cree Territory: The Need to End Government Marginalization of the James Bay Crees. A Brief prepared by the Grand Council of the Crees (Eeyou Istchee) to Québec's Commission on Water Management. http://www.gcc.ca/pdf/ENV000000002.pdf.
–. n.d. "About the Grand Council of the Crees (Eeyou Istchee)." Grand Council of the Crees, http://www.gcc.ca/gcc/.
Haida Nation v. BC (Minister of Forests), [2004] 3 S.C.R. 511; [2002] 2 C.N.L.R. 121 (BCC.A.); [2001] 2 C.N.L.R. 83.
Haley, D., and M. Luckert. 1998. "Tenures as Economic Instruments for Achieving Objectives of Public Forest Policy in British Columbia." In *The Wealth of Forests: Markets, Regulation and Sustainable Forestry,* edited by C. Tollefson, 123-51. Vancouver: UBC Press.
Hessing, M., M. Howlett, and T. Summerville. 2005. *Canadian Natural Resource and Environmental Policy: Political Economy and Public Policy*. 2nd ed. Vancouver: UBC Press.
James Bay and Northern Quebec Agreement. 1975. http://www.gcc.ca/.
La Forest, Hon. G.V. 2007. "Reminiscences of Aboriginal Rights at the Time of the Calder Case and Its Aftermath." In *Let Right Be Done: Aboriginal Title, the Calder Case and the Future of Indigenous Rights.* edited by H. Foster, H. Raven, and J. Webber, 54-60. Vancouver: UBC Press.
May, E. 1998. *At the Cutting Edge: The Crisis in Canada's Forests*. Toronto: Key Porter.
McGregor, D. 2000. "From Exclusion to Co-Existence: Aboriginal Participation in Ontario Forest Management Planning." PhD diss., University of Toronto.
–. 2011. "Aboriginal/Non-Aboriginal Relations and Sustainable Forest Management in Canada: The Influence of the Royal Commission on Aboriginal Peoples." *Journal of Environmental Management* 92:300-10.
Michano, J. 2009. Committee Transcripts: Standing Committee on General Government, August 11. Bill 173, Mining Amendment Act, 2009. http://www.ontla.on.ca/.
Moses, T. 2005. "A Modern Treaty That Opens the Door to the Duty to Consult and Accommodate: An Overview of the James Bay and Northern Quebec Agreement and the Paix des Braves." Paper presented at the New Duties for the Crown and Aboriginal Peoples, Pacific Business and Law Institute conference, Ottawa, April 26-27.
NAFA and IOG (National Aboriginal Forestry Association and Institute on Governance). 2000. *Aboriginal-Forest Sector Partnerships: Lessons for Future Collaboration.* Ottawa: NAFA and IOG. http://www.nafaforestry.org/.
NFSC (National Forest Strategy Coalition). 2003. *National Forest Strategy 2003-2008: A Sustainable Forest; The Canadian Commitment.* Ottawa: NFSC.
Nisga'a Final Agreement. 1999. http://www.gov.bc.ca/.
Nisga'a Lisims Government, British Columbia and Canada. n.d. Nisga'a Final Agreement Implementation Report 2006-2007, 2007-2008, http://www.gov.bc.ca/arr/firstnation/nisgaa/implement/down/nisgaa_fa_implementation_report_2006-08.pdf.
Notzke, C. 1994. Aboriginal Peoples and Natural Resources in Canada. Concord, ON: Captus Press.
OMNR (Ontario Ministry of Natural Resources). 2004. *Forest Management Planning Manual.* http://www.mnr.gov.on.ca/.
–. 2009. *Forest Management Planning Manual.* http://www.mnr.gov.on.ca/.

Pousette, J., and C. Hawkins. 2006. "An Assessment of Critical Assumptions Supporting the Timber Supply Modelling for Mountain-Pine-Beetle-Induced Allowable Annual Cut Uplift in the Prince George Timber Supply Area." *BC Journal of Ecosystems and Management* 7(2): 93-104. http://www.forrex.org/.

RCAP (Royal Commission on Aboriginal Peoples). 1996. *Report of the Royal Commission on Aboriginal Peoples: Restructuring the Relationship.* Vol. 2, Part 2. Ottawa: Communications Group. http://www.collectionscanada.gc.ca.

Scott, C.H. 2005. "Co-Management and the Politics of Aboriginal Consent to Resource Development: The Agreement Concerning a New Relationship between the Government of Québec and the Crees of Québec (2002)." In *Canada: The State of the Federation 2003; Re-Configuring Aboriginal-State Relations,* edited by M. Murphy, 133-63. Montreal and Kingston: McGill-Queen's University Press.

Senate Subcommittee on the Boreal Forest. 1999. *Competing Realities: The Boreal Forest at Risk.* http://www.parl.gc.ca/.

Sustainable Forest Development Act. 2010. R.S.Q., c. A-18.1.

Tobias, T.N. 2000. *Chief Kerry's Moose: A Guidebook to Land Use and Occupancy Mapping, Research Design and Data Collection.* Vancouver: Ecotrust Canada and Union of BC Indian Chiefs. http://www.ecotrust.org/.

–. 2010. *Living Proof: The Essential Data Collection Guide for Indigenous Use and Occupancy Map Surveys.* Vancouver: Ecotrust Canada and Union of BC Indian Chiefs.

Wyatt, S., J.-F. Fortier, and M. Hébert. 2009. "Multiple Forms of Engagement: Classifying Aboriginal Roles in Contemporary Canadian Forestry." In *Changing the Culture of Forestry in Canada: Building Effective Institutions in Sustainable Forest Management,* edited by M.G. Steven and D.C. Natcher, 163-80. Edmonton: CCI Press.

Wyatt, S., D.C. Natcher, P. Smith, and J.-F. Fortier. 2010. "Aboriginal Land Use Mapping: What Have We Learned from 30 Years of Experience?" In *Planning Co-Existence: Aboriginal Issues in Forest and Land Use Planning,* edited by M.G. Stevenson and D.C. Natcher, 185-98. Edmonton: CCI Press.

Part 4
Traditional Ecological Knowledge and Use

9
Early Occupation and Forest Resource Use in Prehistoric British Columbia

Brian Chisholm

For the most part, this volume considers the contemporary context of First Nations' involvement in the forest resource sector of the economy. However, the use of forest resources in British Columbia is not just a recent phenomenon. As the Elders often say – their people have been here since the beginning, since time immemorial. Archaeologists may disagree on the length of time that represents, but it is perfectly clear that the First Nations are indeed the first people in British Columbia, and that they have been reliant on forest resources for a very long time. Ethnographic reports have outlined the use of wood and other plant materials for construction of shelters, manufacture of containers, making fish weirs, making tools, and so on. Lithic materials, used to make wood-working tools and for other purposes, have also been discussed in detail. The question that an archaeologist would ask and that this chapter addresses is, how far back in time do we have evidence of the use of such resources? Here the focus is on time depth and evidence of *early* settlement and forest resource use in British Columbia; it is not exhaustive, however.

Coastal Settlement

As discussed by Heusser (1960) and Fladmark (1975) over thirty years ago, the Northwest Coast was most likely accessible, at least to some degree, to humans and human occupation throughout most, if not all, of the last ice age, and certainly by at least 10,200 calendar years before present (BP) (Josenhans et al. 1997).

At On Your Knees Cave, Prince of Wales Island, Alaska, just north of Haida Gwaii, human remains have been dated to 9730 ± 60 and 9880 ± 50 years BP (Heaton and Grady 2003). Evidence of early human occupation dates to around 9400 BP at the Arrow Creek site in the Gwaii Haanas Archipelago, Haida Gwaii. A basaltic artifact with a barnacle attached produced a date for the barnacle of 9210 ± 190 BP; it was recovered from about two and a half metres above sea level (Josenhans et al. 1995). Also, a piece of sea otter bone showing cut marks and associated with a number of lithic artifacts was recovered from the Echo Bay site in the Gwaii Haanas Archipelago, and dated to 9270 ± 120 BP. An artifact of *Ursus americanus* (black bear) bone from the same cave was dated to 10,300 ± 50 BP. Nearby in British Columbia, McLaren (2008) reports a series of uncalibrated dates for anthropogenic deposits from sites in the Dundas Islands that date back over 9,500 years.

Dates for the Namu site, on the central coast of the province, extend back to 9720 BP (R. Carlson 1996). There is a variety of lithic artifacts present in the early layers of the site, as well as some preservation of faunal species. People at Namu were clearly making use of the marine environment for their subsistence from the earliest times.

Located under the parking lot of the BC Ferries terminal at Port Hardy, the Bear Cove site was occupied over eight thousand years ago. Faunal materials from the site reflect a subsistence base of fishing, sea mammal hunting, and terrestrial mammal hunting (C. Carlson 1979).

The Glenrose site sits on the south bank of the Fraser River at the western end of and below the Surrey uplands. At the time the site was first occupied, 8150 ± 250 BP (Matson 1976), much of the present Fraser Delta lands did not exist; they were still far upstream in the Fraser River drainage.

The Milliken site is located on the east side of the Fraser River a few kilometres north and east of the town of Yale. Dates given for the earliest phase of occupation are 9000 ± 150 BP and 8150 ± 310 BP (Borden 1975; Pokotylo and Mitchell 1998). Included in the recovered materials were numerous wild cherry seeds.

Interior Settlement

In the Peace River area of the province, overlooking Charlie Lake, on the western outskirts of Fort St. John, is a small cave. When it was first occupied it would have had a view of glacial Lake Peace. The lowest cultural level of the Charlie Lake Cave site has yielded radiocarbon dates ranging from 10,100 BP to 10,770 BP (Fladmark 1996). In southeastern British Columbia, Fedje and White (1988) have reported dates between 10,700 BP and 9600 BP, similar to those for Charlie Lake Cave and for the Vermillion Lakes sites in the Crowsnest Pass region.

Although the Interior Plateau was ice-free and habitable by at least 11,500 years ago, evidence for early settlements is scarce. Early finds include the Landels, Drynoch Slide, and Gore Creek sites, as well as several potentially old surface finds. The surface finds are undated but are suspected to be old on the basis of typological similarities with similar finds elsewhere that are known to be old (see Rousseau 1993; Stryd and Rousseau 1996). The Landels site, near Ashcroft, was investigated and reported by Rousseau (1991) and Rousseau et al. (1991) and dated to 7700 BP and 8400 BP. The Drynoch Slide site, located south of Spences Bridge on the Trans-Canada Highway, was dated to 7530 BP by Sanger (1967); however, this is not generally considered to be a secure date. The Gore Creek site provides the earliest evidence of human occupation in the South Thompson River drainage. Human remains of an individual apparently trapped by a mudslide were found below the Mazama ash layer and dated to 8250 BP (Cybulski et al. 1981).

From these dates it is clear that there has been settlement of the coast and interior of British Columbia for over ten thousand years and some time beyond that. However, the main concern here is the time depth for evidence of the use of forest resources.

The Use of Plant Resources for Food

Evidence for food resource use is available from various sources, including faunal and floral analyses, tools and containers, and bone chemistry. Ethnographic evidence has allowed compilation of long lists of plant species used and their various uses (see Appendix 9.1); however, it is difficult to link such evidence to the archaeological record. Floral samples do not preserve as well as bone and are more difficult to recover from deposits, in part because of their small sizes. Thus, our knowledge of the use of plants for food, medicine, or other purposes lags behind our knowledge of animals. This has led to an undervaluing of plants as food resources in the past, a situation that is slowly being rectified.

Ethnographic evidence suggests that species such as mountain potatoes, balsamroot, and wild lily bulbs, among others, were being used along the coast and in the interior of the province. Berries were a major source of plant food for both coastal and interior people but, unfortunately, most excavations have not yielded their remains. Work in progress on the Pitt River in the Fraser Valley has recovered a number of plant specimens from wet site environments. At the Scowlitz site near the mouth of the Harrison River, Lepofsky and Lyons (2003) identified forty-two plant taxa in deposits dating around 1000 BP to 800 BP, based primarily on seed and charcoal samples. Among the oldest evidence for the use of plants for food are the carbonized wild cherry pits recovered from the basal level of the Milliken site, dating to around 9000 BP.

Evidence for plant use in the interior regions of British Columbia has been based largely on the presence of roasting pits. Most roasting pits date to the last few hundred years; however, at the Parker site in Oregon Jack Creek Valley, near Ashcroft, there is a date of 3130 BP for a cooking pit feature (Rousseau et al. 1991). Similar roasting/cooking pits are found in the Upper Hat Creek Valley, dating back to about 2300 BP (Pokotylo and Froese 1983). More recent pits have been found in a number of interior sites.

But not all of the plant remains come from food species; some are utilized in the acquisition of other foods. For example, fish traps and fish weirs are sometimes preserved for long periods. They were made from a variety of wood species, depending on what was locally available. Early dates for fish weirs are reported from Glenrose at 4590 ± 50 BP and 4370 ± 50 BP (Matson and Coupland 1995).

Plant materials were also used for making containers for food storage. Ethnographically, we know that bent wooden boxes and other wooden items for food storage and preparation were common along the coast. Texts also mention the use of cedar and spruce roots and withes and bark for the manufacture of basketry, hats, clothing, and other items. However, good preservation of such materials and items is restricted to wet site situations, which are few. Therefore we do not have a good idea of how long they have been in use along the coast or in the interior.

Unfortunately, the analysis of archaeological basketry has been carried out by only a small number of researchers, and the results are far from complete. Basketry materials are commonly wood splints from species such as western red cedar, with occasional use of yew, fir, and spruce. Roots, when used, are nearly always from spruce, while other species provide withes. Bark from western red cedar is used, with bitter cherry and maple sometimes being added.

Basketry is recovered from archaeological wet sites where ground water has provided anaerobic conditions that prevent bacterial action and hence preserve deposited organic material. One useful aspect of studying basketry is that styles of weaving and decoration may vary from one maker to another, or from one tribal group to another. Thus, baskets may provide useful information on cultural distinctions within the early populations of British Columbia. Those distinctions may not be visible through any other means. Early dates for basketry include specimens from the Glenrose site at about 4300 BP, and at St. Mungo at about 4000 BP, and Musqueam NE at about 3000 BP (Bernick 1998). Bernick (1998) has shown that styles of basketry dating back about three thousand years in the Fraser Delta region are essentially the same as they are now. This lends credence to the idea that the cultural attributes that we see in recent populations in British Columbia have been in place with little modification for a relatively long period, certainly prior to European contact on the coast.

Wood in Structures

As wood does not preserve for long periods in the BC environments, it is hard to find good evidence of early wooden items. However, tools, believed to have been used for woodworking, have been found in the earliest sites. Cobbles with sharp edges may have been used for chopping down trees, while long wedge-shaped tools of wood, antler, and stone may have been used for splitting tree trunks into planks or hollowing them out to make watercraft. Traces of early house structures have been found in a number of sites.

In the lower Fraser Valley region, not far from the mouth of the Harrison River, are two sites that have provided information on early house structures. The Xa:ytem (Hatzic Rock) site just west of Mission was once situated on the banks of the Fraser River and has produced a date of *circa* 4800 BP for a house structure (Matson and Coupland 1995), and the site may have been occupied as early as around 9000 BP. The Maurer site is located near Agassiz in the lower Fraser Valley. It contains the remains of what appear to be a large rectangular wooden house structure that has been dated to between 5500 BP and 3500 BP (LeClair 1973; Schaepe 2003). Reevaluation of the site (Schaepe 2003) produced a date of 4230 BP and confirmed that there was a house structure at that time. At both sites, it appears that posts were arranged in a rectangular pattern, suggesting a form similar to the more recent plank houses of the Northwest Coast settlements, although it is not clear if planks were used in the two houses (Schaepe 1998). They were semi-subterranean to some extent; the Xa:ytem house seems to have been cut into the riverbank, with the excavated dirt being used to form a platform area in front. Other sites along the British Columbia coast that contain the remains of house structures are not as old as Maurer and Xa:ytem. Hence, these two sites represent, at least for the time being, the earliest known use of what appear to be plank house structures on BC's Northwest Coast. The sizes of these two houses are not as large as similar ethnographically recorded structures, but they do indicate that the people of the lower Fraser Valley had a settlement pattern that was at least semi-sedentary, and that they had the appropriate skills and tools for the building of larger plank houses.

Houses at the Paul Mason site in Kitselas Canyon on the Skeena River were large multifamily rectangular structures. Coupland (1985, 1988) reported dates of 3230 ± 160 BP (SFU 134) and 2750 BP (WSU 2922). Again, these houses are not as old as the Maurer and Xa:ytem structures, but they are similar in form to the ethnographically reported structures of the coast, and to some degree to the two early lower Fraser Valley houses. Clearly, the house structure pattern of the ethnographic Northwest Coast societies, using large timbers and planks, has a deep time depth. Similar house patterns with similar dates have been identified in other sites on the north coast – in the Dundas Islands, for example (Ruggles 2007; McLaren 2008).

In the interior of the province, house styles were somewhat different from those on the coast. Ethnographically, we know that people utilized semi-subterranean earth-covered pit house structures in the winter, and in some areas they used mat lodges – lean-to and A-frame shelters – at other times of the year. Thousands of remnant pit house depressions have been recorded along the various rivers of the interior. These houses utilized large timbers to frame a strong roof structure over a dug-out area. The roof was covered with smaller timbers and then overlaid with sod and earth to provide an insulating layer. The warmth thus provided made the pit houses warm and livable during the winter.

Early examples of pit houses are found in the nearby southern plateau region in Washington, Oregon, and Idaho. A date of 5,550 ± 120 BP has been reported for House 1 at the Hatwai site in central Idaho (Ames 1991). Ames suggests that there are at least six houses in the plateau dating from five to six thousand years ago.

Although the Plateau Pithouse Tradition seems to have begun about 2550 BC (Pokotylo and Mitchell 1998), the majority of pit houses on the plateau date to around 2000 BP or more recently. Among the oldest semi-subterranean mat lodge houses known in the BC interior are those reported from the Baker site near Monte Creek, with dates from 4500 BP to 4000 BP (Wilson et al. 1992). Early pit house remnants were found at the Deer Park site on Lower Arrow Lake, near Castlegar, dating to between 3250 BP and 2400 BP (Turnbull 1977). The appearance of this house style has been linked to environmental changes that accompanied the moist and cool climate that existed from about 2050 BC to 1350 BC (Kuijt 1989; Rousseau 1990). Prior to about 2000 BP, most pit houses were around seven to thirteen metres in diameter, but after that there seems to have been more variation in size, with some houses being up to twenty-two metres in diameter, particularly in the Lillooet and Chilcotin River areas (Pokotylo and Mitchell 1998). Kuijt (1989) and Rousseau (1990) have suggested that changes in climate and resource availability were responsible for changes in pit house form.

In addition to house structures, there are other notable uses for timbers in prehistoric British Columbia. Although totem poles and various carved and decorated objects, such as masks, boxes, and also canoes, are of immense interest to archaeologists as well as to the growing audience for Northwest Coast art, because of preservation conditions, there is little evidence of such items older than two to three hundred years or so. It is not clear how old some of the artistic designs and patterns are, though we do see evidence of some design elements in artifacts dating to over four thousand years.

Of significant concern to current archaeologists are those trees that provide evidence of the traditional use of tree bark and planks for various purposes. Culturally modified living trees, usually cedar, retain evidence of the removal of long strips of bark or of planks. Ethnographic evidence exists for this use, and it is clear from the number of trees thus modified that such activities were widespread along the coast and throughout the interior. It may be possible to obtain dates for such uses from the tree rings, but unfortunately, since trees do not live long enough to give a complete record of such usage over time, we cannot infer it for more than a few hundred years, although the continuity of cultural evidence in other areas of activity leads one to believe that use of bark extended far back in time throughout the province.

Summary

The earliest archaeological evidence for settlement in British Columbia predates 10,000 years ago in the eastern interior areas of Crowsnest Pass and the Peace River. On the coast, the earliest dates extend back to between 9,500 and 10,000 years ago.

Plant foods are also not as visible as animal foods; however, wild cherry pits at the Milliken site date possibly as far back as nine thousand years and have provided useful information about both foods and season of occupation of the site. Other evidence for food plant use is more recent.

Another major resource of interest is the timber and wood used for building houses and items of daily use, such as canoes, fish weirs, and storage boxes. Wood doesn't preserve as well as stone or sometimes as bone, though there are traces of its use that date back around five thousand years. It is clear that, in the Fraser Valley, wooden house structures were present similar in form to the ethnographic period structures and dating back about five thousand years. Evidence for house structures in the interior is not quite as old but does date to over four thousand years ago. Other wooden items are less visible archaeologically, though handles for fish knives and other small items occasionally appear in site deposits. Remnants of fish traps and fish weirs are occasionally found in sites along the coast but are usually relatively recent. Other plant materials – used in basketry, for example – have been preserved in wet site conditions for over three thousand years. Culturally modified trees, which are important resources in the discussion of traditional usage and traditional territories, do not live long enough or preserve well enough to give great time depth to their history.

Although the above discussion is limited in scope, it is clear that the First Peoples of British Columbia have been dependent on and have been making extensive use of the natural resources surrounding them, eating a variety of foods from both the marine and terrestrial environments, using timber and other wood materials to make their houses and domestic materials, and making their tools out of both local and traded raw materials. The patterns that were observed by the early European explorers, traders, and missionaries extend far back in time, long before the first Europeans set eyes on this part of the world and its inhabitants.

Appendix 9.1

Plant resources harvested by Aboriginal peoples in northwestern North America

Type of resource	Species examples
Fibrous tree bark	western red cedar, *Thuja plicata;* birch, *Betula papyrifera*
Wooden planks	western red cedar
Bark for medicinal use	red alder, *Alnus rubra;* cascara, *Rhamnus purshiana*
Roots for basketry	red cedar; Sitka spruce, *Picea sitchensis*
Fibrous stems and leaves for mats, cordage, baskets	cattail, *Typha latifolia;* tule, *Scirpus acutus;* stinging nettle, *Urtica dioica;* Indian-hemp, *Apocynum cannabinum;* slough sedge, *Carex obnupta*
Withes and branches for basketry, rope, fish traps	saskatoon berry, *Amelanchier alnifolia;* hazelnut, *Corylus cornuta;* red cedar; willow, *Salix* spp.
Pitch for medicine, adhesives	western hemlock, *Tsuga heterophylla;* lodgepole pine, *Pinus contorta;* Sitka spruce; subalpine fir, *Abies lasiocarpa;* and other conifers
Medicinal plants and roots	mountain valerian, *Valeriana sitchensis;* Indian hellebore, *Veratrum viride*
Edible berries, fruits, nuts	salmonberry, *Rubus spectabilis;* highbush cranberry, *Viburnum edule;* salal, *Gaultheria shallon;* beaked hazelnut, *Corylus cornuta;* huckleberries, *Vaccinium* spp.; soapberries, *Shepherdia canadensis*
Green leaves, shoots as vegetables	cow parsnip, *Heracleum lanatum;* fireweed, *Epilobium angustifolium;* Indian celery, *Lomatium nudicaule*
Root vegetables	blue camas, *Camassia* spp.; yellow avalanche lily, *Erythronium grandiflorum;* spring beauty, *Claytonia lanceolata*; balsamroot, *Balsamorhiza sagittata;* rice-root, *Fritillaria* spp.; springbank clover, *Trifolium wormskioldii;* silverweed, *Potentilla anserina* ssp. *pacifica*
Edible tree cambium	western hemlock; Sitka spruce; black cottonwood; *Populus balsamifera* ssp.; *trichocarpa;* pines, *Pinus* spp.
Edible mushrooms	pine mushroom, *Tricholoma magnivelare*; cottonwood mushroom, *Tricholoma populinum*

Note: Adapted from Turner, Ignace, and Ignace (2000).

References

Ames, Kenneth M. 1991. "Sedentism: A Temporal Shift or a Transitional Change in Hunter-Gatherer Mobility Patterns?" In *Between Bands and States,* edited by Susan A. Gregg, 103-33. Center for Archaeological Investigations, Occasional Paper no. 9. Carbondale: Southern Illinois University.

Bernick, Kathryn. 1998. "Stylistic Characteristics of Basketry from Coast Salish Area Wet Sites." In *Hidden Dimensions: The Cultural Significance of Wetland Archaeology,* edited by Kathryn Bernick, 139-56. Vancouver: UBC Press.

Borden, Charles. 1975. "Origins and Development of Early Northwest Coast Culture to about 3000 BC." National Museum of Man, Archaeological Survey of Canada, Mercury Series, Paper 45. Ottawa: National Museums of Canada.

Carlson, Catherine. 1979. "The Early Component at Bear Cove." *Canadian Journal of Archaeology* 3:177-94.

Carlson, Roy L. 1996. "Introduction to Early Human Occupation in British Columbia." In *Early Human Occupation of British Columbia,* edited by Roy L. Carlson and Luke Dalla Bona, 3-10. Vancouver: UBC Press.

Coupland, Gary. 1985. "Prehistoric Cultural Change at Kitselas Canyon," 214. PhD diss., University of British Columbia.

–. 1988. "Prehistoric Economic and Social Change in the Tsimshian Area." In *Research in Economic Anthropology, Supplement 3: Prehistoric Economies of the Pacific Northwest Coast,* edited by Barry L. Isaac, 211-43. Greenwich: JAI Press.

Cybulski, Jerome S., Donald E. Howes, James C. Haggerty, and Morley Eldridge. 1981. "An Early Human Skeleton from Southern Central British Columbia: Dating and Bioarchaeological Inferences." *Canadian Journal of Archaeology* 5:49-59.

Fedje, Daryl W., and James White. 1988. "Vermillion Lakes Archaeology and Paleoecology: Trans Canada Highway Mitigation in Banff National Park." Microfiche Series 463. Ottawa: Environment Canada Parks Service.

Fladmark, Knut R. 1975. "A Paleoecological Model for Northwest Coast Prehistory." Archaeological Survey of Canada, Paper no. 43, National Museum of Man Mercury Series, Ottawa.

–. 1996. "The Prehistory of Charlie Lake Cave." In *Early Human Occupation of British Columbia,* edited by Roy L. Carlson and Luke Dalla Bona, 11-20. Vancouver: UBC Press.

Heaton, Timothy H., and F. Grady. 2003. "The Late Wisconsin Vertebrate History of Prince of Wales Island, Southeast Alaska." In *Ice Age Cave Faunas of North America,* edited by B.W. Schubert, J.I. Mead, and R.W. Graham, chap. 2. Bloomington: Indiana University Press.

Heusser, C.J. 1960. *Late-Pleistocene Environments of North Pacific North America: An Elaboration of Late-Glacial and Post-Glacial Climatic, Physiographic, and Biotic Changes*. Special Publication no. 35, American Geographical Society.

Josenhans, Heiner, Daryl Fedje, Kim W. Conway, and J. Vaughn Barrie. 1995. "Post Glacial Sea Levels on the Western Canadian Continental Shelf: Evidence for Rapid Change, Extensive Subaerial Exposure, and Early Human Habitation." *Marine Geology* 125:73-94.

Josenhans, Heiner, Daryl Fedje, Reinhard Pienitz, and John Southon. 1997. "Early Humans and Rapidly Changing Holocene Sea Levels in the Haida Gwaii – Hecate Strait, British Columbia, Canada." *Science* 277:71-74.

Kuijt, Ian. 1989. "Subsistence Resource Variability and Culture Change during the Middle-Late Prehistoric Cultural Transition on the Canadian Plateau." *Canadian Journal of Archaeology* 13:97-118.

LeClair, Ron. 1973. "Investigations at the Maurer Site, an Early Pithouse Manifestation in the Upper Fraser Valley: Preliminary Report." Unpublished report, Archaeological Sites Advisory Board Department of the Provincial Secretary Victoria.

Lepofsky, Dana, and Natasha Lyons. 2003. "Modeling Ancient Plant Use on the Northwest Coast: Towards an Understanding of Mobility and Sedentism." *Journal of Archaeological Science* 30(11):1357-71.

Matson, R.G. 1976. "The Glenrose Cannery Site." National Museum of Man, Archaeological Survey of Canada, Mercury Series, Paper no. 52. Ottawa: National Museums of Canada.

Matson, R.G., and Gary Coupland. 1995. *The Prehistory of the Northwest Coast.* San Diego: Academic Press.

McLaren, Duncan. 2008. "Sea Level Change and Archaeological Site Location on the Dundas Island Archipelago of North Coastal British Columbia." PhD diss., University of Victoria.
Pokotylo, David L., and Patricia Froese. 1983. "Archaeological Evidence for Prehistoric Root Gathering in the Southern Interior Plateau of British Columbia: A Case Study from Upper Hat Creek Valley." *Canadian Journal of Archaeology* 7(2):127-58.
Pokotylo, David L., and Donald Mitchell. 1998. "Prehistory of the Northern Plateau." In *Handbook of North American Indians,* vol. 12, *Plateau,* edited by Deward E. Walker Jr., 81-102. Washington, DC: Smithsonian Institution.
Rousseau, Mike K. 1990. "Changes in Human Sedentism, Mobility, and Subsistence during the Plateau Pithouse Tradition on the Canadian Plateau." Paper presented at the 1990 meeting of the American Anthropological Association, Las Vegas.
–. 1991. "Landels: An 8500 Year Old Deer Hunting Camp." *The Midden* 23(4):6-9.
–. 1993. "Early Prehistoric Occupation of South-Central British Columbia: A Review of the Evidence and Recommendations for Future Research." In "Changing Times: British Columbia Archaeology in the 1980's," edited by K. Fladmark, special issue, *BC Studies* 99:140-83.
Rousseau, M.K., R. Muir, D. Alexander, J. Breffit, S. Woods, K. Berry, and T. Van Gaalen. 1991. "Results of the 1989 Archaeological Investigations Conducted in the Oregon Jack Creek Locality, Thompson River Region, South-Central British Columbia." Unpublished report, Ministry of Tourism and Ministry Responsible for Culture, Victoria.
Ruggles, Angela. 2007. "Is Home Where the Hearth Is? Evidence for an Early Non-Domestic Structure on the Dundas Island of North Coastal British Columbia." MA thesis, University of British Columbia.
Sanger, David. 1967. "Prehistory of the Pacific Northwest Plateau as Seen from the Interior of British Columbia." *American Antiquity* 32(2):186-97.
Schaepe, David. 1998. "Recycling Archaeology: Analysis of Material from the 1973 Excavation of an Ancient House at the Maurer Site." MA thesis, Simon Fraser University.
–. 2003. "Validating the Maurer House." In *Archaeology of Coastal British Columbia: Essays in Honour of Professor Philip M. Hobler,* edited by Roy L. Carlson, 113-52. Burnaby, BC: Archaeology Press, Simon Fraser University.
Stryd, Arnoud R., and Michael K. Rousseau. 1996. "The Early Prehistory of the Mid-Fraser-Thompson River Area." In *Early Human Occupation of British Columbia,* edited by Roy L. Carlson and Luke Dalla Bona, 177-204. Vancouver: UBC Press.
Turnbull, C. 1977. "Archaeology and Ethnohistory in the Arrow Lakes, Southeastern British Columbia." National Museum of Man, Archaeological Survey of Canada, Mercury Series, Paper no. 85. Ottawa: National Museums of Canada.
Turner, Nancy J., Marianne Boelscher Ignace, and Ronald Ignace. 2000. Traditional Ecological Knowledge and Wisdom of Aboriginal Peoples in British Columbia. *Ecological Applications* 10(5):1275-87.
Wilson, I.R., B. Smart, N. Heap, J. Warner, T. Ryals, S. Woods, and S. MacNab. 1992. "Excavations at the Baker Site, EdQx 43, Monte Creek, Permit 91-107." I.R. Wilson Consultants. Unpublished report, Ministry of Tourism and the Ministry Responsible for Culture, Victoria.

10

Cultural Resource Management in the Context of Forestry in British Columbia

Existing Conditions and New Opportunities

Andrew R. Mason

As stated in the Introduction, the primary objective of this volume is to provide readers with the opportunity to learn about a wide variety of First Nations issues as they relate to forestry practices in British Columbia. By gaining these insights, readers may better understand some of the challenges and opportunities that will present themselves in light of increased movement toward Aboriginal self-government, increased access to resources, and increased Native land management.

With this objective in mind, the purpose of this chapter is to examine the cultural resources aspect of forest land management. Cultural resources in British Columbia include archaeological sites and traditional land use sites (see Chisholm, this volume, for an overview of archaeology in British Columbia). Given this volume's emphasis on forestry, particular attention is paid to culturally modified trees (Figure 10.1) (see Stewart 1984; Stryd and Eldridge 1993; Stryd and Feddema 1998; Archaeology Branch 2001). Within the context of this part of the volume, "Traditional Ecological Knowledge and Use," cultural resource management in British Columbia can be viewed as a state-imposed regulatory framework that is at times at odds with the values of First Nation communities.

The chapter is divided into three sections. Following this introduction, a review of the cultural resource regulatory framework in British Columbia is provided. This section documents existing conditions in the province and is a logical starting point for a discussion of this topic. Within this section are technical definitions that are important for understanding some of the challenges associated with the management of cultural resources in British Columbia (see also Lewis and Sheppard, this volume, who explore First Nations' spiritual perceptions of the forested landscape and the problems created by a general lack of awareness of culturally significant places by public agencies and land managers). The second section provides insights into Aboriginal views on cultural resources and identifies a critical limitation in the existing regulatory framework, particularly its inability to fully address

Figure 10.1 **Post-1846 culturally modified tree, Pacific Spirit Park, Vancouver**

Aboriginal concerns over the protection and management of cultural resources. This section references a select number of Aboriginal heritage policy documents to provide these insights. The information found in the first two sections has been compiled over a number of years from disparate, often unpublished sources and in some instances puts this information "on record," at least in a published context, for the first time. The final section presents an alternate model for cultural resource management in British Columbia that attempts to reconcile the world views of both Aboriginal and non-Aboriginal peoples.

Heritage Statutes and Policy in British Columbia

Three types of cultural resources are discussed in this chapter – archaeological sites, traditional use sites, and culturally modified trees.

Archaeological Sites:

> Archaeological resources consist of the physical remains of past human activity. The scientific study of these remains, through the methods and techniques employed in the discipline of archaeology, is essential to the understanding and appreciation of prehistoric and historic cultural development in British Columbia. These resources may be of regional, provincial, national or international significance." (Archaeology Branch 1990, 3)

Post-contact era heritage sites created by Aboriginal people and non-Aboriginal people can also be called archaeological resources. However, the latter are not discussed within the context of this chapter.

Traditional Use Sites:

> The term Traditional Use site means any geographically defined area that has been customarily used by one or more contemporary groups of aboriginal people for some type of culturally significant activity. These sites may not reveal physical evidence of use. Traditional use sites are usually documented through oral, historical, and archival sources. Examples of Traditional Use Sites include: ritual bathing pools, resource gathering areas, locations of culturally significant events etc. (Archaeology Branch 1996a, 3)

Culturally Modified Trees:

> A CMT [culturally modified tree] is a tree that has been altered by aboriginal people as part of their traditional use of the forest. Non-aboriginal people also have altered trees, and it is sometimes difficult to determine if an alteration (modification) is of aboriginal or non-aboriginal origin. There are no

> reasons why the term "CMT" could not be applied to a tree altered by non-aboriginal people. However, the term is commonly used to refer to trees modified by aboriginal people in the course of traditional tree utilization ... (Archaeology Branch 2001, 1)

As noted by Klimko, Moon, and Glaum (1998), culturally modified trees (CMTs) are somewhat unique, as they may be considered an archaeological site and/or a traditional use site. CMTs are important to Aboriginal groups, since the year of the modification(s) can often be determined using basic dendrochronological methods and represent continuous use and occupation of the landscape. For these reasons, CMTs are particularly important to First Nations in asserting Aboriginal rights and title to a given area (see Bell 2001).

In British Columbia, a wide range of provincial and federal statutes, policies, guidelines, and protocols exist to safeguard cultural resources. The following discussion examines several of these safeguards in some detail. Understanding the current cultural resource protection environment in British Columbia, and the opportunities and challenges it provides, is important to later sections of this chapter.

Depending on the jurisdiction of the land in question, the cultural resource statutes, policies, and management strategies that apply vary. For example, in British Columbia, cultural resources on provincial Crown land and private property are managed and regulated differently from the same resources found on federal Crown land. The following discussion examines the management structure and regulatory framework for cultural resources on provincial Crown land and private property in British Columbia, then looks at the same issues at the federal level.

Provincial Crown Land and Private Property

For the management of cultural resources on provincial Crown land and private property, the Ministry of Forests, Lands and Natural Resource Operations (MFLNRO) is the lead agency, although the Ministry of Aboriginal Relations and Reconciliation also plays an important role. Within the MFLNRO is the Archaeology Branch. This branch includes both the Archaeological Permitting and Assessment Section and the Archaeological Site Inventory Section. It is the latter that is responsible for the recording, maintenance, and distribution of cultural resource information, as well as Provincial Heritage Register information.

The Archaeological Permitting and Assessment Section is responsible for administering the Heritage Conservation Act (HCA) permit process, as well as representing archaeological resource interests on project review committees established under the British Columbia Environmental Assessment Act (BCEAA).[1] The early development of this branch of government has been well documented by Apland (1993).

The province began to address issues surrounding traditional use sites in 1994, under the Cultural Heritage Resources Inventory System (CHRIS). From the outset, and during most of its existence, the program was located within the Heritage Conservation Branch, though it was briefly resident within the Archaeology Branch (both branches being part of the then Ministry of Small Business, Tourism and Culture) with the hope of conjoining these data with the Provincial Heritage Register (Deloitte and Touche 1997). This program grew over the next few years, as major Land and Resource Management Plan (LRMP) studies were underway and the treaty process was high on the government's agenda (see Archaeology Branch 1996a). In November 1995, the renamed Traditional Use Study Program was transferred to the Ministry of Forests (MoF) (see BC MoF 2000, 2001). The program became part of a new ministry in 2001 when it was assigned to the Ministry of Small Business, Tourism and Culture. In 2002, the BC government decided to terminate the program; it officially ended in March 2003.

As alluded to in the previous paragraphs, the primary piece of legislation regarding cultural resources on private property and provincial Crown lands in British Columbia is the HCA. The HCA provides the basis of heritage protection in British Columbia, but there are also a number of complementary statutes, agreements, policies, operational procedures, guidelines, and information bulletins.[2]

Given that this chapter explores the management of cultural resources in light of forestry in British Columbia, the following discussion summarizes salient elements of the HCA, the Forest and Range Practices Act,[3] and the protocol agreement between the Archaeology Branch and the Ministry of Forests (Archaeology Branch 1996b). These three instruments have the greatest bearing on cultural resources and forest practices in British Columbia.

Heritage Conservation Act

All archaeological sites, whether on provincial Crown or private land, that predate AD 1846 are automatically protected under the HCA.[4] Certain sites, including burials and rock art sites with historical or archaeological value, are protected regardless of their antiquity. The types of cultural resources that are protected under the HCA are defined in section 13 (Heritage Protection) and are largely limited to "physical evidence of human habitation or use before 1846." The 1846 threshold date is not culturally relevant other than being the date that the courts have identified as the assertion of British sovereignty in British Columbia, corresponding with the signing of the Oregon Boundary Treaty, 1846.

In addition to cultural site types protected under section 13 of the HCA, section 4 (Agreements with First Nations) provides an opportunity for the lieutenant-governor in council to protect additional types of sites that are not listed in section 13 (e.g., post-1846 CMTs, transformer sites[5]). However, a section 4 agreement has never been signed, raising questions concerning the effectiveness of this aspect of the legislation. In practice, under the HCA there is no legal protection for many post-1846 archaeological sites and traditional use sites on private property or provincial Crown land in British Columbia.

Forest and Range Practices Act

The recently introduced Forest and Range Practices Act (FRPA) seeks to provide a simplified planning and approval process for proposed forestry developments in British Columbia. Under the FRPA, forest practice standards are set in regulation rather than statute, and more emphasis is placed on professional accountability ("results-based"). With the FRPA, agreement holders are required to prepare forest stewardship plans (FSPs) for approval by the minister prior to timber harvest or road construction on land to which the licence applies. FSPs must adhere to the FRPA value regime, which includes a series of resource values and provincially set objectives. Under section 149(1) of the FRPA, the province has the ability to make regulations prescribing objectives. Accordingly, agreement holders must recognize these values and carry out actions (i.e., obtain results or implement strategies) to achieve the objectives. Within the context of this chapter, those objectives that relate to cultural heritage resources are most relevant.

As defined by the FRPA, "cultural heritage resources are the traditional uses of a First Nation. They should not be confused with archaeological resources ..." (BC MoFR 2006, 59). In essence, cultural heritage resources, not otherwise protected by the HCA, may be managed under the FRPA and, given their association with the realm of Aboriginal rights and title, they may be both challenging and highly dependent on individual circumstance. Government objectives for cultural heritage resources are "to conserve, or, if necessary, protect cultural heritage resources that are; (a) the focus of a traditional use by an aboriginal people that is of continuing importance to that people, and (b) not regulated under the *Heritage Conservation Act*" (BC MoFR 2006, 111). In effect, the FRPA ventures into the territory of the Heritage Conservation Act (section 4) and seeks to manage traditional use sites. As for archaeological resources, they remain protected under the act, and individuals responsible for the preparation of FSPs are still advised to "ensure archaeological resources are protected as per the requirements of the *Heritage Conservation Act*" (BC MoFR 2006, 59). Although the FRPA may not address archaeological resources directly, at the operational level some forest districts still require archaeological studies to ensure due diligence before issuing a cutting permit or road building permit (Glaum 2006).

To assist foresters and to provide them with the information they need to make informed decisions for the effective and efficient management of archaeological resources in forestry operations, the Archaeology Branch has prepared the *British Columbia Archaeological Resource Management Handbook for Foresters* (Archaeology Branch 2007).

A tangible outcome of the FRPA was the use of the Government Actions Regulation provision under the statute to establish the Kweh-Kwuch-Hum (Mt. Woodside) Cultural Heritage Resource, Resource Feature in the Chilliwack Forest District, which created a 1,072-hectare management area that contains sacred sites, historic sites, archaeological sites, and resource procurement sites of great spiritual, personal, cultural, and economic importance to the Sts'ailes (Chehalis) people (BC MoFR 2008); Chehalis Indian Band and Chilliwack Forest District 2008).

Protocol Agreement between the Archaeology Branch and the Ministry of Forests

The Archaeology Branch and Ministry of Forests' Protocol Agreement on the Management of Cultural Heritage Resources (Archaeology Branch 1996b) states that "cultural heritage resources will be managed so that their inherent values are protected, maintained, or enhanced according to the principles of integrated resource management." The protocol agreement was established to define the roles and responsibilities of both the Archaeology Branch and the MoF, and to integrate cultural heritage resources within the MoF's land and resource management planning and operations. This agreement conforms with, and is subject to, the HCA, the Forest Act, and the Ministry of Forests Act. With the adoption of the Forest and Range Practices Act there have been changes, but portions of the protocol remain in effect, particularly those related to data sharing. However, some responsibilities are no longer mandated by the FRPA and are beyond the authority of the MFLNRO (Glaum 2006). Further, under this agreement, cultural heritage resources are to be protected and managed in accordance with existing or future policies and operational procedures.

As originally conceived, practice under this agreement obliges the MFLNRO district manager to (1) initiate and fund archaeological overview assessments (AOAs) that have not been addressed by other government agencies; (2) incorporate Archaeology Branch requirements in forestry documents; and (3) notify licencees of the need for archaeological assessments prior to development. For its part, the Archaeology Branch is required to (1) set the standards and policies for AOAs and heritage permits; (2) review archaeological assessment reports; (3) issue HCA permits; and (4) provide technical input and advice for appropriate measures to manage impacts to archaeological resources. Licencees fund and commission the archaeological assessments and implement the proposed cultural resource management recommendations (see Klimko, Moon, and Glaum 1998).

Federal Crown Land

For federal Crown land in British Columbia, the legislative framework is unclear. There is no comprehensive federal statute directing how (or whether) a given department is supposed to treat cultural resources on its lands (Canada 1988; Pokotylo and Mason 2010; see also CAA 1986).

Indeed, in this regard, Canada has the dubious distinction of being the only member of the G8 nations lacking comprehensive federal cultural resource management legislation (Cameron 2002; see also Burley 1994). In the absence of comprehensive cultural resource management legislation, federal land managers are left to rely on general policies applicable to all departments and the specific directives, if any, of their own department (Canada, Parks Canada 1993, 1994, 2000; CEAA 1996; see also Burley 1994; Yellowhorn 1999).

Organizationally, the Archaeological Services Branch of the Parks Canada Agency has responsibilities pertaining to cultural resources on Parks Canada lands. In British Columbia, the Archaeological Services Branch also responds to requests for advice from other federal government departments. Questions pertaining to heritage policy and legislation are addressed by the Heritage Branch of the Department of Canadian Heritage. That department, together with Parks Canada, has developed a body of directives governing cultural resources under the control of either Parks Canada or the department (see Canada, Parks Canada 2000). Within Parks Canada, the Cultural Resource Management Policy (1994) and Guidelines for the Management of Archaeological Resources in the Canadian Parks Service (Canada, Parks Canada 1993) are applied.

Additional details concerning federal legislation, guidelines, policies, and operational procedures that address cultural resources in some fashion can be found in DCH (1994); Battiste and Henderson (2000); and Pokotylo and Mason (2010).

Within the context of forestry on federal Crown land in British Columbia, Natural Resources Canada (NRC), operating under the federal Forestry Act R.S.C. 1985, chapter F-30, and the Timber Regulations 1993, SOR/94-118, is the lead agency. Within British Columbia, NRC manages 4,400 hectares of land, with only 230 hectares being production forest. NRC also works for other federal departments, such as the Coast Guard, airport authorities, and Parks Canada. Much of this work for other departments relates to site maintenance (e.g., removal of hazard trees or invasive species), though some timber harvest also occurs. In general, the federal government's forest lands consist of small parcels, with two exceptions – the Chilcotin Military Reserve (40,000 hectares) near Riske Creek and the Dominion Coal Blocks (20,000 hectares) near Fernie, British Columbia. The Chilcotin Military Reserve is owned by the Department of National Defence but is the responsibility of Aboriginal Affairs and Northern Development Canada. Forestry operations on the Chilcotin Military Reserve were managed by the provincial MoF on behalf of the federal government until 1989, when a moratorium on the use of these lands was implemented. With respect to the Dominion Coal Blocks, the management of these NRC lands is delegated to the provincial MFLNRO. For federal lands managed by the MFLNRO, cultural resources are treated as if the lands are within provincial jurisdiction, albeit lacking any enforcement capability. NRC essentially defaults to provincial statutes on its own lands, or works under the specific directives, if any, of other departments when working on those lands (Robinson 2003). NRC does not have a specific operational procedure for cultural resources.

To summarize, within the context of forestry on federal lands in British Columbia, there is limited protection for cultural resources. With the exception of forestry activities on Parks Canada lands, there is no legislative protection for archaeological and traditional use sites unless the limited provisions of another federal act, such as chapter 37 of the Canadian Environmental Assessment Act, 1992 (CEAA), apply. For example, the CEAA states that one of the purposes of the act is "to ensure that projects are considered in a careful and precautionary manner before federal authorities take action in connection with them, in order to ensure that such projects do not cause significant adverse environmental effects" (section 4(1)(a)). Under the CEAA, environmental effects include archaeological resources and the current use of lands and resources for traditional purposes by Aboriginal persons (i.e., traditional use sites). The shortcoming of this "protection" is the absence of statutory directives with respect to how these resources and features are to be "considered" (i.e., managed).

First Nations Perspectives on Cultural Resources

Given the diversity and geographic distribution of First Nations throughout British Columbia, it is difficult to prepare a summary of their views on the definition and management of cultural resources in the province. In all likelihood, the range of responses would be as diverse as the First Nations themselves. Nevertheless, this section attempts to provide some insights in this area. The means for gaining these insights lie in the heritage policies, guidelines, agreements, by-laws, protocols, treaty provisions, and permitting systems established by several First Nations within the province (see Tables 10.1 and 10.2) (see Zacharias and Pokotylo 1997).[6]

These aforementioned documents are useful, as they articulate specific First Nations' views on what cultural resources are and how they should be managed. Each of these documents provides a range of perspectives on cultural resources, but for the purposes of this discussion, the definition of what constitutes a cultural resource is examined. For example, are First Nations' definitions of an archaeological site or cultural resource similar to, or widely divergent from, government definitions of what receives legislated

Table 10.1

BC Aboriginal groups with heritage policies, guidelines, agreements, by-laws, protocols, and treaty provisions

Ahousaht First Nations
Allied Tsimshian Tribes
Boston Bar First Nation
Cariboo Tribal Council
Gitanyow
Haida Nation
Heiltsuk Nation
Hul'qumi'num Treaty Group
Huu-ay-aht First Nations
In-SHUCK-ch Nation
Ka:'yu:'k't'h'/Che:k:tles7et'h' First Nations
Kamloops Indian Band
Ktunaxa Kinbasket Treaty Council
Lil'wat Nation/Mount Currie Band (Creekside Resources Inc.)
Lillooet Tribal Council
Musqueam Indian Band
Neskonlith Indian Band
Nisga'a
Okanagan Indian Band
Seyem' Qwantlen Resources Ltd.
Shuswap Nation Tribal Council
Skeetchestn Indian Band
Spallumcheen First Nation
Squamish Nation
Stó:lō
Sts'ailes Indian Band
Tahltan Central Council
Taku River Tlingit First Nation
Toquaht Nation
Treaty 8
Tsawwassen First Nation
Tsilhqot'in National Government
Tsleil-Waututh First Nation
Uchucklesaht Tribe
Ucluelet First Nation
Union of British Columbia Indian Chiefs
Upper Nicola Band
Upper Similkameen Indian Band
Westbank First Nation

Table 10.2

BC Aboriginal groups with cultural resource permitting systems or equivalent

Adams Lake Indian Band
Heiltsuk Nation
Huu-ay-aht First Nations
In-SHUCK-ch Nation
Kamloops Indian Band
Ktunaxa Kinbasket Treaty Council
Musqueam Indian Band
Neskonlith Indian Band
Okanagan Indian Band
Seyem' Qwantlen Resources Ltd.
Shíshálh Nation
Skeetchestn Indian Band
Squamish Nation
Stó:lō
Sts'ailes Indian Band
T'Sou-ke Nation
Toosey Indian Band
Treaty 8
Tsilhqot'in National Government
Upper Similkameen Indian Band
Westbank First Nation

protection? Three examples, representing different levels of political organization, are considered. These are the Upper Similkameen Indian Band, the Nisga'a, and the Union of BC Indian Chiefs.

Upper Similkameen Indian Band

Based in Princeton, British Columbia, the Upper Similkameen Indian Band maintains its own heritage investigation permitting system and has a policy document concerning heritage resources within its traditional territory (Upper Similkameen Indian Band, n.d.). In terms of what constitutes a cultural resource, the policy document is clear:

> Upper Similkameen definitions of heritage and heritage sites are not always appreciated or immediately apparent to mainstream Canadians. Upper Similkameen history is anchored in antiquity and is intimately connected with the cultural and physical landscape. Our people believe it is artificial to separate matters of spiritual, social, heritage and economic significance. (Upper Similkameen Indian Band, n.d., 1)

The Upper Similkameen Indian Band policy uses four general categories of heritage sites: (1) plant and animal resource use areas, (2) sacred and spiritual places, (3) areas of historical significance, and (4) archaeological sites. The definition of heritage resources is further expanded to include artifacts, documents, and other media pertaining to the Upper Similkameen Indian Band that are held by various individuals and agencies. Clearly, the definition of a heritage site requiring management and protection is much broader than as defined under the HCA. Within a provincial Crown land forestry context, this category of site could potentially be managed under the FRPA using the Government Actions Regulation mechanism.

Nisga'a

In the Nisga'a example, cultural resources are addressed in a comprehensive fashion. In chapter 17, Protection of Heritage Sites (sections 36 to 39), of the Nisga'a Final Agreement (1998), the treaty allows the Nisga'a Nation to develop their own processes to manage heritage sites on Nisga'a lands. Further, the treaty states that British Columbia will develop or continue processes to manage heritage sites until the Nisga'a government establishes its own policies. In other words, the treaty allows the Nisga'a to develop their own heritage protection processes, and until such time that this has been implemented, the provincial status quo is maintained.

Of particular interest to this chapter is the treaty document's definition of a heritage site. Under this agreement, heritage sites include archaeological, burial, historical, and sacred sites. The Nisga'a document is not clear on how traditional use sites are to be treated, but it could certainly be argued that many traditional use sites are also sacred sites. By including sacred sites, the definition of what constitutes a heritage site is more expansive than what is identified under section 13 of the HCA. As with the preceding example for sites of this nature within a provincial Crown land forestry context, this category of site could potentially be managed under the FRPA using the Government Actions Regulation mechanism.

Union of BC Indian Chiefs

The Union of BC Indian Chiefs (UBCIC) is a political organization representing a large number of BC First Nations that have opted not to participate in the BC treaty process; its example provides yet another perspective on cultural resources for consideration.

In 1992, the UBCIC membership met for a graveyards claim conference. One outcome of this conference was a draft position paper that claimed unextinguished ownership of their respective homeland territories and their unbroken spiritual connection to their ancient, sacred lands (UBCIC 1992). A resolution (no. 3) arising from the same meeting called for the preparation of draft First Nation protective legislation covering sacred sites, burial areas, artifacts, and places of spiritual significance. In this draft legislation, a "heritage area" was defined as "an area or site which contains heritage objects and includes an area or site which has special spiritual significance to the _______ Nation." A "heritage object" was defined as "material, artifact, object or feature of human origin or other physical characteristics that is evidence or may be evidence of aboriginal occupation or use, found in a heritage area" (ibid., Appendix B, 1).

Similar to the Upper Similkameen Indian Band and Nisga'a examples, the UBCIC greatly expands the provincial definition of a cultural site that warrants protection and management to include spiritual and sacred sites. Again, within the context of forestry on provincial Crown land, this expanded definition of a cultural site could potentially be managed under the FRPA using the Government Actions Regulation mechanism.

For each of these organizations, the definition of what constitutes a cultural resource warranting protection differs from the current definition provided by government. These First Nations groups do not differentiate between archaeological sites, traditional use sites, or sacred sites. In these three examples, which can be described as generally representative of most First Nations policy documents that were reviewed, the definition of cultural sites that are significant and warrant some form of legislative or other protection are far broader than provincial definitions and pose a significant resource management challenge to government, industry, and First Nations. With the introduction of the FRPA there is now a process to manage, but not necessarily protect, "non-archaeological cultural sites" in the context of forestry on provincial Crown lands. This suggests the positions of the two parties are coming closer together and should be viewed as a positive development. However, statutory protection is still absent, and the cultural resource management provisions under the FRPA appear cumbersome, time-consuming, and not suited to broad adoption. Further, the FRPA does not extend to private property or non-forestry developments on provincial Crown land. An example of the challenges posed by this difference in perspective and the regulatory gap is illustrated by the following case study.

Ahousaht First Nations Post-1846 CMTs

North of Tofino, on the west coast of Vancouver Island, within the traditional territory of the Ahousaht First Nation, the author completed an archaeological impact assessment study on an eighty-five-hectare parcel of largely forested private land (Golder Associates 2000). As part of this study, archaeologists and representatives of the Ahousaht and Hesquiaht First Nations revisited four previously recorded archaeological sites and identified another twelve locations containing cultural resources. These latter twelve sites were recent CMTs, some only a few years old, and as a result, did not qualify for automatic protection under the HCA. Using government definitions, these sites are traditional use sites.

These post-1846 features were recorded by the archaeologists and added to the Provincial Heritage Register – a practice since discontinued (see Archaeology Branch 2001). Given the age of these features, their documentation and subsequent placement in the register did not provide any form of protection. Following the field component of the study, the draft results and management recommendations were provided to the Ahousaht First Nation for comment. In its reply (Atleo 2000), the Ahousaht First Nation requested that all cultural resources be protected from development regardless of age. However, under the HCA, the province was obligated only to protect the four previously recorded archaeological sites, as they were the only locations considered to meet the criteria identified in section 13 of the HCA.

This typical example illustrates the conflicting definition of what is culturally important to First Nations and what is protected by statute. This is a difficult problem to address, and it is unlikely that current statutes and agreements could be easily amended to broaden the range of cultural sites afforded protection. As stated earlier in this chapter, section 4 of the HCA does provide the mechanism to protect these cultural site types. However, given the absence of a single instance where this section was implemented and an agreement to protect a cultural site was successful, it is apparent that the HCA, at least in its current form, is not the vehicle to protect sites of this nature.

The Forest and Range Practices Act does have a mechanism to manage resources similar to those of concern to the Ahousaht First Nation. However, the potential to manage does not necessarily mean protection, and this management tool is limited to forestry developments on provincial Crown land and in all likelihood is not meant for broad application but more for managing "exceptional areas" of cultural significance to First Nations. Further, as the Ahousaht case study took place on private land, an effective management or protective framework for sites of this nature remains elusive.

An alternative model for cultural resource management in British Columbia is provided in the last section of this chapter.

Reconciling World Views: Finding the Third Way

> From the standpoint of sustainable development, how can harmonious management of natural resources be envisaged if the special features and diversity that characterize tradition and previous forms of resource exploitation are ignored? (Ofoumon 1997, 63)

The preceding discussion illustrates the divergent views between government and First Nations of the definition of a cultural resource that warrants management and protection measures. Given the culturally relative definition of significance and what a cultural resource is, an alternative model is presented that uses the traditional territories of First Nations as the management unit. This model would balance the needs of Aboriginal people, government, and industry while seeing that cultural resources (including archaeological sites, traditional use sites, and sacred sites) receive adequate consideration and are managed in a culturally responsible fashion (see also Marc Stevenson, this volume, for a model to integrate traditional knowledge, and holders of this knowledge, into sustainable forest management and a proposal for policy reform and the creation of new institutional arrangements).

The model that follows is not without its own challenges. The model effectively sidesteps the issue of overlapping First Nations territories, divergent priorities, and the capacity of some First Nations to manage or participate in the system that is proposed. Despite these challenges, the model serves to present another way to consider the delivery of cultural resource management in British Columbia and to stimulate further discussion so we may create a cultural resource management framework for British Columbia that is fair, culturally relevant, and effective.

The cultural resource management model proposed below for British Columbia uses as the basis for this new approach the UK example of an archaeological trust (e.g., York Archaeological Trust) (see Hunter, Ralston, and Hamlin 1993).

To summarize, beginning in the early 1970s, in response to the pace of development and its impact on cultural resources, the UK Department of Environment (DoE) provided operating grants to establish a network of approximately eighty regional archaeological units that were intended to provide comprehensive archaeological coverage across the United Kingdom. The role of these units was to respond to development threats (mostly transportation related), employ professional archaeologists, and mount excavations. Some of these units were independent trusts, others were based in local authorities, and some were based out of museums or universities.

Initially, each archaeological unit limited its activities to its hometown (e.g., York or Winchester), county (e.g., West Yorkshire), or region (e.g., Wessex). Once this network was established, it was the intent of the DoE to have local authorities assume responsibility for the costs associated with these units, thus creating a comprehensive archaeological service. As this latter objective was never met, the DoE discontinued all but project-specific funding in the early 1980s. However, with the introduction of new legislation at that time that required archaeological resources to be considered in the planning and development process, the archaeological units have evolved into self-sustaining entities funded by development (Hunter, Ralston, and Hamlin 1993).

In the context of British Columbia, through enabling legislation, networks of archaeological advisory bodies, or archaeological trusts, could be established within First Nations' traditional territories and given the responsibility for the protection and management of cultural resources in that area. The trust could be structured as an incorporated entity, comprising a chair and board of directors with representatives from the First Nation, industry, and local government. This diverse composition is critical, as the trust would be required to meet the cultural resource management objectives of society as a whole, not just of one segment.

The specific role of the trusts would include defining the types of cultural resource sites that require protection and/or some form of management and reviewing of development plans to determine which projects require cultural resource reviews or assessment. Under these locally based trusts, board members would seek to accommodate cultural resource values while finding ways for development proponents to continue with their work. In essence, the establishment of archaeological trusts would bring an element of stability and surety to development in the province, including forestry operations. Further, by managing the resource at the level of a traditional territory, it will be easier to ensure that the resource base is managed using sustainable practices, in turn ensuring that the full range and diversity of these sites remain for future generations.

The trust could either maintain its own qualified staff to undertake the required cultural resource assessments or subcontract this work to qualified consultants. As in the UK DoE example, initial funding for the trusts would be provided by the provincial government, with subsequent funding generated by revenues from cultural resource assessments that development proponents would be required to undertake.

Under this model, the mandate of the Archaeology Branch would be refocused to include a coordinating role for the archaeological trusts, maintenance of the Provincial Heritage Register, enforcement, and quality-assurance audits.

Conclusions

This chapter has provided provincial government definitions of archaeological sites, traditional use sites, and culturally modified trees. Within the context of these cultural resources, the scope and nature of provincial and federal heritage legislation were presented. In contrast to government, the heritage policies and guidelines of several BC Aboriginal organizations were reviewed, providing an alternative view of what cultural resources are and what cultural resources warrant statutory protection and management. This discussion illustrated that what government statute protects and what Aboriginal groups believe warrants protection and management differ significantly, although management provisions under the FRPA suggest the views of the two parties may be converging. This is a positive development, and steps to simplify the process for broader application and adoption for all lands in British Columbia should be encouraged.

The final section of this chapter provided an alternative to the current cultural resource management regulatory framework in British Columbia. Under this system, the management of cultural resources would become the duty of a number of archaeological trusts, based on First Nations' traditional territories. The strength of this decentralized management model includes more effective and sustainable cultural resource management (e.g., based on meaningful landscape units and/or traditional territories), a culturally relative definition of cultural resources, and an increased ability to manage and protect site types not currently being effectively protected by government statute (e.g., transformation sites). This model is scalable and could easily be applied across Canada based on treaty settlement areas or First Nations' traditional territories.

Acknowledgments

Several individuals shared their time and expertise to provide background information on the nature of statutes and government programs, Aboriginal policy and permitting systems, their evolution, and their current status. These individuals include Susan Rivet, Art Robinson, Eldon Yellowhorn, Allison Cronin, and Margaret Mason. From the BC Archaeology Branch, Ray Kenny, Justine Batten, Doug Glaum, and Al Mackie shared their knowledge of the regulatory environment in British Columbia. Several colleagues from Golder Associates provided valuable insights and comments, including Jeff Bailey, Ben Hjermstad, Kirsty Miskovich, and Natasha Thorpe. Finally, special thanks are due to Milt Wright, Marty Magne, and Kim Lawson.

Notes

1 Heritage Conservation Act, R.S.B.C. 1996, c. 187; British Columbia Environmental Assessment Act, R.S.B.C. 1996, c. 119.
2 http://www.for.gov.bc.ca/.
3 Forest and Range Practices Act, S.B.C. 2002, c. 69 (FRPA).
4 The Heritage Conservation Act supersedes all conflicting legislation.
5 For example, within Stó:lō culture, transformer sites are locations attributed to, or associated with, the deeds or actions of Xa:ls and/or other "transformers" (see Mohs 1994). According to Stó:lō traditions, when Xa:ls travelled through Stó:lō territory, a number of transformations were made. Evidence of these transformations are today associated with many geographical features (e.g., large boulders, bedrock outcroppings) within Stó:lō territory. Transformation sites are not exclusively a Stó:lō phenomenon. (See Lewis and Sheppard, this volume.)
6 Examples from American Tribal organizations also exist (see Battiste and Henderson 2000; Stapp and Burney 2002).

References

Apland, Brian. 1993. "The Roles of the Provincial Government in British Columbia Archaeology." In "Changing Times: British Columbia Archaeology in the 1980's," edited by K. Fladmark, special issue, *BC Studies* 99:7-24.

Archaeology Branch (British Columbia, Ministry of Tourism, Culture and the Arts, Archaeology Branch). 1990. *BC Archaeological Resource Management Handbook*. Victoria: Archaeology Branch, Ministry of Tourism, Culture and the Arts.

–. 1996a. *British Columbia Traditional Use Site Recording Guide*. Victoria: Archaeology Branch, Ministry of Tourism, Culture and the Arts.

–. 1996b. Protocol Agreement with the Ministry of Forests. Victoria: Archaeology Branch, Ministry of Tourism, Culture and the Arts.

–. 2001. *Culturally Modified Trees of British Columbia: A Handbook for the Identification and Recording of Culturally Modified Trees*. Version 2.0. Victoria: Archaeology Branch, Ministry of Tourism, Culture and the Arts.

–. 2007. *British Columbia Archaeological Resource Management Handbook for Foresters*. Victoria: Archaeology Branch, Ministry of Tourism, Culture and the Arts.

Atleo, Ann. 2000. Letter to Golder Associates dated September 1, 2000. "Re: Archaeological Assessment – Openit Peninsula." Appendix 2 in Golder Associates 2000.

Battiste, Marie, and James (Sa'ke'j) Youngblood Henderson. 2000. *Protecting Indigenous Knowledge and Heritage: A Global Challenge*. Saskatoon: Purich Publishing.

BC MoF (British Columbia, Ministry of Forests). 2000. *Traditional Use Study Field Training Manual – Draft for Discussion Purposes Only*. Victoria: Aboriginal Affairs Branch, Ministry of Forests.

–. 2001. *Traditional Use Studies: Data Capture Specifications*. Victoria: Aboriginal Affairs Branch for the Resources Inventory Committee, Ministry of Forests.

BC MoFR (British Columbia, Ministry of Forests and Range). 2006. *Administration Guide for Forest Stewardship Plans*. Version 1.05. Victoria: Resource Tenures and Engineering Branch, BC Ministry of Forests and Range.

–. 2008. "Groundbreaking Agreement Protects Chehalis Heritage." Ministry of Forests and Range news release 2008FOR0100-000978. Victoria.

Bell, Catherine. 2001. "Protecting Indigenous Heritage Resources in Canada: A Comment on *Kitkatla v. British Columbia*." *International Journal of Cultural Property* 10(2):246-63.

Burley, David V. 1994. "A Never Ending Story: Historical Developments in Canadian Archaeology and the Quest for Federal Heritage Legislation." *Canadian Journal of Archaeology* 18:77-98.

CAA (Canadian Archaeological Association). 1986. "The Need for Canadian Legislation to Protect and Manage Heritage Resources on Federal Lands." *Canadian Archaeological Association Newsletter* 6(1):1-8.
Cameron, Christina. 2002. "Canada's Historic Places: Towards the Canadian Register; Implications for Heritage Conservation Policies," lecture at City Program, Simon Fraser University, February 22.
Canada. 1988. *Federal Archaeological Heritage Protection and Management: A Discussion Paper.* Ottawa: Department of Communications.
–. Parks Canada. 1993. *Guidelines for the Management of Archaeological Resources in the Canadian Parks Service*. Ottawa: Parks Canada Agency.
–. 1994. *Parks Canada, Guiding Principles and Operational Policies*. Ottawa: Minister of Supply and Services Canada.
–. 2000. *Unearthing the Law: Archaeological Legislation on Lands in Canada*. Ottawa: Archaeological Services Branch, Parks Canada Agency.
CEAA (Canadian Environmental Assessment Agency). 1996. *A Reference Guide for the Canadian Environmental Assessment Act: Assessing Environmental Effects on Physical and Cultural Heritage Resources*. Hull: Minister of Supply and Services Canada.
Chehalis Indian Band and Chilliwack Forest District. 2008. "Kweh-Kwuch-Hum (Mt. Woodside) Spiritual Areas and Forest Management: A Policy Pilot by the Chehalis Indian Band and Chilliwack Forest District." Unpublished report on file with the Ministry of Forests and Range, Chilliwack, BC.
DCH (Canada, Department of Canadian Heritage). 1994. *Parks Canada Guiding Principles and Operational Policies*. Ottawa: Minister of Supply and Services Canada.
Deloitte and Touche. 1997. "Evaluation of the Traditional Use Study Program." Unpublished report on file with the Ministry of Aboriginal Relations and Reconciliation, Vancouver, BC.
Glaum, Doug. 2006. pers. comm. October 2.
Golder Associates. 2000. "Report on Archaeological Impact Assessment of D.L. 1371, Openit Peninsula, Clayoquot Sound, BC." Unpublished report.
Hunter, John, Ian Ralston, and Ann Hamlin. 1993. "The Structure of British Archaeology." In *Archaeological Resource Management in the UK: An Introduction,* edited by John Hunter and Ian Ralston, 30-43. Reprint, Sparkford, UK: Sutton Publishing, 2001.
Klimko, Olga, Heather Moon, and Doug Glaum. 1998. "Archaeological Resource Management and Forestry in British Columbia." *Canadian Journal of Archaeology* 22:31-42.
Mohs, Gordon. 1994. "Sto:lo Sacred Ground." In *Sacred Sites, Sacred Places,* edited by D.L. Carmichael, J. Hubert, B. Reeves, and A. Schanche, 184-208. New York: Routledge.
Nisga'a Final Agreement. 1998. *Nisga'a Final Agreement*. Victoria: Queen's Printer.
Ofoumon, David. 1997. Sacred Forests in Africa. *Ecodecision* 23:61-63.
Pokotylo, David, and Andrew R. Mason. 2010. "Archaeological Heritage Resource Protection in Canada: The Legal Basis." In *Cultural Heritage Management: A Global Perspective,* edited by Phyllis Mauch Messenger and George S. Smith, 48-69. Gainesville: University Press of Florida.
Robinson, Art. 2003. pers. comm. to Andrew Mason, April 29.
Stapp, Darby C., and Michael S. Burney. 2002. *Tribal Cultural Resource Management: The Full Circle to Stewardship*. Walnut Creek, CA: AltaMira Press.
Stewart, Hilary. 1984. *Cedar: Tree of Life to the Northwest Coast Indians*. Vancouver: Douglas and McIntyre.
Stryd, Arnoud H., and Morley Eldridge. 1993. "CMT Archaeology in British Columbia: The Meares Island Studies." In "Changing Times: British Columbia Archaeology in the 1980's," edited by K. Fladmark, special issue, *BC Studies* 99:184-234.
Stryd, Arnoud H., and Vicki Feddema. 1998. *Sacred Cedar: The Cultural and Archaeological Significance of Culturally Modified Trees*. Vancouver: David Suzuki Foundation.
UBCIC (Union of BC Indian Chiefs). 1992. "Ownership, Jurisdiction, Repatriation: Draft Position Paper on First Nation Graveyards, Burial Areas, Sacred Sites and Heritage Objects." Unpublished report.
Upper Similkameen Indian Band. n.d. "Policy on Heritage Resources within the Jurisdiction and Management Area of the Upper Similkameen Indian Band." Unpublished document.

Yellowhorn, Eldon. 1999. "Heritage Protection on Indian Reserve Lands in Canada." *Plains Anthropologist* 44(170), Memoir 31:107-16.

Zacharias, Sandra, and David L. Pokotylo. 1997. "Cultural Heritage Policy Overview." Unpublished report prepared for the BC Ministry of Forests, Nelson Forest District, Nelson, BC.

11

Blue Ecology
A Cross-Cultural Ecological Vision for Freshwater

Michael D. Blackstock

> The blood that runs through the veins are the rivers that run through Mother Earth. People need to understand that, without life giving water, there will be nothing.
>
> – Chief Gwininitxw, Yvonne Lattie, Gitxsan Nation

In this chapter, I present a probe of the question, what is water? My purpose is to reveal cross-cultural assumptions and definitions of freshwater, thereby helping to reconcile forest-related conflicts between First Nations and government agencies. This approach is my local contribution to the emerging global recognition that freshwater has the potential to become a catalyst for cooperation rather than a source of conflict. Endorsed by the United Nations, this movement is characterized as the "blue revolution" (Hinrichsen, Robey, and Upadhyay 1997; Cosgrove 2003). As a mediator and forester, I have come to see the value of foundational research that examines cross-cultural perceptions of forest values such as water. When mediating a dispute, there usually isn't enough time for a detailed exploration of underlying perspectives and values. Since water is such a common part of life, it is easy to assume that we must all have the same perception of water's nature. Assumptions, attitudes, and resulting behaviour can create barriers to resolving conflict and disputes. Our attitude, how closely we choose to hold onto our world views, or how we relate to others' is manifested in our behaviour as participants in conflict resolution.

Over the past fifteen years, I have heard First Nation communities' concerns and conducted research with Elders on their urgent fears about freshwater. I first began to hear the Elders speak about deteriorating water quality and availability while mediating disputes over whether and how tree harvesting should occur. Key players in these disputes were usually First Nations Elders, chief, and council; foresters from the forest company; and Ministry of Forests representatives. Over time, I noticed that the Elders started to speak of concerns about water. Typically, foresters would seem a bit puzzled by the Elders' apparent tangential stories, which told of how water was drying up or how beaver had left a watershed. The foresters' replies usually explained that they were following British Columbia's Forest Practices Code (and now the Forest and Range Practices Act) and, therefore, *all should be well.* But the Elders were not reassured by these words or the code requirements. Differences exist between how First Nations people and those trained in Western science (e.g., foresters) perceive freshwater. Miscommunication and misunderstandings over an apparently simple concept like water can lead to mistrust and potentially to conflict. Each disputing party may assume that the other "understands," rather than probing for the other's perceptions of water and other similar keystone concepts.

In this chapter, I explore first the concept of blue ecology as a proposed theoretical foundation for a cross-cultural cooperative approach to forest management. I begin with a review of blue ecology, and then focus on a dispute resolution case study of Mt. Ida. This case study juxtaposes First Nations' and Western science's perspectives on freshwater and its management in forested ecosystems by investigating three questions about water relations in trees and ecosystems. Next, the two world views are compared. I then examine how these world views can be reconciled using a blue ecology vision and principles. My central argument is that Western science has an incomplete understanding of water (Blackstock 2002), and thus an in-depth exploration of First Nations' ecological perspectives on water can contribute a more intuitive layer of knowledge, and furthermore propose a vision for reconciling conflicts over how to manage forest lands and water.

Blue Ecology Background

Blue ecology is defined as "an ecological philosophy, which emerged from interweaving First Nations and Western thought, that acknowledges water's (i.e., fresh and salt) essential rhythmical life-spirit and central functional role in generating, sustaining, receiving and ultimately unifying life on Earth Mother" (Blackstock 2009, 308-9).[1] Water is believed to have a spirit by many cultures, and in particular indigenous peoples, and it is commonly characterized as the lifeblood of the planet (Blackstock 2005b; UNESCO-IHP 2008; TURKKAD 2009).[2] Indigenous people commonly use the descriptor "water is lifeblood," since water is seen as the blood flowing through the terrestrial veins of Earth Mother.[3]

Elsewhere in this volume, Lewis and Sheppard describe First Nations' spiritual conception of forests. Secwepemc, Syilx, St'at'imc, and Nlaka'pamux Elders (see Figures 11.1 to 11.4) inspired the interweaving of cross-cultural perspectives on ecology.[4] Water is a meditative medium, a purifier, a source

Figure 11.1 (clockwise from top) **Elders Mary Thomas, Mary Louie, Millie Michell, and Albert Joseph**

of power. Most importantly, it has a spirit. Water is alive – biotic: "Water flows, it is 'living,' it moves: it inspires, it heals, it prophesies" (Eliade 1958, 199-200; Anderson 2010, 21-22).[5] Water the transformer is always moving. The amount of water in the world is relatively constant; however, its form, availability, and quality is not. Nlaka'pamux (Thompson) Elder Millie Michell passed a torch to me on October 2, 2000, during an interview presented in "Water: A First Nations Spiritual and Ecological Perspective" (Blackstock

2001b). At the age of eighty-six, she was passionate about water and wanted to share her deep concern about freshwater being available in sufficient quality and quantity for future generations – she was worried that it was drying up and becoming polluted. Water is all around us in an ordinary sense; its ubiquity tends us to ambivalence. But what is water? How would a First Nation Elder or a Western scientist answer this question? Through the use of ethnographic research methods and interviews with four highly respected Elders from the southern Interior of British Columbia, I document First Nations' perspectives on water (Blackstock 2001b). The Elders were very concerned about whether enough clean drinking water would exist for future generations. Here, I highlight the Elders' vision of the relationships between water, land, and animals and encourage a shift toward *water-based* ecosystem management, thus proposing to repair the definition of forest ecosystems in a way that interweaves First Nations' philosophy with Western science's ecosystem-based management approach. This results in a "blue ecology" redefinition of a forest ecosystem as "a segment of the *sacred and profane* landscape, composed of relatively uniform climate, soil, plants, animals, and micro-organisms, *which is a community complexly interconnected through a network of freshwater hydrological systems* (Blackstock 2012, 10)."

First Nations-like reverence for water was once common across cultures in the Western world.[6] Inuit Elder Paingut "Annie" Peterloosie (2011) shares her ancestors' poignant and timely (i.e. from a climate change perspective) teaching related to reverence and showing respect to the Arctic's icebergs (translated from Inuktitut):

The icebergs are very big.
They are good drinking water from the sea salt.
They are not to be played with.
They are not to be shot at.
They are not to be broken down, when played with.

Our Inuit ancestors
Told us not to break them.
And near the edge of the ice
were told not to camp there
because it might break
Inuit were told not to sleep overnight.

And they were told not to urinate
because it would soil them and are vulnerable.
It [iceberg] does not want to get dirty
because it is believed that there are people in it.

It is true.

Elder Annie York, of the Nlaka'pamux Nation, shared her reverence for water in *They Write Their Dreams on the Rock Forever* (York, Daly, and Arnett 1993, 129): "There was a special reverence for the water. You come to a stream. You don't just go there and put your hand in and drink. You cross yourself to the four directions, and you say a prayer. That came with the creation of the first person." Many First Nations' creation oral history cycles begin when there was just water on earth – it is the primal substance from which coyote or duck, for instance, emerged to create earth. Sanderson (2008) retells how the Creator gave the Cree trickster Wesakachak the power to bring earth from beneath the floodwater, to remake the world. Syilx Elder Harry Robinson (1989, 31) says: "God made the sun ... then after that and he could see – all water. Nothing but water. No trees. No nothing but sun way up high in the sky." Later, Coyote created earth by diving into the water to get a grain of dirt, which expanded into earth as we know it today. The Gitxsan creation oral tradition speaks of the flood, which describes how a people were brought by flood to live in their traditional territory. After the flood, Raven brought sunlight and then freshwater. Txamsem (Raven) tricked the chief who controlled water into believing that he had soiled himself. Txamsem said he would help clean the chief, but first he would need some of his water. The chief agreed and, in the meantime, Txamsem "took the water bag and began to saturate his blanket with water and then he ran out and threw the water to all the different directions which then caused many rivers to flow and now there was water all over the world" (Cove and Macdonald 1987, 16). Gitxsan Elder Chief Mary Mackenzie shared during the *Delgamuukw* court case her knowledge of how the *halayt* (shaman or healer) gained power and learned their songs from the waterfall:

> Now, to get that – those songs, these people *[halayts]*, they're taken out where there's a waterfall and they stay there for some time. And this person would go to this waterfall and sit by this waterfall by the hours and then you hear the echo of the waterfall and they listen to that and they say little by little there's words coming out from that waterfall and this they have to – they have to remember.

Kennedy et al. (1993, 43) reported a similar example: "In former times people trained for guardian spirit power in the rocky area where the Carmanah Point Lighthouse (Vancouver Island, British Columbia) is situated today. Those who were training, scrubbed themselves with huckleberry bush branches and bathed in natural rock-depression basins at this point." Grand Chief Dr. Gordon Antoine (Nlaka'pamux) recalls how the old timers [Elders] talked about water with such fervour that he characterized water "as a kin to religion" (Antoine 2003). Water is powerful and yet it can be so gentle;

water is more powerful than fire: fire you can stop, but water you can't (Thomas 2000; Joseph 2002). Blackstock (2001b) investigates the *going to the water* ritual described by Elders and Kilpatrick (1991, 51). Elder Mary Thomas explains the ritual like this:

> Without the water we can't survive. And I can remember our Elders talking about it. Therefore, when we're weighted down with a lot of grief, your life is becoming unmanageable or you're going through a lot of pain, the first thing our grandmother and my aunt and my mother would say, "Go to the water."

The Mohawk word for water is *ohnekanos,* which means that spirit that brought us here through the universe, through the spirit world, and that will always take care of us (Anderson 2010, 8-11). Anderson describes a poignant Mohawk "dew" ceremony in which fresh morning dew is rubbed on the skin to attain spiritual growth. Many Aboriginal cultures believe there is a "distinct relationship between women and water" because "women's bodies have the capacity to host and sustain the life force that water represents" (ibid., 9).[7]

Now that I have reviewed the sacred nature of blue ecology, let's examine the concept within the Mt. Ida case study context.

Mt. Ida (Kela7scen) Case Study

During the summer of 1998, the Silver Creek forest fire burnt the front face of Mt. Ida, which is adjacent to the town of Salmon Arm, British Columbia (see Figure 11.2). The mountain is known as Kela7scen to the Secwepemc people, who consider the mountain sacred. *Kela7scen* is the Secwepemc word for volcano. Hot springs emerge from the basaltic fissures on the mountain. This area is in the Interior Cedar Hemlock (ICH) biogeoclimatic zone. After the Silver Creek fire was doused, a dispute ignited between First Nations (specifically, the Adams Lake and Neskonlith bands of the Secwepemc Nation), the Ministry of Forests, and two forest companies over whether previously approved harvesting should occur on the mountain. In December 1998, I became involved in the dispute as a mediator. A technical working group, with membership from the bands and the Ministry of Forests, was established to explore solutions to how burnt timber could be harvested off the front face of the mountain. The group found innovative ways to modify harvesting plans that would preserve freshwater, honour the sacredness of Mt. Ida, allow time for the mountain to heal, and permit the forest company to modestly proceed with modified harvesting plans.

During a tour of Mt. Ida, Elder Mary Thomas became deeply concerned about how the creeks appeared to be drying up and also expressed her desire

Figure 11.2 **Mt. Ida (Kela7scen), near Salmon Arm, British Columbia**

to let the mountain heal for five years. Her teachings and questions frame this case study on how blue ecology can be used as an alternative ecosystems-based approach to forest management. Mary asked me three seemingly simple questions about water relations in trees and ecosystems that probe these interests:

1 How much water is stored in a tree?
2 Does the tree enrich the soil with the sun's energy?
3 What happens to creeks and springs when the trees are harvested?

Her questions arose from decades of observations of ecological changes on Mt. Ida, teachings from her Elders, and her ethnobotanical research with Dr. Nancy Turner (Turner 2000).

As a layperson, I can offer a good insight into *how* to approach answering the questions using a cross-cultural approach. The questions are raised from a First Nations traditional ecological framework; the answers will arise from interweaving the strengths of each knowledge system. A good question raised by Elders, along with their long-term reliable observations, has great value to science. Exploration of each question below begins with a science-related primer; question three is complex and thus involves an in-depth exploration of what each knowledge base has to offer toward finding an answer.

Question 1: How Much Water Is Stored in a Tree?

The study of water relations in trees is a forest biology topic dedicated to the study of the nature and function of the hydraulic architecture of a tree (Aitken and Guy 2005). Water has four functions in plants.

1 As a constituent, it accounts for 80 to 90 percent of the fresh weight of most herbaceous plants (approximately 50 percent of woody plants).
2 As a solvent, it aids in the movement of gases, minerals, and other solutes through and between cells.
3 As a reactant, it is a constituent of hydrolytic processes.
4 As a maintainer of cell turgor, it aids in cell enlargement and growth (Kramer 1983).

The water content of plants fluctuates both seasonally and diurnally. Internal tree storage compartments are recharged at night or during low transpiration demand periods (Meinzer et al. 2004). Water storage capacity (WCS) is defined as the quantity of water that can be lost without irreversible wilting (Ewers and Cruiziat 1991). Significant water storage capacity exists in the trunk of conifers (e.g., for *Pseudosuga menziesii,* 74.8 percent of the tree's water is stored in the trunk, 19.8 percent in the roots, 5.2 percent in the branches, and 0.1 percent in the leaves). Water coming out of storage in the sapwood represents about 17 percent of the total daily transpiration stream (Pallardy et al. 1994). For short-term or diurnal needs, the available water storage capacity is in the inner bark (cambium); the sapwood contains water that can be accessed over a longer term (days or weeks) (Ewers and Cruiziat 1991). Water loss through leaf stomata in the transpiration process is a major pathway for the discharge of soil and groundwater (O'Grady, Eamus, and Hutley 1999). Waring and Schlesinger (1985) estimated that sapwood storage of water may approach 300 tonnes/ha, which is equivalent to the amount of water transpired by the forest in a five- to ten-day period.

To determine the actual water content of certain trees, one could use Pallardy et al.'s approximations (1994). The theoretical maximum water content for a one-cubic-metre tree is a thousand litres. The estimates shown in Table 11.1 provide a rough guide. Water stored within trees is significantly less than the amount stored in the soil or transpired by the tree. However, to understand water relations in trees is to understand the importance of timing. Where is the available (unbound) water when the plant is under water stress and near wilting point? And at the stand level, where is the available water stored for short-term access by organisms in an ecosystem under water stress? Although the amount of water stored in a tree is relatively small compared with the amount transpired each day, the stored water plays an important role when the tree's demand for water is high during

Table 11.1

Estimates (± approximately 25 litres) of the amount of water stored in trees (roots, branches, leaves, and trunk) by species, for trees found on or near Mt. Ida

	Capacitance		Conductance	Diameter
Tree species	Per tree (L)[a]	Per hectare (L)[b]	Per tree (L)[c]	(cm)
Douglas fir	152[d]	76,111	64	38[e]
			530	134
			16.4-57.3	30-35[f]
Lodgepole pine	220	110,000[g]	25	25
			44	20-26[h]
Red cedar	80	40,000	n/a	
Birch	250	125,000	42-70	n/a[i]
Aspen	400	200,000	n/a	
Cottonwood	500	250,000	51	15[j]
	640	400,000[k]	n/a	

Notes:

a Volume of water in litres stored in a *whole* tree (roots, branches, foliage, and stem) with a hypothetical stem volume of 1 cubic metre.

b Number of litres stored per hectare assumes 500 stems per hectare. This number is subject to a number of scaling errors. As a comparative benchmark, a typical in-ground swimming pool (16 by 30 feet) holds about 83,000 litres.

c Volume of water in litres used by a tree in one day (diameter of tree at breast height [dbh]). O'Grady, Eamus, and Hutley (1999) assert that a high correlation exists between diameter at breast height and tree water use.

d Based on the studies of David Simpson (research scientist, BC Ministry of Forests, pers. comm., November 10, 2002), the sapwood would have 0.11 cubic metres of water (for a tree of 0.9 cubic metres), assuming a 37 percent moisture content. This figure (0.137 cubic metres) would represent 75 percent of the water in a tree, so it was adjusted accordingly.

e Averages reported by Wullschleger, Meinzer, and Vertessy (1998).

f Early summer average daily water use by interior Douglas fir trees in plots A and B is estimated to be 1.08 and 1.59 millimetres, which is 10,800 to 15,900 litres per day per hectare (Simpson 2000).

g The capacitance for lodgepole pine, red cedar, birch, aspen, and cottonwood were based on the green weight (kg/m^3) and oven dry weight (12 percent) (kg/m^3) reported by Alden (1995, 1997). A 12 percent adjustment was made to account for oven-drying; Pallardy et al.'s proportions (1994) were then applied to get whole tree water storage.

h Averages reported by Wullschleger, Meinzer, and Vertessy (1998).

i Averages reported by ibid.

j Averages reported by ibid. For cottonwood in the four- to five-metre height range, they report daily use of 109 litres. These data are weak in scope. In general, water use by mature cottonwood is probably higher.

k Based on the average initial moisture content of black cottonwood as published by the Northern Hardwood Initiative in Quesnel, British Columbia, www.cfquesnel.com.nhi/Content/Section5/5_4.htm. Pallardy et al.'s proportions (1994) were applied to estimate whole tree moisture content.

transpiration or times of drought. The question of how much water is in a tree is important because it causes one to reflect on a forest as an interaction of two spirits *in situ:* the spirit of the tree and of water. It was curious to me that the topic of water capacitance in a tree was not typically addressed in the research literature; this dearth of information may offer insight into how the two cultures have differing perspectives of the forest.

Question 2: Does the Tree Enrich the Soil with the Sun's Energy?

Water sticks, or coheres, to itself or to substances to which it comes into contact. In combination with water potential gradients, the water surface tension created by cohesion is for the most part responsible for the rise of water from the roots to the leaves. The leaves create the lifting action through transpiration and the cohesive property of water prevents cavitation and enables capillarity action (Davis and Day 1961; Aitken and Guy 2005). Capillary rise of water through the soil profile, especially in coarse-textured soils, can occur at rates of about one millimetre per day (ten thousand litres per hectare per day). Water travels from the water table through the soil pores to the roots, up through the trunk (xylem and sapwood) to the leaves, and then out the leaf stomata. In saplings and small trees, the mass flow rate of water through the stem is essentially equivalent to the canopy transpiration; however, a considerable lag may occur between fluctuations in transpiration and stem water flux. In the morning, the crown water flux may be greater than the basal water flux. In the afternoon and evening, the converse is true, which indicates a recharge of stem water storage reserves (Wullschleger, Meinzer, and Vertessy 1998). Water is redistributed within the tree and also passively from the tree back into the soil. This emerging concept is called hydraulic redistribution (Brooks et al. 2002; Meinzer et al. 2004); formerly described as hydraulic lift (Richards and Caldwell 1987).

When the tree is transpiring near midday, using deep roots it lifts water from the moist soil to where it is needed for transpiration. Wullschleger, Meinzer, and Vertessy (1998) assumed that stomatal conductance and transpiration was positively correlated with the hydraulic conductance of the soil–root–leaf pathway. As transpiration slows in the evening, the tree begins to replenish the short- and long-term reserves in the trunk that it may have used to supplement the water main flow from the roots. Reliance on stored water increases with tree size (Phillips et al. 2003). At night, the water efflux (i.e., passive leaking of water from the roots into the rooting zone) typically occurs in the upper layer of denser and younger roots, in about the top sixty centimetres of soil (Meinzer et al. 2004). Trees will allocate more resources to root growth during dry periods to overcome soil drying (Eamus 2001).[8] The hydraulic redistribution of water in the upper soil layer demonstrates how deeper roots can have an effect at night as well. Rectification, a property of a barrier that allows flow in one direction and not in the other, increases as the roots age, so the water passes more easily across the cell wall barrier of younger roots to the lower water potential in the dry upper layer of soil. Essentially, the tree provides not only available water to neighbouring plants, which is easily accessed for short-term use the next day, but also a water solution that may be nutrient enriched (Caldwell, Dawson, and Richards 1998; Brooks et al. 2002), confirming that it is possible for trees to enrich the soil.

Davis and Day (1961) expressed wonder at the remarkable ability of water to act as a universal chemical solvent; they described the sea from which evolutionary life emerged as *Magna Mater* – the mother of life. Biologically significant is water's inertness: it has the ability to deliver this aqueous solution, without being modified itself. Over one-half of the world's chemical elements, as well as an abundance of micro- and macro-organisms, are found in rivers, streams, lakes, and oceans. Naturally occurring water is an aqueous solution of suspended micro-organisms and chemicals, ever flowing, connecting, and serving tissue fluids, blood plasma, and the liquids that flow in cell interiors.

Water, in its naturally occurring form, is biotic. Elders believe that all beings, including trees and water, have a spirit (aura). From a scientific perspective, the Elders' teaching that tree sap can enrich the soil seems to be supported through the emerging research on hydraulic redistribution. Hydraulically redistributed water may be available for reabsorption by the same plant or neighbouring plants of the same or other species that have active roots (Meinzer et al. 2004, 919). Plants, as individuals and communities, seem able to self-manage the flow of water, depending on their demands and water availability. The emerging field of hydraulic redistribution seems to support Mary Thomas's belief that a tree can enrich the soil.

Question 3: What Happens to Creeks and Springs When the Trees Are Harvested?

This is a surprisingly complex question requiring a background review of groundwater hydrology, springs, and the sponge effect. These concepts are examined first from a Western science perspective and then from the point of view of traditional ecological knowledge. Mary's question is answered in the concluding subsection, "Ecosystem Water Balance and Health."

Background for Question 3

Western Science

In the preface of *Basic Ground-Water Hydrology* (Heath 1987), Ralph Heath states that groundwater is one of the United States' (and equally so for Canada's) most valuable resources, and yet surprisingly, the occurrence of groundwater is poorly understood and the subject of common misconceptions. Two-thirds of the world's freshwater is found underground in aquifers and appears at the surface in springs. A natural spring occurs when the water becomes trapped under a less porous soil layer. The resulting pressure forces the water to the surface, usually where a more porous layer meets the surface. All water that occurs beneath the land's surface is referred to as groundwater (ibid.). Groundwater, though difficult to visualize, occurs in the pores between soil particles (silt, sand, and gravel) or in cracks in bedrock, much like a sponge holds water. Rain or snowmelt percolates into the soil in permeable groundwater recharge areas. Excess water, which remains after requirements for plants and the unsaturated soil are satisfied, percolates down to the water table. The most significant function of groundwater is perhaps its gradual discharge into streams, rivers, and springs to maintain stream flows during dry weather. The water table is the level where water stays. It is the very top of the zone of saturation.

Visualize the soil layer as consisting of two sponge-like layers – an unsaturated layer on top of a saturated one. The unsaturated layer is further divided into the soil, or rooting zone, and a lower intermediate zone. The saturated layer is divided into the transitional capillary fringe and groundwater (see Heath [1987] for excellent diagrams illustrating this concept). Groundwater moves very slowly at about 1/9000th of the rate of river water (ibid.). If you were to squeeze the unsaturated sponge, it would feel wet because of a mix of air and water in the pores; however, it would not drip water like the saturated sponge. Starting at the top surface, the unsaturated sponge dries out as water is used by plants and evaporates from the soil. The water table is a separating line between the unsaturated and saturated zones; the transition between them is called the capillary fringe. Simply, the water table is the level of water in a well. All water that occurs beneath the land's surface is referred to as groundwater (ibid.). An aquifer is a specific kind of saturated sponge that consists of a geologic formation (e.g., sandstone, carbonate rock, basaltic, volcanic, unconsolidated sand, and gravel), which is porous enough to retain, transmit, and yield significant amounts of groundwater when saturated. Aquifers are not underground lakes or rivers. Recharge areas with permeable unsaturated soil in forests or grasslands allow water to infiltrate the aquifer. The aquifer, in turn, supplies significant quantities of groundwater to springs and wells through a system of interconnecting pores and fractures within the host formation.

First Nations Traditional Ecological Knowledge[9]

Springs provide very pure water for medicinal plants and localized sources of water for small animals. These waters are used for making medicinal tinctures, are curative, and may *give* spiritual power (Blackstock 2001b; Klubnikin et al. 2000). Haida Elder Agnes Russ (at the age of 105, in 1962) described how, after a great flood, her grandmother's grandmother brought forth a freshwater spring to bring water to her village people, located on the west coast of Haida Gwaii, British Columbia. On the highest peak behind her flooded village she stuck her cane in the ground and sang ten songs belonging to her family, and then she pulled it out and a freshwater spring shot up (Russ 1962).

The Hopi call springs *paahu* – "natural water" or "spring" – and they are absolutely central in Hopi social and environmental thought because they are the prototypical water sources (Whiteley and Masayesva 1998, 13). The Hopi consider the area around the spring sacred; clay, reeds, and spruce branches are collected from the area to bring the power of the spring to the village. Spring water is used on a daily basis, to bless fields and to welcome babies into the world (Sierra Club 2002, 510). Vernon Masayesva, executive director of Black Mesa Trust, former chair of the Hopi tribe and member of the Water Clan, compares the Hopi view of underground water with that of Western science:

> Western science describes neat but unconnected layers of aquifers. Hopi see the water underneath us as a living, breathing world we call Patuwaaqatsi, or "water-life." Plants breathe in moisture from the sky, and cloud people reciprocate by pulling the moisture to the plants' roots. Hopi believe that when we die we join the cloud people and join in their journey home to Patuwaaqatsi; and so all Hopi ceremonies are tied to the water world, and all the springs along the southern cliffs of Black Mesa serve as religious shrines or passageways to water-life. (Sierra Club 2002, 510)

The Hopi hydrological cycle is quite different from that of Western science: the Hopi believe that the spruce tree has the ability to attract clouds and moisture to the land (Waters 1963, 51, 200). A very sacred carbonate-spring is located mid-slope amid the beautiful Hat Creek Valley's grasslands, in British Columbia. Local Aboriginal people continue to visit, pray, and leave offerings at this site, as shown in Figure 11.3.

Elder Mary Louie believes that Earth Mother filters or cleanses groundwater through the soil by expanding and contracting herself like a sponge. In addition, she may "push the water away" in retaliation to people who disrespect water or Earth Mother. Mary believes that we must show respect to water by giving offerings of prayer, coins, or tobacco.[10] The dialogue below provides a glimpse into this complex concept:

> Mary Louie: The water on top is still [quiet or without movement], but underneath you get a fast current, and that current goes all over the place. It's like [blood in] your veins. And when it comes out in the spring time, springs come down. In May, the land goes down [that is, Earth Mother is

Figure 11.3 **Soda Spring, Hat Creek Valley, British Columbia**

> contracting, or breathing in, causing groundwater to be pushed out onto the earth's surface], but in June or July it comes back up again [that is, Earth Mother is expanding, or breathing out, causing surface water to be sucked back into the earth], then it dries out. The whole land goes down; that's why you get the high water – all the springs are running out. The land filters the water ... and [when] the water goes in the lakes, the mud sucks it down. That's why the water gets polluted. But when the soil starts getting polluted, it can't filter the water anymore. (Louie 2000)

First Nations' Focus on Sponge Effect

What are the cumulative effects, at all scales, on the ecosystem's sponge-like functions (e.g., groundwater recharge areas, wetlands, plants, moss, beaver ponds)? Furthermore, what happens if the function of these sponges become impaired, as postulated in Mary Thomas's third question? Based on his observations of the land during his thirty years of experience in forestry and guide outfitting, Elder Albert Joseph suggests that forestry practitioners be mindful of the sponge effect when contemplating interventions such as harvesting:

> Albert Joseph: They should have knowledge enough to understand how water ... stays in the forests, through the moss and the swamps and the trees. Like, see if moss will hold water for four or five months, then the creek will

> run all summer. So moss is like a sponge. Squeeze a sponge, it's dry, it's the same with moss. That's how a certain amount of water comes out of the hillside, especially on the shady side. And you know those creeks on the mountain, they run almost all summer ... they were fast-running streams four or five years ago. And then you burn or log ... and now the [streams] are all dry, because the moss is all gone. (Joseph 2002)

Other indicators, for Joseph, of the existence of springs or seeps include the feel of the spongy ground underfoot or the way his horses avoid wetter ground because "mud is their worst enemy"; "lots of rabbit tracks" also can indicate springs. Joseph also remembered a place called Beaver Meadows, near the Paradise Valley, where there were dams every couple of hundred yards. He thought that the beaver created reservoirs (which could be thought of as saturated sponges) to provide a controlled release of water during the low-flow periods in late summer.

I have heard Elders on numerous occasions worry about the disappearance of beaver from valleys and the subsequent effects on the local hydrological cycle. Olson and Hubert (1994) outline the following benefits of beaver activity in an aquatic ecosystem:

- elevated water tables that enhance riparian areas
- reduced stream velocity
- improved interception of runoff
- improved habitat for birds, waterfowl, and animals
- improved water availability and storage for summer drought.

Wetlands (i.e., swamps, bogs, muskeg, pothole marshes, and spring-fed seeps on side hills) are transitional zones between dry uplands and open water. These areas are important sponges created by emerging springs, by beavers, and by seasonal pooling of precipitation or streams. They are defined as ecosystems dominated by water-loving plants and having saturated soils (Banner and Mackenzie 2000). Wetland environments have had an important role in human affairs around the globe and throughout antiquity. Cattail (*Typha* spp.), tule (*Scirpus* spp.), reed (*Phragmites* spp.), and Indian hemp *(Apocynum cannabinum)* are important wetland plants used for baskets, mats, sandals, rope, and building construction (Nicholas 1998). Albert Joseph believes that horsetails *(Equisetaceae)* have a purifying or cleansing function in the wetlands. Joseph has noticed that sometimes water, in wetlands or creek edges, has an oily "rainbow" film on the surface, and if you drink it, "you will get diarrhea awfully bad" (Joseph 2002).

Elder Mary Thomas is concerned about the springs and streams drying up on Mt. Ida. She remembers that there used to be lots of water on Kela7scen but, after the fire and logging on the mountain, she fears "the snow is going to melt and it's going to swish right out of there ... it's not going to saturate the mountain."

Ecosystem Water Balance and Health

To answer Mary Thomas's third question, we must think of the integral effects of water relations in plants and the mysteries of groundwater hydrology. I provide a more detailed analysis of ecosystem water balance and low-flow hydrology in Blackstock (2005b). Essentially, though, water balance is an accounting of the freshwater inputs (gains) and outputs (losses). During the summer, and sometimes winter, streams, creeks, rivers, and so on, have low-flow periods, which can be striking examples of water balance, as was observed in the summer of 2006 in British Columbia, when river levels were at an all-time record-low water level. The Elders notice unseasonable low water levels. Researchers such as Hetherington (1987), Pomeroy et al. (1997), Johnson (1998), Smakhtin (2001), and Pike and Scherer (2003) provide useful literature surveys. In general, they see higher post-harvest stream flows, which begin to subside about six years after site disturbance. There is "diverse and apparently conflicting evidence" on the effects of forestry on low flows (Johnson 1998, 1). The effects on low flows seem to depend on the pattern of disturbance in a catchment, intensity of harvesting (e.g., clear-cut or selection), and the seral stage of the stands (Johnson 1998; Smakhtin 2001). Smakhtin (2001, 152) cautions: "At the same time, both increase and decrease in low flows are theoretically possible [after deforestation]." Elders teach, and science confirms, that groundwater discharges or sponge effect discharges from lakes, marshes, beaver-dammed ponds, mossy seeps, melting snow, or glaciers are the main sources of water during low flows (Smakhtin 2001). Pomeroy et al. (1997) introduce the notion that boreal forests have a "self-control" ability as water managers, similar to what is discussed above in the section on hydraulic redistribution. The resilience of this ability to create climate and water supply is as of yet undefined.

After harvesting a forest, significant changes may occur to runoff, stream channel location, interception of rainfall, summer low flows, snowfall melt rates, infiltration of water into soils, and water availability for plants in the rooting zone. Most importantly, the ecosystem's sponge effect may be affected, which potentially results in impaired hydrologic self-management abilities. In addition, global climate warming could also factor into observations of declining river flows, lake and wetland levels, and adverse changes in water renewal times (Schindler 2001). As the average ambient air temperature rises due to climate change, so too does air's ability to hold water (humidity) (see Blackstock [2008] for a discussion on water and clouds as arbiters of climate).

In answer to Mary's third question, it is difficult to measure and assess the Elders' general observation that creeks and springs are drying up. I respect the Elders' connection to the land and their observations of changes. Elders' keen interest in water suggests that forestry professionals and researchers should think of a healthy ecosystem as one in which freshwater, of sufficient quality and quantity, is delivered in a functional rhythm.

Case Study Teachings

Elder Mary Thomas asked wise questions. What can be learned from them and this case study? First, I compare traditional ecological knowledge and Western science perspectives using freshwater as an example. I then discuss how these world views can be reconciled in forest-related conflict by proposing a vision and five guiding principles.

Comparing Blue Ecology with Western Science Ecology

If freshwater is the most mistreated and ignored natural resource, then groundwater is the least understood, simply because we can't observe its connecting function (Schindler 2001). Water is the lifeblood as it traces through the labyrinth of pathways connecting organisms in the ecosystem. In a poignant essay about Gitxsan ecological philosophy, Elder Marie Wilson (1989, 10-11) tells of how the ancient ones believed that all created life was equal, necessary, and a vital part of the interconnected whole that we now call Planet Earth (Anderson 2010). They also believed that this interconnected whole was created to be in perfect balance and must remain so if all parts are to survive in comfort and harmony. A comparison of these two ecological perspectives is presented in Table 11.2; also see Sanderson (2008) and Tipa (2009).

Reconciliation

Water, especially spring water, is a metaphor for healing and reconciliation. Peter Warshall (2001, 56), through his experience in watershed planning and governance in the United States, has come to believe that springs can bring humans back to the basics of life. Water has a unifying role at the ecosystem and human level. Water, without fail, is recognized throughout the globe as crucial to human life, or, in Albert Joseph's words, "Without water, you might as well give up 'cause that's a slow way to die" (humans begin to feel thirst after a loss of only 1 percent of bodily fluids and risk death if fluid loss nears 10 percent [Hinrichsen, Robey, and Upadhyay 1997]).

I see opportunity, in this unifying human interest, to build cross-cultural consensus. Water is a common interest to people because it is essential to the everyday life in a plethora of ways (UN 2003). I previously proposed a conflict-resolution process based on trust building and cross-cultural communication (see Blackstock 2001c). This process recognized that differences

Table 11.2

Comparison of Western science and First Nations traditional ecological knowledge using freshwater as an example

Characteristic	Western science	Traditional ecological knowledge
Epistemology	Analytical	Intuitive
	Cause and effect	Relationship-oriented
	Rely on observations over short term (e.g., 100 years)	Rely on very long-term observations (thousands of years) and communication with spirit world
	Taxonomic	Integrative
	Researchers are experts	Community are experts
	Requires empirical proof	Ancient focus on the prime importance of water
	Dichotomous (e.g., man versus nature; mind over matter; abiotic versus biotic)	Earth Mother is an interconnected whole, a unifying approach; no dichotomy between living and dead
Definition of ecology	Water is implicit to the definition	Water is explicit to the definition
Water	Important, but abiotic; services the biotic world	Biotic, with spirit and will; the lifeblood of Earth Mother
	Forest hydrology focus on snowmelt, surface flow, low flow, and soil moisture regime	Focus on groundwater, springs, and sponge effect
	A secular hydrological cycle	A sacred hydrological cycle
	A resource that provides services to humans	A living being with which we coexist
		Ability to form relationships with and between life forms
Water relations	Water flux is a focus of water relations in plants	Water capacitance, and the ability of the tree's aura to instill energy from the sun into sap, which is shared underground
Water cycle	Hydrologic	Blue ecology

Notes: Readers are referred to Battiste and Henderson (2000, 117-25) for a First Nations perspective on why "Western or Eurocentric thought diverted away from the doctrine of intelligible essences (spiritual forces, life giving energy or having will)." Theodor Schwenk (1996, 9-12), in *Sensitive Chaos*, provides a good introduction to the sacred meandering nature of freshwater: "The more people learned to understand the physical nature of water and to use it technically, the more their knowledge of the soul and spirit of this element faded." Tipa (2009, 107) describes how Maori protect the meandering pattern of rivers to preserve balance.

between parties' perspectives on fundamental concepts, such as freshwater, can aggravate conflict and instill mistrust. For instance, a forester may prematurely discount claims by Elders that the streams are drying up. Trudy Govier (1997), a Canadian sociologist and philosopher, emphasized that trust is fundamentally an attitude. Our trusting attitude toward another person or institution reflects our expectations of benign behaviour based on beliefs about the trusted person's motivation, integrity, and competence. Miscommunication can result in the erosion of trust. The trust-based mediation proposal suggests that the preparation stage of a mediation process is critical. It is a chance to build trust by sharing perspectives on keystone world view concepts, and to then commit to respecting these perspectives throughout the mediation.

A mediator can reframe conflict by encouraging disputing parties to reflect on and identify their common interest to share and preserve freshwater. Dimitrov (2002, 685) discusses "ecological security" as a means to address anthropocentric assumptions about water: "Rivers divide nations, and rivers connect nations. Whether shared water is a uniting force or a divisive force is a matter of attitude." Blue ecology provides a common ground, a new attitude, for mediators and foresters to assist disputing parties to understand each other and then build and monitor solutions based on a set of agreed-on principles. The set of guiding principles introduced below lays an important foundation to reconcile future forest-related conflict. A forester or resource manager can become curious of other world views and open to the opportunities of improving on Western science's understanding of ecology: *it costs you nothing to change your attitude towards water.*

The Vision and Guiding Principles of Blue Ecology

I propose a "water first" vision for human-planned interventions in the ecosystem. I encourage acknowledgement of water's sacred and profane unifying role in the ecosystem (Blackstock 2012).[11] Forest land managers who adopt the blue ecology vision strive to produce clean freshwater in a reliable seasonal rhythm, thus asking themselves, how will this human-planned ecosystem intervention affect the rhythm and quality of water flows? I propose five blue ecology principles to support the implementation of this vision:

1 *Spirit:* Water is biotic, with spirit and will. Water is the lifeblood of the ecosystem. Water is continuously moving and shape-shifting. Sanderson (2008) explains the Cree concept of water: the physical manifestation of spirit, and that free-flowing water that keeps moving is healthy water. Tipa (2009, 106, 99-100) describes the Maori's indicators of healthy flowing water and that they, like British Columbia's First Nations, do not make a distinction between abiotic and biotic; thus, "all things possess a spiritual essence." Anderson (2010, 9) shares a key point expressed by eleven First

Nations grandmothers: "Water is life because water is spirit, and without spirit we have no life ... Without water in our bodies we are dead; not only because of the dehydration that happens in the physical domain, but because of a lack of the spirit energy that signifies life." The Altaians believe their Katun River is a "living being" (Klubnikin et al. 2000, 1299).

2 *Harmony:* Harmonious sustainability in a functional rhythm engenders healthy bodies and ecosystems. Humans should give back when borrowing freshwater from future generations. Water users should be obligated to maintain water that passes through their area of responsibility in as pristine a state as possible and should be penalized for not doing so. Elder Mary Louie believes that water should not be owned as property – just as humans cannot own an orca or an eagle. She teaches giving back to the water by offering gifts and prayer. Most Elders encourage us to be in sync with the rhythms of nature and water.
3 *Respect:* Unsustainable withdrawals (i.e., when water is withdrawn faster than it is recharged or in a manner that irreparably damages the water) from water systems should be prohibited. Millie Michell (2000) emphasizes the importance of educating children to respect water and especially groundwater: *Respect water by not throwing rocks into it or laughing or yelling at it.*
4 *Unity:* Topologically, water is the connector and unifier of the ecosystem and humanity. Water also has the potential to build consensus. Local watershed citizens who are closest to the water can handle matters at the least centralized component of authority (i.e., principle of subsidiarity) and therefore influence their own destiny in water-related management and conflict resolution (Cosgrove 2003).
5 *Balance:* We should practise restrained and measured water withdrawals in combination with and giving back (i.e., restoration, monitoring, or ceremony) to watersheds and water.[12]

These principles are portrayed in Figure 11.7 of the intuitive blue ecology water cycle (see Blackstock [2009, 309-10] for a full description of the cycle). Western science's analytical hydrologic cycle is the more common depiction (see Oki and Kanae [2006] for details on the Western science global hydrologic cycle). Forest professionals can benefit from a more holistic understanding of water by learning from both water cycle models. Blue ecology is an example of how a cross-cultural framework can arise, as a positive effect, from reconciling conflict. Cooperative approaches for resource managers (e.g., such as the case study in Tipa [2009]) that acknowledge and interweave world views are needed to achieve sustainable survival with dignity. As Anderson (2010, 31) states, "Good health is dependent on how well we manage our relationship with water." We have responsibilities associated with borrowing freshwater from future generations, which include respecting

Figure 11.4 **Blue ecology water cycle**

and responding to Elders' concerns that some of our lakes, springs, and creeks are drying up.[13] In parts of the world, climate change will result in water being the most valued "commodity" produced from forested lands (Thompson 2007), necessitating a change in our management priorities and approaches. The first question asked when contemplating resource management impacts should be, how does it affect the water? The highest environmental assessment test for development planning is the water-first principle: planned development (e.g., real estate, urban planning, hydro power, architecture, forestry, agriculture, fishing, aquaculture, mining, oil and gas extraction) cannot impede the functional delivery of quality water to ecosystems in a healthy rhythm.

Acknowledgments

The author would like to acknowledge Elders Mary Louie, the late Dr. Mary Thomas, Nathan Spinks, the late Paingut "Annie" Peterloosie, Albert Joseph, the late Grand Chief Dr. Gordon Antoine, and the late Mildred "Millie" Michell for sharing their wisdom. The following people assisted with the research or review: Joyce Sam, David Simpson, Charlene Levis, Gordon Sloan, Susan Bannerman, Julie Taylor Schooling, Pamela Perrault, Dr. David Tindall, Jeff Stone, Susanne Barker, Kathy Holland, and Marjorie Serack. Dr. Bill Poser and Dr. Alana Johns assisted with the transliteration and translation of Inuktitut. As the author, I am solely responsible for the content and any potential errors.

Notes

1 Dr. Marcus Barber (2012) describes the Yolngu's (Australian Aborigine) cosmology related to the coastal water cycle where the continual mixing and movement of water occurs in the intertidal, underground, and estuarine ecotypes. Furthermore, Barber (2012, 124) says: "The mixture of salt water and fresh water is a symbol of fertility, a source of power and meaning rather than confusion, and the dynamism of water expresses the productive flows of Yolngu social relationships." In blue ecology, water is also viewed as a spectrum of salinity rather than an arbitrary dichotomy (i.e. fresh and salt) artificially separating the wholeness and intermixed nature of water. This chapter focuses on fresh(er)water, i.e., "not of the sea," since this book is about "forests." The blue ecology vision applies to all water fresh and salt.

2 The author is a member of the UNESCO-IHP expert advisory group on water and cultural diversity.

3 Interestingly, Leonardo da Vinci's water theory centred on his comparison of the movement of water on earth to the flow of blood in the human body (Pfister, Savenije, and Fenicia 2009, 85).

4 I explore the concept of blue ecology from a poetic point of view in my books *Salmon Run* (2005a, 24) and *Oceaness* (2010). Also see Jeanette C. Armstrong's poem "Water Is Siwlkw" (UNESCO 2006, 18-19) for an indigenous poetic interpretation of water as our medicine.

5 Pauline (a Cree Grandmother) says that when you acknowledge the spirit of water through prayer or blessing, it becomes medicine (Anderson 2010, 10).

6 Eliade (1958), Schwenk (1996), and TURKKAD (2009) provide a good discussion of cross-cultural reverence for water.

7 "A number of grandmothers drew the equation between life-giving waters carried by women and what occurs with Mother Earth in her life-giving cycles and abilities" (Anderson 2010, 11).

8 Inverse flow of water from roots in a recently wetted desert soil to roots in the deeper drier soil allows the deep roots to grow toward a deep water table. Caldwell, Dawson, and Richards (1998) called this process inverse hydraulic lift, and report roots of Kalahari Desert plants reaching a depth of sixty-eight metres.

9 See Blackstock (2001a, 173) for a definition of traditional ecological knowledge.

10 The Altaians indigenous people of Siberia "leave offerings of light-colored cloth or coins at springs" to show a proper respectful attitude toward the spring: "Nobody should laugh and joke when speaking to a sacred spring ... you must be reverent to the spirit of the spring" (Klubnikin et al. 2000, 1299-1300). They know the location of all the springs in their territory since, as well as being sacred, "springs that are naturally high in minerals do not freeze in the winter" (ibid., 1300).

11 Blue ecology encompasses both freshwater and saltwater.

12 A blue ecology approach should also make financial sense. Although the scope of this chapter does not allow for a detailed economic analysis, there are some good references. Refer to Lansing, Lansing, and Erazo's excellent work *The Value of a River* (1998) for an economic analysis of the "free gifts of Nature" within the Skokomish culture context. Also see the United Nations' discussion (UN 2003, 318-19) on "virtual water" and the advantages of reframing conflict over sharing the water allocations to sharing water benefits.

13 See Blackstock and McCallister (2004) for further discussion on water-based ecosystem restoration and Elders' perspectives on the role of water in grassland ecosystems.

References

Aitken, S., and R. Guy. 2005. "Forestry 200: Forest Biology Course Site; Water Relations." Faculty of Forestry, University of British Columbia, http://courses.forestry.ubc.ca/.

Alden, H.A. 1995. *Hardwoods of North America*. General Technical Report FPL-GTR-83. Washington, DC: US Department of Agriculture Forest Service.

–. 1997. *Softwoods of North America*. General Technical Report FPL-GTR-102. Washington, DC: US Department of Agriculture Forest Service.

Anderson, Kim. 2010. *Aboriginal Women, Water and Health: Reflections from Eleven First Nations, Inuit, and Métis Grandmothers*. Paper commissioned by Atlantic Centre of Excellence for Women's Health (ACEWH) and Prairie Women's Health Centre of Excellence (PWHCE). http://www.acewh.dal.ca/.

Antoine, Gordon Grand Chief. 2003. Personal communication, April 25.

Banner, A., and W.H. Mackenzie. 2000. *The Ecology of Wetland Ecosystems*. Extension note 45. Victoria: BC Ministry of Forests.

Barber, Marcus. 2012. "River, Sea, and Sky: Indigenous Water Cosmology and Coastal Ownership amongst the Yolngu People in Australia." In *Water, Cultural Diversity and Global Environmental Change: Emerging Trends, Sustainable Futures,* edited by B.R. Johnston, L. Hiwasaki, I. Klaver, A. Ramos, and V. Strang, 124-25. Dordrecht, The Netherlands: UNESCO and Springer SBM.

Battiste, M., and J.Y. Henderson. 2000. *Protecting Indigenous Knowledge and Heritage: A Global Challenge*. Saskatoon: Purich Publishing.

Blackstock, M.D. 2001a. Faces in the Forest: First Nations Art Created on Living Trees. Montreal and Kingston: McGill-Queen's University Press.

–. 2001b. "Water: A First Nations Spiritual and Ecological Perspective." *BC Journal of Ecosystems and Management* 1(1):54-66.

–. 2001c. "Where Is the Trust? Using Trust-Based Mediation for First Nations Dispute Resolution." *Conflict Resolution Quarterly* 19(1):9-30.

–. 2002. "Water-Based Ecology: A First Nations Proposal to Repair the Definition of an Ecosystem." *BC Journal of Ecosystems and Management* 2(1):7-12.

–. 2005a. *Salmon Run: A Florilegium of Aboriginal Ecological Poetry*. Kamloops, BC: Wyget Books.

–. 2005b. "Blue Ecology: A Cross Cultural Ecological Approach to Reconciling Forest Related Conflicts." *BC Journal of Ecosystem and Management* 6(2): 39-54.

–. 2008. "Blue Ecology and Climate Change." *BC Journal of Ecosystems and Management* 9(1):12-16.

–. 2009. "Blue Ecology and Climate Change: Interweaving Cultural Perspectives on Water; An Indigenous Case Study." *The Role of Hydrology in Water Resources Management*. Proceedings of a symposium held on Capri, Italy, October 2008. IAHS Publication no. 327:306-13. http://iahs.info/redbooks/.

–. 2010. *Oceaness*. Kamloops, BC: Wyget Books.

–. 2012. "Blue Ecology and the Unifying Potential of Water." In *Water, Cultural Diversity and Global Environmental Change: Emerging Trends, Sustainable Futures,* edited by B.R. Johnston, L. Hiwasaki, I. Klaver, A. Ramos, and V. Strang, 132-33. Dordrecht, The Netherlands: UNESCO and Springer SBM.

Blackstock, M., and R. McCallister. 2004. "First Nations Perspectives on the Grasslands of the Interior of British Columbia." *Journal of Ecological Anthropology* 8(1):24-46.

Brooks, J.R., F. Meinzer, R. Coulombe, and J. Gregg. 2002. "Hydraulic Redistribution of Soil Water during Summer Drought in Two Contrasting Pacific Northwest Coniferous Forests." *Tree Physiology* 22:1107-17.

Caldwell, M.M., T.E. Dawson, and J.H. Richards. 1998. "Hydraulic Lift: Consequences of Water Efflux from the Roots of Plants." *Oecologica* 113(2):151-61.

Cosgrove, W.J. 2003. *Water Security and Peace: A Synthesis of Studies Prepared under the PCCP – Water for Peace Process*. Geneva: UNESCO and Green Cross International.

Cove, J.J. and G.F. Macdonald. 1987. Tsimshian Narrative I: Tricksters, Shamans and Heroes. Ottawa: Canadian Museum of Civilization.

Davis, K.S., and J.A. Day. 1961. *Water: The Mirror of Science*. Garden City, NY: Anchor Books.

Dimitrov, Radoslav S. 2002. "Water, Conflict, and Security: A Conceptual Minefield." *Society and Natural Resources* 15:677-91.

Eamus, D. 2001. "How Does Ecosystem Water Balance Influence Net Primary Productivity? A Discussion." In *NEE Workshop Proceedings*. Canberra: Commonwealth of Australia.

Eliade, M. 1958. *Patterns in Comparative Religion*. Lincoln: University of Nebraska Press.

Ewers, F.W., and P. Cruiziat. 1991. "Measuring Water Transport and Storage." In *Techniques and Approaches in Forest Tree Ecophysiology,* edited by J.P. Lassoie and T.M. Hinckley, 91-116. Boca Raton, FL: CRC Press.

Govier, T. 1997. Social Trust and Human Communities. Montreal and Kingston: McGill-Queen's University Press.

Heath, R.C. 1987. *Basic Ground-Water Hydrology.* Water Supply Paper no. 2220. Denver: US Geological Survey.

Hetherington, E.D. 1987. "The Importance of Forest in the Hydrological Regime." In *Canadian Aquatic Resources,* edited by M.C. Healey and R.R. Wallace, 179-211. Ottawa: Department of Fisheries and Oceans.

Hinrichsen, D., B. Robey, and U.D. Upadhyay. 1997. *Solutions for a Water-Short World.* Population Information Program, Johns Hopkins School of Public Health, Baltimore. Population Reports, Series M, no. 14.

Johnson, Richard. 1998. "The Forest Cycle and Low Flow River Flows: A Review of UK and International Studies." *Forest Ecology and Management* 109:1-7.

Joseph, A. 2002. Taped interview on October 2.

Kennedy, Dorothy, Randy Bouchard, Morley Eldridge, and Alexander Mackie. 1993. "Vancouver Island Cultural Resource Inventory." March 31.

Kilpatrick, A.E. 1991. "Going to the Water: A Structural Analysis of Cherokee Purification Rituals." *American Indian Culture and Research Journal* 15(4):49-58.

Klubnikin, Kheryn, C. Annett, M. Cherkasova, M. Shishin, and I. Fotieva. 2000. "The Sacred and the Scientific: Traditional Ecological Knowledge in Siberian River Conservation." *Ecological Applications* 10(5):1296-306.

Kramer, P.J. 1983. *Water Relation of Plants.* Toronto: Academic Press.

Lansing, J.S., P.S. Lansing, and J.S. Erazo. 1998. "The Value of a River." *Journal of Political Ecology* 5:1-21.

Louie, M. 2000. Taped interview on September 6.

Meinzer, F.C., J.R. Brooks, S. Bucci, G. Goldstein, F.G. Scholz, and J.M. Warren. 2004. "Converging Patterns of Uptake and Hydraulic Redistribution of Soil Water in Contrasting Woody Vegetation Types." *Tree Physiology* 24:919-28.

Michell, M. 2000. Taped interview on October 2.

Nicholas, G.P. 1998. "Wetlands and Hunter-Gatherer Land Use in North America." In *Hidden Dimensions: The Cultural Significance of Wetland Archaeology,* edited by K. Bernick, 31-45. Vancouver: UBC Press.

O'Grady A.P., D. Eamus, and L.B. Hutley. 1999. "Transpiration Increases the Dry Season Patterns of Tree Water Use in Eucalypt Open-Forests of Northern Australia." *Tree Physiology* 19:591-97.

Oki, Taikan, and Shinjiro Kanae. 2006. "Global Hydrological Cycles and World Water Resources." *Science* 313:1068-72.

Olson, R., and W.A. Hubert. 1994. *Beaver: Water Resources and Riparian Habitat Manager.* Laramie: University of Wyoming Press.

Pallardy, S.G., J. Cermak, F.W. Ewers, M.R. Kaufmann, W.C. Parker, and J.S. Sperry. 1994. "Water Transport Dynamics." In *Resource Physiology of Conifers: Acquisition, Allocation, and Utilization,* edited by W.K. Smith and T.M. Hinkley, 299-387. Toronto: Academic Press.

Peterloosie, Paingut Annie. 2011. Translated written note of March 11, in Inuktitut syllabics, to author.

Pfister, Laurent, Hubert H.G. Savenije, and F. Fenicia. 2009. *Leonardo da Vinci's Water Theory: On the Origin and Fate of Water.* Oxfordshire, UK: IAHS Press.

Phillips, N.G., M.G. Ryan, B.J. Bond, N.G. McDowell, T.M. Hinckley, and J. Cermak. 2003. "Reliance on Stored Water Increases with Tree Size in Three Species in the Pacific Northwest." *Tree Physiology* 23:237-45.

Pike, Robin G., and Rob Scherer. 2003. "Overview of the Potential Effects of Forest Management on Low Flows in Snowmelt-Dominated Hydrologic Regimes." *BC Journal of Ecosystems and Management* 3(1):44-60.

Pomeroy, J.W., R.J. Granger, A. Pietroniro, J.E. Elliot, B. Toth, and N. Hedstrom. 1997. *Hydrological Pathways in the Prince Albert Model Forest.* Saskatoon: National Hydrology Research Institute, Environment Canada.

Richards, J.H., and M.M. Caldwell. 1987. "Hydraulic Lift: Substantial Nocturnal Water Transport between Soil Layers by *Artemesia Tridentata* Roots." *Oecologica* 73:486-89.

Robinson, H. 1989. *Write it on Your Heart: The Epic World of an Okanagan Storyteller.* Edited by W. Wickwire. Vancouver: Talonbooks/Theytus Press.

Russ, Agnes. 1962. Haida Elder interviewed by Inbert Orchard. http://www.cbc.ca/.
Sanderson, Cheryl Darlene. 2008. "Nipiy Wasekimew/Clearwater: The Meaning of Water, from the Words of Elders; The Interconnections of Health, Education, Law and the Environment." PhD diss., Simon Fraser University.
Schindler, D.W. 2001. "The Cumulative Effects of Climate Warming and Other Human Stresses on Canadian Freshwaters in the New Millennium." *Canadian Journal of Fisheries and Aquatic Sciences* 58:18-29.
Schwenk, Theodor. 1996. *Sensitive Chaos*. 3rd rev. ed. East Sussex, UK: Rudolf Steiner Press.
Sierra Club. 2002. "Comments on Peabody Mining Application Cultural Impacts." In *Native American Sacred Places,* 495-515. Hearing before the Committee on Indian Affairs, United States Senate, 107th Congress, April 27. Washington: US Government Printing Office.
Simpson, D.G. 2000. "Water Use of Interior Douglas-Fir." *Canadian Journal of Forest Research* 20:534-47.
Smakhtin, V.U. 2001. "Low Flow Hydrology: A Review." *Journal of Hydrology* 240:147-86.
Thomas, M. 2000. Taped interview, October 13.
Thompson, J. 2007. "Running Dry: Where Will the West Get Its Water?" *Science Findings* 97:1-6. http://www.fs.fed.us/.
Tipa, Gail. 2009. "Exploring Indigenous Understandings of River Dynamics and River Flows: A Case from New Zealand." *Environmental Communication* 3(1):95-120.
TURKKAD (Turkish Women's Cultural Association). 2009. *Water and Culture: Diverse Water Cultures.* Proceedings of the fifth World Water Forum, Istanbul, March 14-15. Istanbul: NEFES.
Turner, N.J. 2000. *Plant Technology of First Peoples in British Columbia*. Vancouver: UBC Press.
UN (United Nations). 2003. *Water for People, Water for Life*. http://unesdoc.unesco.org/.
UNESCO. 2006. *Water and Indigenous Peoples*. Edited by R. Boelens, M. Chiba, and D. Nakashima. Vol. 2 of Knowledges of Nature 2 series. Paris: UNESCO.
UNESCO-IHP (UNESCO-International Hydrological Programme). 2008. "Water and Cultural Diversity: Towards Sustainability of Water Resources and Cultures." Brochure. Paris: UNESCO-IHP. http://www.anthropoasis.free.fr/.
Waring, R.H., and W.H. Schlesinger. 1985. *Forest Ecosystems: Concepts and Management.* Toronto: Academic Press.
Warshall, P. 2001. "Watershed Governance: Checklists to Encourage Respect for Waterflows and People." In *Writing on Water,* edited by D. Rothenberg and M. Ulvaeus, 40-56. Cambridge, MA: Terra Nova.
Waters, Frank. 1963. *Book of the Hopi*. New York: Viking Press.
Whiteley, Peter, and Vernon Masayesva. 1998. "The Use and Abuse of Aquifers: Can the Hopi Indians Survive Multinational Mining?" In *Water, Culture, and Power: Local Struggles in a Global Context,* edited by John M. Donahue and Barbara Rose Johnston, 9-34. Washington, DC: Island Press.
Wilson, M. 1989. "Ecology: A Native Indian Perspective." *Legal Perspectives* 14:10-11.
Wullschleger, S.D., F.C. Meinzer, and R.A. Vertessy. 1998. "A Review of Water Use Studies in Trees." *Tree Physiology* 18:499-512.
York, Annie, Richard Daly, and Chris Arnett. 1993. *They Write Their Dreams on the Rock Forever: Rock Writings of the Stein Valley of British Columbia*. Vancouver: Talonbooks.

12
First Nations' Spiritual Conceptions of Forests and Forest Management

John Lewis and Stephen R.J. Sheppard

The potential for destruction and alteration of spiritually significant landscapes typically stems from misunderstanding. The problem centres on inadequate definitions of what First Nations peoples mean when they assert that the landscape has ancient and profound spiritual significance. Although public land managers are becoming increasingly aware of cultural land-based values, they are experiencing difficulty fully understanding them and in designing and implementing actions to accommodate them (Redmond 1996; Satterfield 2002; Stoffle, Toupal, and Zedeno 2003). These values are extremely hard to define, yet in British Columbia, the current forest policy framework mandates that the cultural and spiritual interests of First Nations peoples – expressed as "Aboriginal rights" – be explicitly accommodated in forest management planning (British Columbia 2002, 27-28). Therefore, a poor understanding of these values leads both to noncompliance with policy and to conflict and prolonged mistrust between public agencies and First Nations groups.

If public land managers are to be responsive to the needs and values of indigenous cultures, they must work toward a fuller understanding of those needs and values, in a spatial, landscape-specific context that resonates with local communities and can be translated into forest management planning. The broader objective of our research has been to explore the spiritual perceptions of forested landscapes as expressed by First Nations people. What remains of the spiritually significant landscapes that are being used by Aboriginal people in Canada and the United States is being threatened by resource development, by extractive industries such as timber and mining, and by expanding tourism and recreation use (Cummings 1998; Boyd and Williams-Davidson 2000; Nagle 2005). Most culturally significant places are situated on public lands, and thus the responsibility for these areas rests in the hands of public agencies and professional land managers, many of whom may not understand land-based spiritual values and, perhaps because of their lack of comprehension, simply regard them as irrelevant to the technical or commercial interests of resource management (Redmond 1996).

To gain this understanding, we have reviewed the relevant literature and asked members of the Cheam (pronounced *chee-am*) First Nation of British Columbia to explain what land-based spiritual values mean to them. Having spent two years working with the Cheam First Nation, we have heard people say unequivocally, "This land is our home," "This land is sacred to our people," "This stream, this forest, this fish, this tree is essential to our way of life" (Cheam participants, Field Notes). In the process, much has been learned that can help clarify how forest management interacts with spiritual values in the landscape. This understanding results from personal experience coupled with a more formal type of interpretive or qualitative research that integrates several data sources.

The Cheam First Nation: Research Setting and Background

Our field research began on the Cheam Reserve, a five-hundred-hectare community of approximately 150 people located in the Upper Fraser Valley of British Columbia (adjacent to the community of Agassiz between Chilliwack and Hope). The Cheam people are members of the larger Stó:lō (pronounced *staw-lo*) Nation, a Coast Salish group whose traditional territory encompasses one hundred square kilometres and spans the lower length of the Fraser River, from its estuary in Vancouver to the Fraser Canyon northeast of the towns of Hope and Yale. The reserve takes its name from the Salishan word meaning "always wild strawberries," reflecting the abundance of wild berries and other foodstuffs – wild game, fish, medicinal and food plants – that traditionally sustained the Cheam people. Ethnohistorical accounts of the area in the mid- to late eighteenth century describe the landscape in the shadow of Mount Cheam, the Fraser Valley's highest peak and centre of Cheam spirituality, as abundant in plant and animal life (Crosby 1907). Thick forests of cedar and hemlock interspersed with valley bottom stands of cottonwood and aspen provided habitat for deer and elk herds that supplemented the Cheam's staple diet of salmon from the Fraser River.

The contemporary Cheam lifestyle is modern in several ways, and over the past two hundred years the Cheam and their Stó:lō neighbours have lived through many of the changes experienced by other Pacific Northwest Native peoples. The hunters and fishers of the past who travelled through the valley from summer to winter villages in dugout cedar canoes have now settled permanently on reserves to live in government-built houses serviced with electricity and running water. Sources of household income are many: income generated from fishing and hunting, wages earned on and off the reserve through part-time and full-time employment, as well as family allowances, old-age pension benefits, and social welfare provided by different levels of government. All families rely on store-bought food in addition to the substantial amounts of salmon, deer meat, fowl, and berries harvested by both men and women from spring to early autumn. Several visits to Cheam households have brought us face to face with the rich products of contemporary fishing, hunting, and plant harvesting activities.

The persistence of the Cheam subsistence economy in the face of modern cultural influences is a remarkable testament to its capacity to coexist with the material trappings of a society that many outsiders regard as antithetical to a traditional Aboriginal way of life. As regional developments have proliferated in recent years in the Upper Fraser Valley, cultural preservation concerns have become more pronounced. Logging and recreational activities (e.g., hiking, fishing, camping, boating) interfere with Native uses of the forests and mountain highlands as places of spiritual refuge and, for traditional dancers, as questing sites for spiritual visions (Cheam participant, pers. comm., July 1999). In their pronouncement of forests as culturally meaningful and sacred places, the Stó:lō people of the Pacific Northwest are expressing their intimate relationship with the forests because they are *providers*. In essence, forests are sacred and meaningful places because they are much more than a source of commercial wealth; they are meaningful because they are a living source of physical and spiritual sustenance. This is a fundamental source of Native assertions that human beings are integral parts of the land and that they must live responsibly as a harmonious part of what Euro-Canadians and Americans would call the "natural order."

Methods

Research Design

The selection of a particular research strategy should be determined by how the problem is shaped, and by the kinds of questions that it raises. For this study, a qualitative research design was used for several reasons. A key difference between qualitative research and other forms of social scientific methods is the assumption that we must first discover what people really think and the reasons that they give for their perceptions and actions (Spradley 1979). To date, there has been no systematic examination or documentation of the nature-based cultural values of the Cheam or their Stó:lō neighbours in terms of their perceptions of the spirituality of forested landscapes. Given this lack of prior research with the Stó:lō (and many other First Nation communities throughout British Columbia), and the emphasis that qualitative inquiry places on exploration and inductive analysis from which testable hypotheses and theories can be generated, a qualitative approach seemed to be the most appropriate strategy.

Participant Selection

The strategy used to select the research participants can best be described as "purposive" (Lincoln and Guba 1985). Commonly used in social science research, purposive selection is based on the notion that in order to gain the most insight into a particular problem, participants who can provide the most appropriate information need to be selected. In order to understand land-based spiritual values from a First Nations perspective, it was felt that the most obvious and appropriate approach would be to obtain information directly from an Aboriginal community willing to participate in such research. Moreover, we believed that the research participants chosen from that community ought to satisfy particular criteria:

- *Knowledgeable:* having an understanding of and appreciation for traditional cultural and spiritual uses of the forest.
- *Role in the community:* band members who, by virtue of their profile in the community, are exposed to a range of cultural perspectives and perceptions of forests and forest management.
- *Communication ability:* having a relative ease with outsiders and ability to discuss, at length, cultural issues and concepts.

For their part, key informants from the Cheam community provided a list of prospective participants who satisfied our selection criteria, and whom we contacted at a later date to schedule interviews. From an initial list of twelve suggested participants, eleven consented to participate. The group of eleven participants appeared to be appropriate according to McCracken (1988), who recommends eight as the maximum number of participants in a long interview situation. Although we would hesitate to argue that the individuals chosen represent a full cross-section of Cheam society, we can assert that the participant sample was diverse in age (from twenty-five to eighty), represented varying levels of involvement in the cultural life of the community, reflected a wide spectrum of educational achievement (grade eight to undergraduate university), and included both male and female community members. Such active and knowledgeable participants may not be statistically representative of the larger population, but they do tend to represent those who are more influential in local decision-making processes in that they are often solicited by government and industry planners to provide cultural perspectives on planning issues and, therefore, provide an illustrative case study that may help to understand related populations.

Data Collection

As with many qualitative research designs, this study involved multiple data collection methods. Over the course of our working relationship with the Cheam community, we were allowed to:

1 Review documentary material, both internal (e.g., community newsletters, memoranda, and council minutes) and external (e.g., published articles, consulting reports, and deputations to government hearings).
2 Attend community meetings and informal discussions at the band offices, as well as reside briefly within the Cheam community.
3 Accompany community members into the field to visit different sacred forest areas, examine spiritual sites that have been affected by forest management and other land uses, and participate in the harvesting of culturally significant medicines and foods.
4 Conduct semi-structured interviews with community members.

In addition to photocopied documents, photographs, and journal notes compiled from field excursions and site visits, a significant portion of our data came from recorded interviews conducted with each of the eleven community participants. These took the form of semi-structured conversations that took about an hour and a half to complete. We favoured the semi-structured format because, although the interviews were guided by a list of questions to be raised, neither the exact wording nor the order in which they were asked needed to remain constant (Merriam 1988). In this way, we were permitted to respond to ideas as they arose through the course of our discussions, and to explore more fully the perspectives of each participant.

We chose to use a version of the photo-elicitation technique in our discussions with the Cheam participants, based on the following considerations (Ball and Smith 1992). First, photographic materials can be useful aids in semi-structured interviews to the extent that they give participants the opportunity to pull together and discuss perceptions or concepts that may be difficult for them to articulate. In addition, interviews can become stilted exercises in which the interviewer probes for information that the participant may not be able (or willing) to express, which only accentuates further the social disparities that exist between First Nation participants and Euro-Canadian researchers. Visual referents, however, can enhance the participant's comfort with the interview process by providing a common reference point for the discussion that shifts the focus away from the participant as the subject of questioning and engages both the researcher and the participant as co-investigators of the photographic materials. In effect, the strength of the technique rests in the power of imagery to invite people to take the lead in an inquiry and render full use of their expertise (Lewis and Sheppard 2006).

After the introductions and review of a statement of consent, the interviews began by showing the participants maps, photographs, and photo-realistic simulations of alternative land management scenarios at two different but spiritually significant locations in the Cheam community's traditional territory. The first set of images depicted hypothetical timber management options on the lower slopes of Mount Cheam (Figure 12.1). The simulations were based on a series of three management conditions developed in collaboration with forest operations experts from the University of British Columbia: preservation (or the existing condition), partial cut harvesting (75 percent removal), and clear-cut harvesting (100 percent removal). While reviewing these simulations, the participants were asked a range of questions drawn from the interview schedule. When it appeared that the participant had no further insights to provide about the visual information, the forest images were replaced by maps, photographs, and visual simulations depicting a series of riparian restoration scenarios along a stream segment that had formerly been used by the Cheam community as a meditation area and travel corridor (Figure 12.2). The questions were repeated from the interview guide, and discussions with the participants proceeded until it became evident that no additional information was forthcoming.

Analysis

The data collection process was completed after eleven interviews were conducted, recorded, and transcribed, and a set of field notes and documentary materials were assembled. Glaser and Strauss's constant comparative method (1967) was used to analyze the data. The technique begins by organizing the transcribed interview data into a research database. According to the approach prescribed by Glaser and Strauss (1967) and Goetz and LeCompte (1984), as the database is read through, a running list of the themes or regular categories is developed and maintained. After this initial review, the data is combed again and culled into distinct themes, identifying "units of information that [eventually] serve as the basis for defining categories" (Lincoln and Guba 1985, 344). To refine the themes, units are sorted into piles by comparing the information in one unit with that in the next. Thus, through a process of recursive review, each pile becomes a conceptual category or theme that describes the participants' conceptions of land-based spirituality. This inductive process is integral to eliciting valid constructs, as the patterns emerge mainly from the data rather than being predetermined or overly biased by the researchers' preexisting notions.

Results

Mythic Themes: Commitments to the Creator Across the Landscape

The Cheam people have an extremely strong relationship with their land. Its character – the legends and oral traditions that are ascribed to it – determines to what uses certain places are put and how the land is to be treated. Several participants referred to the Cheam's relationship with the Creator and notions embedded in Stó:lō spirituality that underpin their relationship to the natural

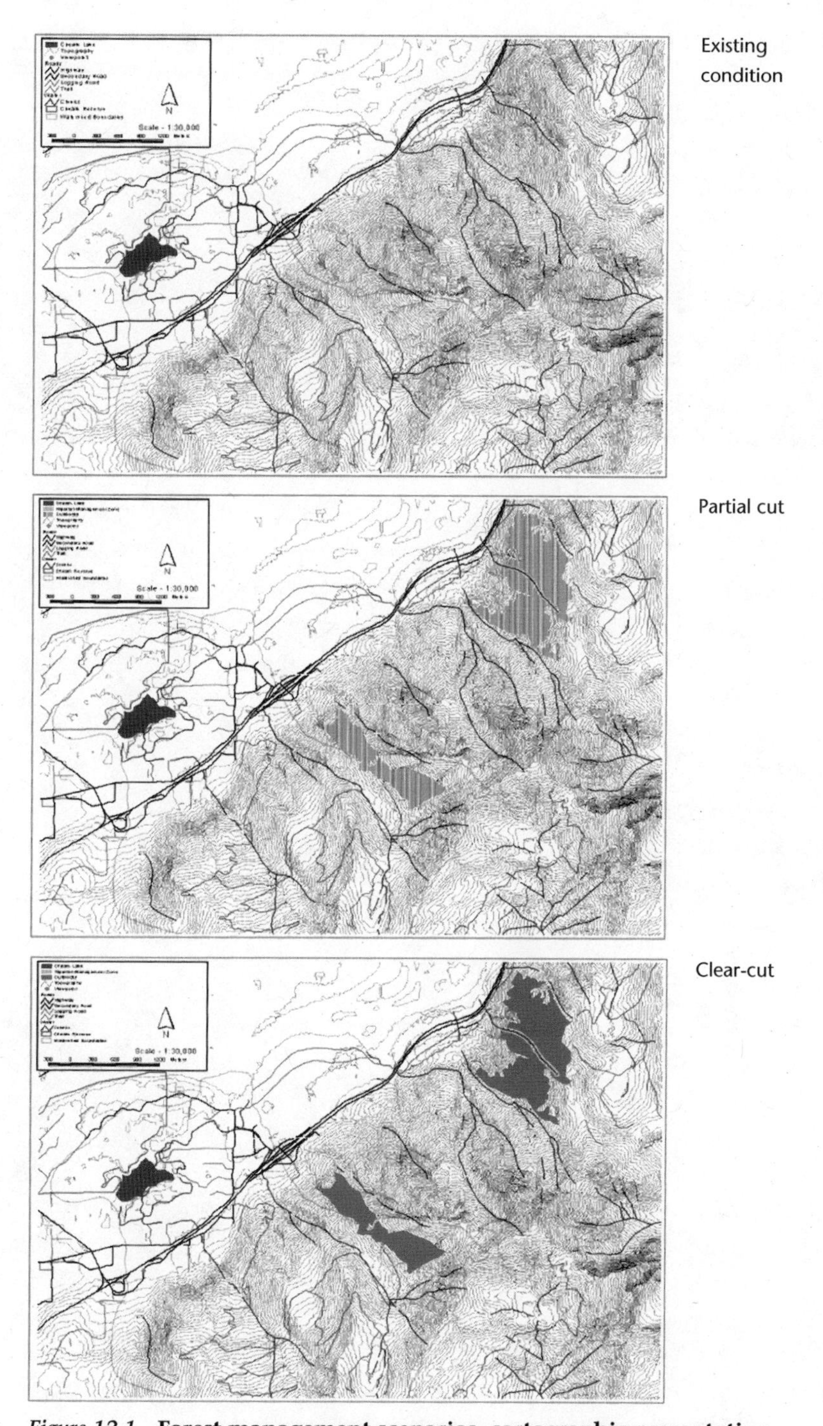

Figure 12.1 **Forest management scenarios, cartographic presentation**

Existing condition

Partial cut

Clear-cut

Figure 12.2 **Forest management scenarios, landscape simulation presentation**

world. For instance, participants often characterized the uses that the Cheam make of the land as a right that flows directly from the Creator:

> We know that we have been given the right to do, not by the Supreme Court of Canada, but by the Creator to go out and celebrate, to go out there and maintain our connections to the forest and to utilize it ... (Cheam participant in Lewis 2000, 210)

The Stó:lō narratives relating to the Great Spirit (Chichelh Si:yam; pronounced *chi-chel see-yam*) and other spiritual beings (e.g., Xa:ls; pronounced *hāls*) oblige the Cheam to make productive use of their environment without compromising the land's productive capacity. The spiritual consequences of disrespectful land management were particularly evident in the participants' appraisal of the clear-cut harvesting that has occurred throughout their traditional territory. In this instance, there were intense statements of collective responsibility, which suggested that the Cheam had failed in their spiritual duty to protect the land that had been given to them:

> Xa:ls was this person who taught us how to live with the land, and have a, how do you call it, like an ecological relationship with the land ... looking at this [clear-cut], it's been destroyed, and what the Creator provided for us now we as mortals have to replace it, we have to restore the naturalness. (Cheam participant in Lewis 2000, 211)

The relationship the Cheam have with their environment involves a form of covenant responsibility to utilize and protect the resources that have been given to them, which ultimately has significant implications for landscapes generally. In other words, even apparently unremarkable settings that are lacking in legendary or historical significance, or landscapes that are devoid of spirit beings, can retain a kind of sacred significance:

> When you get right down to it and look at the land itself, the water, I mean everything here has got a certain sacredness to it, and we've got to take care of them ... (Cheam participant in Lewis 2000, 212)

Showing Respect for the Land: Spiritual Conditions of the Landscape

What flows from this generalized spiritual imperative toward all landscapes is a responsibility to use the land according to traditional notions of respect. That Aboriginal peoples utilize resources in a respectful manner is a fairly common but often poorly understood concept. For many contemporary environmental writers, notions of respect are inaccurately couched in terms of a pristine existence that existed prior to contact with European colonists, when Aboriginal peoples trod lightly on the land, with no manifest effect on landscape patterns or ecological processes (White 1996). However, in their most basic expression, notions of respect are rooted in the twofold conception that the wealth of the land has been fashioned for the benefit of all living things by the Creator, and that this relationship of shared dependence, often expressed in terms of humans, animals, and plants (e.g., cedar trees) being kin (Turner 1997), requires consideration of others' needs when resources are taken from the environment. In this sense, notions of respect posit that sufficient resources must be left behind to ensure the continuing survival of other living communities:

> All the mountains, the water, the trees, the plants, the game, the ones that fly, the ones that crawl are very sacred to us because ... the Great Spirit put it there for everyone. So we know that if something like this [timber harvesting] is to be done it should be done in a very sacred and respectful way and some, maybe some of this logging can take some of the resources away but still leave a lot behind. (Cheam participant in Lewis 2000, 186)

Among the activities that demand a respectful treatment of the land's resources, the Cheam participants identified several that are focal points of Stó:lō spiritual life. The most commonly mentioned activity in our discussions with the Cheam related to the personal and ritualized uses of upper watershed streams and the restorative benefits of a more natural forest environment in which the Cheam people can meditate and find spiritual renewal:

> Like our elders said a long time ago, if you're feeling sad and something's bothering you, just go walk through the forest and let the branches and things brush against you to take away your bad feelings and things like that. And be by yourself so that you are able to meditate, if you want to, and do some praying to the Great Spirit to help you in whatever is bothering you. (Cheam participant in Lewis 2000, 208)

Such statements are consistent with other ethnohistorical accounts of Stó:lō spirituality describing the forests and thick stands of riparian vegetation as refugia and places of spiritual cleansing (Suttles 1987). However, two points are worth considering in light of such comments. First, participants often made reference to *particular* streams as the locus of spiritual cleansing and meditation activities:

> Some of these creeks that we're looking at, we use them during the winter season for traditional purposes ... *Like these two here, this one here looks like it's above the Popcum Reserve, a few of our people use them for bathing in the*

> *winter season*. It's a sort of, ceremonial conditioning that they use them for. (Cheam participant in Lewis 2000, 182, emphasis in original)

Localized sites of ritual activity are important because they represent a connection with the place in which they occur and, even more importantly, with the community that has practised these sacred activities for generations. In effect, when place-specific customs are passed from one generation to the next, they have the power to connect people with the places where those rituals are performed, and to the people who share those localized forms of ritual custom. Embedded as it is in a traditional form of spiritual practice, site-specific ritual activity becomes a part of cultural identity as well:

> These sites are part of my Indian identity and these sites are an important aspect of our spirituality. (Stó:lō participant in Mohs 1987, 108)

Second, we learned that identifying and protecting such locales through some form of special-use designation is only a first step in cultural resource management. When asked to identify the factors that make these particular sites unique, in addition to an historical legacy of use by the Cheam community, several participants described the physical conditions of ritual cleansing areas and, in some cases, referred to the aesthetic attributes of ritual sites as an important factor in achieving meditative experiences.

Supporting Spiritual Activities: Physical Conditions of Specific Places and Landscape Types

Among the aesthetic cues the Cheam participants described, visual and auditory privacy for creating a transcendent experience was of paramount importance. They recounted the affective or expressive feelings that are a product of an environment not only visually appealing but rich with life and spirit:

> The way the trees come right to the river, you know they're sort of right in the river ... it's the most beautiful sight there is, it sort of looks untouched. (Cheam participant in Lewis 2000, 185)

This statement is neither unusual nor unexpected to the extent that individuals of many cultures seek the restorative psychological benefits of a natural setting that affords tranquility, respite from hard work, and sensory pleasure (e.g., visual, auditory, olfactory). Thus, whether forest streams, for instance, are used for higher spiritual purposes such as cleansing the body and spirit or simply as a quiet place by which to meditate and contemplate life, the physical qualities of a forest setting are relevant to the Cheam participants' appreciation of environmental condition (Lowenthal 2007).

However, of greater interest were the participants' evaluations of forest settings in terms of symptomatic cues, which reflected the setting's suitability for particular spiritual uses. In the following statement, the participant suggests that one of the stream simulations would provide suitable conditions for ritual meditation or cleansing because of the presence of medicinally important plants that were absent from the other pictures:

> There's more life, more interaction of different life spirits. There's the cedar trees, there's the different plants in here, some of these devil's club plants will only grow because these trees cover them ... (Cheam participant in Lewis 2000, 201)

Statements such as this reflect the specialized environmental knowledge that some First Nations retain. Traditional ecological knowledge or indigenous knowledge is based on observations, interactions, and repeated feedback from the environment, and from this is built a set of observations and classifications about the local environment (Berkes 1999; Marc Stevenson, this volume). The knowledge obtained through purposive encounters with a setting, whether to procure sustenance or to fulfill ritual obligations, affects where the observer looks and what perceptual cues are sought – for example, a hunter looking for prime ungulate browse or, as in the case of the participant quoted above, particular plants that are an important part of spiritual cleansing practices. In effect, traditional ecological knowledge is structured according to particular activities such that each activity has its own set of environmental requirements, and the possibility of each activity is dependent on the availability and recognition of these requirements. For instance, First Nations become successful hunters by learning how to read the landscape, by looking for tangible indications of the presence of valued species, and by identifying the plants and forest structures that different species use for sustenance and refuge (Wiersum 1997; Lowenthal 2007). In a similar fashion, Cheam ritual cleansing depends on the presence of specialized medicinal plants and forest associations that provide the necessary habitat for those plants to thrive. Recognizing and accommodating this kind of knowledge in forest management or ecological restoration may permit land managers to manage (i.e., modify or conserve) a forest landscape in a manner consistent with or, more appropriately, respectful of Aboriginal spiritual uses and their particular physical requirements in the landscape.

Discussion: Implications for Forest Management

Although the spiritual aspects of the landscape will increase in importance for natural resource management, in the absence of working guidelines or, as Andrew Mason posits in his chapter, an expanded definition of cultural significance in government regulation that encompasses sacred or spiritual sites, they will continue to be difficult for resource managers to incorporate into planning practice. Throughout the 1980s and 1990s, the planning techniques for biophysical and ecological inputs grew in sophistication and, with varying degrees of success, have been integrated into the standard practice of environmental planning and design – for example, comprehensive planning, integrated resource management, and multi-criteria analysis. However, despite considerable discussion of and research into the spiritual, social, and cultural aspects of the environment, the incorporation of spiritual inputs within landscape and forest management planning continues to be outside the realm of ordinary practice (Mason, this volume; Satterfield 2002).

Certain caveats must be stated before interpreting these exploratory study results. The use of a particular limited set of photographs and realistic simulations, although typical of resource management issues on Cheam territory, may have biased a proportion of the comments. Different scenarios, images, and even camera viewpoints could have led to different stories and perceptions. Moreover, the imagery presented focused on landscape effects. Full scenarios with more complete information on issues such as logging costs and measurable ecological impacts could have altered some opinions. Whether some comments arose more as a reaction to the relatively recent history of land clearing than from truly ancient views of the precontact forested landscape is unknown. Clearly, much more work needs to be done before we can claim an in-depth understanding of how Cheam or Stó:lō spiritual values are expressed in the landscape.

However, some patterns of indigenous spiritual values in the landscape seem to be clear and broadly consistent with other cultural landscape studies (e.g., Stoffle, Toupal, and Zedeno 2003; de Pater, Scherer-Rath, and Mertens 2008; Dewsbury and Cloke 2009). From our own investigation of indigenous spiritual conceptions of forested landscapes, we have learned that, as a culture that has occupied and sustained itself on forest resources, the Cheam First Nation derives from them not only material sustenance and a sense of identity but also a feeling of inward, spiritual security. For this reason, forests are something the Cheam feel they must cherish and protect, something they can return to for sustenance and support, both physically and spiritually. Cheam expressions of the forests' importance are replete with mythic themes, many of which are deeply rooted in legend and oral tradition. Several of the legends that the Cheam continue to recount describe the gift of specific creatures and places by the Great Spirit to the Stó:lō people for their benefit and use. From these same narratives, the gift of the land and its abundant resource wealth carries with it the associated responsibility to utilize the land with respect, which ultimately means that the physical requirements of other users (both human and animal) need to be recognized and accommodated in any resource use activity. According to the Cheam participants, land-based spiritual activities require similar considerations of respect to the extent that there are identifiable forest locations and conditions conducive to ritual activities, and these conditions ought to be maintained to ensure the viability of spiritual traditions. Thus, in any effort to develop a framework that accommodates indigenous cultural inputs within landscape and forest management, a necessary starting point is to understand the salient physical requirements associated with particular land-based activities, ranging from hunting to ritual cleansing.

A theoretical foundation for such a framework may be found in recent thinking in landscape aesthetics. To most people, aesthetic perception implies trivial decoration and a superficial appreciation for the beautiful and picturesque. However, philosophers have convincingly argued that aesthetic perception has a fundamental effect on how we see the world. Eaton's definition of aesthetic experience is fundamental to an understanding of human perception, in part because it reflects the purposeful basis of environmental valuation:

> Aesthetic perception is marked by perception of and reflection upon *intrinsic properties of objects ... that a community considers worthy of sustained attention.* (Eaton 1997, 88; emphasis added)

Similarly, theorists such as Nassauer (1995) and Sheppard (2001) posit that the power of landscape aesthetics rests in its ability to deliver personally or culturally salient *knowledge* to the perceiver. For instance, although aesthetic knowledge may include a basic appreciation for the scenic attributes of an environment, there are other categories of aesthetic experience that invoke higher levels of cognition:

- *Affective:* the common affective association of settings with human emotions or sensations – e.g., dark forests and mystery or fear.
- *Symbolic:* the representation of abstract ideas or beliefs by physical objects and features – e.g., Mount Sinai and God, Mount Cheam and creation narratives.
- *Symptomatic:* the apprehension of physical signs or cues that manifest or reflect environmental condition – e.g., foliage colouration as an indication of disease or pestilence.

The affective or emotional responses that are commonly associated with aesthetic experiences are essential to all three categories. Symptomatic perception, however, is often the most elusive to obtain, particularly for those of us who are accustomed to living an urban or suburban way of life that is, in varying degrees, detached from the environments from which we all subsist. In this higher form of cognition, it is hypothesized, affective aesthetic responses vary according to the recognition of either favourable or unfavourable environmental cues. Moreover, to attain this level of aesthetic cognition, some understanding or knowledge must be acquired, often through prolonged and active involvement with the landscape via activities that are specific to a cultural group. Tapping into this form of knowledge may provide one crucial means for forest managers and researchers to work with indigenous communities to identify the conditions that are conducive to particular spiritual traditions, and the thresholds that will tell managers when modifications have reached critical limits.

Recent work in the broader area of traditional ecological knowledge has documented indigenous landscape patterns and conditions pertaining to forests, wildlife, soils, and water, and evaluated their use by local communities as resource management standards and models (Berkes and Folke 2001; Houde 2007). On the one hand, this research has initiated a shift in the way that Aboriginal knowledge is regarded: from viewing Native systems of thought and classification as static, anecdotal, unscientific, and subjective, to a recognition that local cultures know their plant, animal, and other biophysical resources intimately and are experts at evolving and fine tuning their land use strategies to environmental opportunities and constraints (Nazarea 1998). The examples of indigenous cultures in British Columbia, and indeed throughout much of the world, demonstrate that sustainable ecosystems include active human use and management, and have done so for thousands of years (Doolittle 2000; Vale 2002; Deur and Turner 2005). In our current research among Aboriginal communities along the remote north coast of British Columbia, we are attempting to underscore the importance of culturally salient and visually recognizable patterns in local perceptions of landscape condition and forest management. In particular, we hypothesize that in settings that have been actively utilized or managed to satisfy particular human requirements (e.g., sustenance, habitation, spiritual, symbolic), culturally rooted conventions of "visible stewardship" (Sheppard 2001) may be found that have a direct bearing on local perceptions of forest conditions that are conducive to these requirements. In light of a critical need for more context-sensitive and culturally relevant indicators of sustainable forest management, the practical objective of this research is to present a case study that suggests a way for government agencies and resource industries to accommodate locally defined conceptions of acceptable landscape change. These culturally defined standards may include the kind of symptomatic perception described earlier that has a direct bearing on perceptions of culturally appropriate forest management.

We believe that, beyond developing more sophisticated environmental inventories, there are additional advantages associated with enhanced understanding of the cultural requirements and perceptions that traditional communities possess, particularly in the area of planned landscape change. If acknowledged and integrated into forest management, differences in environmental cognition between communities may be sources of creativity and discussion. Interaction among people with different cognitive models through shared decision-making frameworks such as co-management or joint venture initiatives may facilitate learning and temper the kind of culturally biased thinking and single-minded management assumptions that Marc Stevenson describes in his chapter in this volume on traditional knowledge. As a method of resource sharing that is primarily applicable to First Nation communities, shared decision making is founded on the basic premise that all cultures have, at their core, fundamental and defining values that are not easily grasped, conceptualized, or appreciated by people outside the culture (Stevenson 1999). Shared decision making is conceived as one means of bringing the holders of different knowledge sets together to facilitate cross-cultural communication, build awareness of the values that define a community and their relationship with the land, and develop constructive solutions that satisfy the needs of both Native and non-Native communities (Robinson and Kassam 1998). As in many other respects of environmental planning, recognition of diversity of cognition among members of a society seems an important step toward fostering socially acceptable forest management and reducing conflicts that hinder the attainment of this goal. Moreover, acknowledging the plurality of perspectives, both within and between communities, may facilitate the search for common ground in the pursuit of broader goals such as environmental conservation and sustainable forest management.

References

Ball, M.S., and G.W.H. Smith. 1992. *Analysing Visual Data*. Vol. 24 of Qualitative Research Methods series. Newbury Park, CA: Sage.

Berkes, F. 1999. *Sacred Ecology*. Philadelphia: Taylor and Francis.

Berkes, F., and C. Folke. 2001. "Back to the Future: Ecosystem Dynamics and Local Knowledge." In *Theories of Sustainable Futures*, edited by L.H. Gunderson and C.S. Holling, 121-46. Washington, DC: Island Press.

Boyd, D., and T. Williams-Davidson. 2000. "Forest People: First Nations Lead the Way toward a Sustainable Future." In *Sustaining the Forests of the Pacific Coast: Forging Truces in the War in the Woods*, edited by D. Salazar and D. Alper, 123-47. Vancouver: UBC Press.

British Columbia. 2002. *Provincial Policy for Consultation with First Nations*. Victoria: Government of British Columbia.

Crosby, T. 1907. *Among the An-ko-me-nums or Flathead Tribes of Indians of the Pacific Coast*. Toronto: William Briggs and Department of Agriculture.

Cummings, C. 1998. "Sacred Lands from a Legal Perspective: Examples from the United States." In *Sacred Lands: Aboriginal World Views, Claims and Conflicts*, edited by J. Oakes, R. Riewe, K. Kinew, and E. Maloney, 277-. Edmonton: Canadian Circumpolar Institute.

de Pater, C., M. Scherer-Rath, and F. Mertens. 2008. "Forest Managers' Spiritual Concerns." *Journal of Empirical Theology* 21:109-32.

Deur, D., and N. Turner. 2005. *Keeping It Living: Traditions of Plant Use and Cultivation on the Northwest Coast of North America*. Seattle: University of Washington Press.

Dewsbury, J., and P. Cloke. 2009. "Spiritual Landscapes: Existence, Performance and Imminence." *Social and Cultural Geography* 10(6):695-711.

Doolittle, W. 2000. *Cultivated Landscapes of Native North America*. Oxford: Oxford University Press.

Eaton, M.M. 1997. "The Beauty That Requires Health." In *Placing Nature: Culture and Landscape Ecology*, edited by J. Nassauer, 85-106. Washington, DC: Island Press.

Glaser, B., and A. Strauss. 1967. *The Discovery of Grounded Theory: Strategies for Qualitative Research*. Chicago: Aldine.

Goetz, J., and M. LeCompte. 1984. *Ethnography and Qualitative Design in Educational Research*. Orlando: Academic Press.

Houde, N. 2007. "The Six Faces of Traditional Ecological Knowledge: Challenges and Opportunities for Canadian Co-Management Arrangements." *Ecology and Society* 12(2):34-51.

Lewis, J.L. 2000. "Ancient Values, New Technology: Emerging Methods for Integrating Cultural Values in Forest Management." MA thesis, University of British Columbia.

Lewis, J.L., and S.R.J. Sheppard. 2006. "Culture and Communication: Can Landscape Visualization Improve Forest Management Consultation with Indigenous Communities?" *Landscape and Urban Planning* 77:291-313.

Lincoln, Y., and E. Guba. 1985. *Naturalistic Inquiry*. Beverly Hills: Sage.

Lowenthal, D. 2007. "Living with and Looking at Landscape." *Landscape Research* 32(5):635-56.

McCracken, G. 1988. *The Long Interview*. London: Sage.

Merriam, S. 1988. *Case Study Research in Education: A Qualitative Approach*. San Francisco: Jossey-Bass.

Mohs, G. 1987. "Spiritual Sites, Ethnic Significance and Native Spirituality: The Heritage and Heritage Sites of the Sto:lo Indians of British Columbia." MA thesis, Simon Fraser University.

Nagle, J. 2005. "The Spiritual Value of Wilderness." *Environmental Law* 35:955-1003.

Nassauer, J.I. 1995. "Messy Ecosystems, Orderly Frames." *Landscape Journal* 14(2):161-70.

Nazarea, V. 1998. *Cultural Memory and Biodiversity*. Tucson: University of Arizona Press.

Redmond, L. 1996. "Diverse Native American Perspectives on the Use of Sacred Areas on Public Lands." In *Nature and the Human Spirit: Toward an Expanded Land Management Ethic*, edited by B.L. Driver, 127-34. State College, PA: Venture Publishing.

Robinson, M., and K. Kassam. 1998. *Sami Potatoes: Living with Reindeer and Perestroika*. Calgary: Bayeux Arts.

Satterfield, T. 2002. *Anatomy of a Conflict: Identity, Knowledge and Emotion in Old-Growth Forests*. Vancouver: UBC Press.

Sheppard, S.R.J. 2001. "Beyond Visual Resource Management: Emerging Theories of an Ecological Aesthetic and Visible Stewardship." In *Forests and Landscapes: Linking Ecology, Sustainability, and Aesthetics*, edited by S. Sheppard and H. Harshaw, 149-72. Wallingford, UK: IUFRO Research Series.

Spradley, J. 1979. *The Ethnographic Interview*. New York: Holt, Rinehart and Winston.

Stevenson, M. 1999. *Traditional Knowledge in Environmental Management? From Commodity to Process*. Edmonton: Sustainable Forest Management Network, University of Alberta.

Stoffle, R.W., R. Toupal, and N. Zedeno. 2003. "Landscape, Nature and Culture: A Diachronic Model of Human-Nature Adaptations." In *Nature Across Cultures: Views of Nature and the Environment in Non-Western Cultures*, edited by H. Selin, 97-114. Dordrecht: Kluwer Academic Publishers, 97-114.

Suttles, W. 1987. *Coast Salish Essays*. Seattle: University of Washington Press.

Turner, N. 1997. "Traditional Ecological Knowledge." In *The Rainforests of Home: Profile of a North American Bioregion*, edited by P. Schoonmaker, B. von Hagen, and E. Wolf, 275-98. Washington, DC: Island Press.

Vale, T. 2002. *Fire, Native Peoples and the Natural Landscape*. Washington, DC: Island Press.
White, R. 1996. "Are You an Environmentalist or Do You Work for a Living? Work and Nature." In *Uncommon Ground: Rethinking the Human Place in Nature,* edited by W. Cronon, 171-85. New York: W.W. Norton.
Wiersum, K.F. 1997. "Indigenous Exploitation and Management of Tropical Forest Resources: An Evolutionary Continuum in Forest-People Interactions." *Agriculture, Ecosystems and Environment* 63:1-16.

Part 5
Collaborative Endeavours

13
Progress and Limits to Collaborative Resolution of the BC Indian Forestry Wars

Norman Dale

> You thought it was over,
> but it's just like before.
> Will there never be an end
> to the Indian wars?
>
> – Bruce Cockburn, "Indian Wars"

Those involved in the use and management of BC coastal forests participated in what seemed a remarkable shift during the past fifteen years: from an almost fabled arena of deadlock – "the war of the woods," as it was often called – to what seemed successful exercises in multiparty consensus building. An impasse was broken – but was it the impasse that most mattered? Was partial concurrence about a collage of coloured polygons on large-scale maps, purporting to represent how land was now to be used and regulated, really the depth of the peace that was needed, especially if one considers the duration and pain associated with what I have elsewhere called "North America's oldest public policy conflict" (Dale 1996)?

The argument in this chapter is not that the widespread application of conflict resolution approaches to the Native – non-Native forestry issues in British Columbia have failed or succeeded – it is too early and the arena too diverse to render such summary judgments. But, rather, it illustrates that caution needs to be taken before concluding that outcomes meet indigenous community needs arising from a legacy of colonialism.

To look at this, I provide perspective from two different "altitudes." The first is a broad regional view of a multiparty planning initiative, the Central Coast Land and Resource Management Plan process, covering its first phase, roughly 1997-2001.[1] This unfolds as a narrative largely about settling that "war in the woods" through bargaining tables, negotiated plans, and widely publicized signed agreements. I then narrow the observational field to the "inside story" of the forestry-related struggles of one coastal First Nation for whom I worked over that five-year period. At that level, "success" becomes a more elusive claim, and signed consensus documents seem less compelling demonstrations of progress away from the cross-cultural deadlocks of post-colonial British Columbia.

Modern Disputes, Historic Conflict

It may be helpful to begin by making a brief distinction between manifest disputes and underlying conflicts (Deutsch 1973). Across the spectrum of conflict settings – from the international down to so-called intra-psychic conflict – humans tend to fight about matters that are short term and readily visible, even superficial, but often quite wilfully neglect and deny the deeper causes of the conflict through which they suffer.

My experience as a mediator has been that frequently in consensus-building processes, once representatives assemble to consider their differences, they start by denying the very existence of a conflict. It can be quite disorienting as a mediator to be challenged with nods of approval all around the table – "Conflict? What conflict?" (see Dale 1999). Reluctance to admit the existence of conflict is consistent with at least some First Nations' culturally pervasive disinclination to openly state disapproval of others (Briggs 1970; Ross 1992). Regrettably, there is no surer way to perpetuate a deadlocked conflict than by refusing to admit its existence.

Underlying the large and small skirmishes between Natives and others over BC forest land use is unresolved title and rights, "the land question" (Cassidy and Dale 1988; Tennant 1990). Had differences over who owns and controls the land base been settled through treaty or surrender when non-indigenous interests first sought to harvest timber, today's disputes would have been different and probably far more restricted (Coates and Carlson, this volume).

In the past twenty years, a steady series of high-profile forestry conflicts flared in British Columbia in which First Nations played a central role. The famous struggles of more recent times, such as those of Clayoquot Sound and the Great Bear Rainforest, were but part of a legacy that goes back to renowned confrontations in the 1980s – Meares Island, the Carmanah, the Tsitika, the Stein Valley, and, probably most famously, South Moresby/Gwaii Haanas (Wilkes and Ibrahim, this volume). Often these controversies were characterized initially as "environment versus development," but they evolved quickly to centre as much on Native land claims (Pinkerton 1983). Until the early 1980s, the Native role in relation to commercial forestry was a subtle and subservient one (Knight 1978; Tennant 1990), for the most part presenting no major obstacle to the business interests of tenure holders who were rapidly using up the most merchantable timber stands in coastal British Columbia. Then, loose alliances between First Nations and environmental nongovernmental organizations started to coalesce against logging that ostensibly threatened ancient rainforests. Native people and environmentalists began to stand side by side in blockades and through consequent legal battles, threatening, some even thought, to "shut the province down" (Blomley 1996; Wilkes and Ibrahim, this volume).

The provincial government's early responses entailed a mix of seeking injunctions against the blockades, denying the validity of the land question, and assembling blue ribbon commissions to seek objective truths where the problem often really lay in conflicting subjectivities. Several important lessons emerged from these highly publicized conflicts at Meares Island, South Moresby, the Stein Valley, and beyond. One was that these internationally known controversies could strongly detract investment in forest and other resource development in British Columbia. Another was that Native-environmentalist coalitions could be very effective in delaying development – at a considerable cost to all, including the protestors. It was also realized that appointing committees and commissions to simply document the problem offered little in the way of breaking deadlock. The context was ripe for an alternative paradigm, one that would put at least some authority back into the hands of Native people and others with a stake in forestry management.

Advent of Consensus-Seeking Approaches to Forestry Conflicts in the 1990s

By the early 1990s, parties from all sides spoke not just of battles but of the wider "war in the woods." Soon, more advanced weaponry, notably well-orchestrated international boycotts, would be brought in. But even before this, a newly elected provincial government and most other parties began to consider collaborative approaches to deal with the chronic forestry conflict. A paradigm had recently emerged of convening all the major parties to environmental conflicts for face-to-face consensus seeking (Susskind and Cruikshank 1987). A *vade mecum* for "principled negotiations" had also come in to good currency with the publication of the best-selling guidebook *Getting to Yes* (Fisher and Ury 1981), which instructed disputants on how to get away from win/lose dynamics and invent mutual gains for all.

The idea of having all parties sit down relatively amicably and talk through logging controversies had precedents in British Columbia. Some of the first constructive responses of government to the aforementioned marquee forestry battles had been the formation of select technical advisory committees to report back on management options. In processes such as the South Moresby Resource Planning Team and the Meares Island Planning Team, there was, however, no obligation to reach consensus but rather just to present a range of alternatives from which public decision makers could then choose (or vary, or even disregard altogether). This limited purpose meant that the key incentive for achieving environmental consensus was missing.

The early 1990s saw several concurrent commissions wrestle with seemingly endless conflict over forest management and especially over old-growth areas. These included the Forest Resources Commission, the BC Round Table on the Environment and Economy, and the Old Growth Strategy Project. In these lay seeds of awareness that merely having good technical understanding does not spontaneously make for conflict-resolving decisions. A more concerted effort to find common ground was needed if bitterly divided parties were to break deadlock. The Forest Resources Commission recommended a follow-up land use commission, with "a core of trained mediators" (BC Forest Resources Commission 1991, 116). On a similar theme, the BC Round Table on the Environment and the Economy recommended that consensus procedures be adopted in addressing the chronically difficult environmental resource decisions key to the province's sustainability (BC Round Table on the Environment and the Economy 1991).

Recognizing the need to have direct face-to-face discussions between disputants gave rise to two broad initiatives of lasting significance: the Commission on Resources and Environment (CORE) process (BC Commission on Resources and Environment 1995) and then the Land and Resource Management Planning (LRMP) approach (BC Integrated Resource Planning Committee 1993). At almost the same time, the paradigm of negotiated settlements for so-called modern treaty making became formalized through the implementation of the recommendations of the BC Claims Task Force (1991). Thus, rather suddenly, multiple forums came into existence as means for parties to forestry conflicts to seek consensus.

The LRMP processes adapted the CORE approach to "higher level" planning areas. By 2001, twenty-six LRMP processes had been initiated across British Columbia. Many of the initial LRMPs proceeded with little or no Aboriginal participation, but on the central and north coast, First Nations came to play a pivotal role in this planning. This was in spite of their routinely declaring reluctance to participate in a process designed and driven by the provincial government. The Heiltsuk Nation, for example, repeatedly expressed quite specifically its deep-rooted dismay while still remaining in the Central Coast LRMP process:

> The process is not what the Heiltsuk Tribal council wants. When we objected to this process we were told it would happen with or without our participation. The Council decided to participate in this process to safeguard Heiltsuk interests in Heiltsuk lands. We feel that we are in this process under duress. (Heiltsuk Tribal Council, n.d.)

Despite these reservations, the Central Coast LRMP process, which began formal sessions in June 1997, proved to be by far the most successful such process in engaging First Nations' participation. Reluctantly or not, most First Nations with traditional territories in the plan area participated. Not only were the experiences of earlier exercises there to learn from but there was also a pressing need to succeed in attracting such participation, since, unlike all other LRMP areas, the Native population on the central coast constitutes a large proportion of the overall population. In the seven years that followed, the Central Coast LRMP involved the Kwakiutl District Council (a tribal council working on behalf of many of Kwak'wala-speaking bands), as well as the separate Heiltsuk, Nuxalk, Oweekeno, and Tlowitsis Nations.[2]

Originally scheduled to conclude by 1999, the process came to be seen as consisting of two phases. Phase 1 came to a much-publicized conclusion in April 2001 in the so-called Great Bear Rainforest Agreement.[3] Despite high rhetoric and claims about this being a turning point and of global significance (British Columbia 2001), the First Nations of the Central Coast LRMP, in fact, did not formally ratify the Phase 1 outcome. Instead, as part of what became Phase 2, it entered a general protocol agreement to engage in further negotiations with British Columbia aimed at a separate agreement. At last, in 2006, seven years behind the original schedule for completion, the BC government was able to announce a conclusion in the form of a combined north and central coast land use decision (British Columbia 2006) supported by most First Nations. Over the years, the Central Coast LRMP process had evolved significantly in allowing the parties to co-invent a role for First Nations consonant with both their and the BC government's position on land jurisdiction. The old "war in the woods" appeared to have ended and, at least on the surface, substantial progress had been made in accommodating Aboriginal aspirations and provincial government mandates.

Although not a story of unequivocal success, the evolution of BC forest land planning in the late twentieth and early twenty-first century has been one of considerable progress compared with the blockades and stalemate of the preceding decade. Amid this advance it is not surprising to encounter overstatements of what was accomplished, especially as it relates to the inclusion of and support from Aboriginal interests and values. Riddell (2005, 65) speaks of "the Great Bear Rainforest and the environmentalists working for its protection" as a "unique model for resolving complex environmental conflicts." Jackson and Curry (2004, 40) likewise not only herald the "democratization of land use planning" achieved in LRMPs but commingle this with the Aboriginal land question: "The new rural planning regime also offers a consensual management system capable of assimilating the tribal holdings created by the settlement of First Nation claims" (ibid., 27). Yet, in at least one key First Nation, the Nuxalk of Bella Coola, there was nothing like a sense of dramatic transformation in the underlying conflict about land. Instead, a community already profoundly disrupted by the broad postcolonial legacy became split along the bitter lines of the forestry conflict, including over whether and how to participate in a planning process firmly under the province's control.

We turn now to how and why the Central Coast LRMP – as well as several contemporary attempts by non-Natives to reach consensus with the Nuxalk on forest land use issues – fell short of being such a turning point in cross-cultural understanding.

One Nation's Involvements in Forestry Consensus Processes

When one focuses on one coastal Nation and how it used or benefited from opportunities for forestry consensus building, the situation becomes more complex and ambiguous, and less compellingly optimistic. From 1997 to 2002 I was the senior planner for the Oweekeno-Kitasoo-Nuxalk Tribal Council (OKNTC). Based in Bella Coola, the heart of Nuxalk traditional territory, I developed an especially close (which is not to say untroubled) working relationship with that First Nation as represented by its chief and council and the young Nuxalk professional staff who made up its resource planning team.

The Nuxalk are the northernmost nation from the Salishan linguistic group. Their traditional territory extends into the watersheds of the Bella Coola River system and along Dean and South Bentinck Channels. Although the Nuxalk have essentially no political and social connections with other Salishan speakers and, in contrast, have long associated more with neighbouring Central Coast First Nations, their traditional and modern structures are also quite distinct from their neighbours. From the perspective of adjacent nations and ethnographers, the Nuxalk are less hierarchically structured. McIlwraith [1948] 1992) comparing leadership forms to neighbouring tribes, concluded that Bella Coola Chiefs had "an extreme paucity of functions ... and prerogatives ... likewise scanty"(176). Today's Nuxalk mostly live in one central village at Bella Coola, a mix of the descendants and moieties of many separate communities, over forty of which were spread out over the coast of the traditional territory well into the twentieth century.

Another relevant distinction is that the Nuxalk, alone among Central Coast First Nations, have adamantly refused to take part in the BC Treaty Commission process. For most nations, participation in modern treaty making has led to guardedness about alternative consensus processes like the LRMP. In contrast, not being in the treaty process initially meant less Nuxalk reluctance to participate, but, because of deep schisms in the community, discussed below, isolationism eventually spilled over to impede full involvement in any process that to them smacked of collaboration.

The Nuxalk also differed from their neighbours in having a far more extensive original endowment of merchantable timber in their traditional territory. Bella Coola became an epicentre of logging much earlier than many coastal areas, and Nuxalk workers adapted to this nontraditional extractive economy and held important jobs within an area of great interest to forestry companies (Knight 1978). Industry's mechanization and centralization of logging operations significantly eroded that workforce participation in the 1990s. This exacerbated long-standing resentment of the unauthorized intrusion of non-Nuxalk into the coastal rainforest. Nonetheless, given the presence within the community of intergenerational logging families, it was perhaps inevitable that the advent of environmental controversies would lead to bitter internal divisions.

From the early 1990s on, struggles about the care and use of forests in their territory were constantly at issue for the Nuxalk leadership and community. In 1995, environmental groups and some of the hereditary leadership of the Nuxalk Nation joined in a well-publicized blockade of logging operations on King Island (Capozza 1998). As in the prominent forest battles of the 1980s by the Haida, Nuu-chah-nulth, and others, a sequence of injunctions, protests, arrests, and boycotts unfolded. This coincided with a time of great internal dissension within the Nuxalk leadership, a factionalism that was to last for many years, intensified by the alliances that different sides made with either pro- or anti-logging interests. A past chief councillor of the Nuxalk was among the hereditary chiefs who stood with such groups as the Forest Action Network and Greenpeace against the proponent, International Forest Products (Interfor). Meanwhile, the then current chief councillor and the majority of elected council sided with the company. The blockade and Interfor's constrained attempts to log went on intermittently for almost two years.

This case saw no conclusive ending analogous to the earlier Meares and South Moresby stories. Convictions and injunctions enabled Interfor to proceed with logging for several years, during which time environmentalists greatly broadened their fight against logging throughout British Columbia's Central Coast. This became known as the Great Bear Rainforest campaign, a highly sophisticated international effort to persuade prospective consumers, especially in Europe and the United States, to boycott Central Coast forest products. Several of the hereditary Nuxalk chiefs were enlisted to travel to the United States and Europe as part of this initiative. Yet, the elected chief and council enthusiastically participated in ongoing efforts to countervail the boycotts, appearing at sessions organized by Interfor where European and American clients were brought on promotional tours to the Bella Coola area. "We will speak for our environment, and do not want outside groups speaking for us or attempting to represent what they believe to be our interests," asserted then elected chief councillor Archie Pootlass (Brett 1997, B8).

By the late 1990s, factions had become so divided within the Nuxalk Nation as to preclude day-to-day social contact. The internal conflicts involved a very conspicuous antagonism between two factional leaders who were also brothers, one the elected chief, and the other a hereditary chief. Even death could not suspend the open hostility.[4] On the surface, the fight was about logging but in fact was underlain by a more deep-rooted clash between the traditional hereditary system and the so-called Indian Act government. At its height this conflict boiled over into verbal abuse, physical threats, lawsuits, and family estrangement. Although the strife eased somewhat after a new chief and council were elected in 2001, during much of this period and beyond, community support for any forestry initiative tended to be divided and hamstrung. Attempts to build any Nuxalk community involvement in externally driven processes like the LRMP failed, as most band members avoided being identified with either side of the chronic and bitter internal struggle.

In 1997 when the Central Coast LRMP was launched, the Nuxalk chief and council decided to become involved, for several reasons distinct from the BC government's program intent. Only to a very limited degree did the chief and council subscribe to the underlying rationale of broad strategic land use planning. The more frequent rationale given was that the Nuxalk had better get involved because the Heiltsuk were in and would make commitments about land inside the overlapping claims areas of the two nations. The Central Coast LRMP also was seen as an opportunity to fund the newly established Nuxalk Integrated Resource Office at a time when band finances were in poor condition. By entering a contribution agreement, the Nuxalk elected council gained access to approximately $80,000 per year for capacity building. Predictably, traditionalists viewed this as selling out to non-Native governments whose occupation and claimed jurisdiction over Nuxalk territory they deemed unlawful.

As time passed, the tangible gains from the Central Coast LRMP were less than had been hoped. A promising young Nuxalk who was recruited to be the LRMP coordinator ran into personal and familial problems that seriously limited his discharge of his duties. It may be argued that such unforeseen personal issues can afflict any planning process, including ones aimed at developing consensus about forest land use. However, we need to red flag the incidence and severity of such "realities of Aboriginal life" (Frideres, this volume) in any community with unusually high rates of social dysfunction. For First Nations afflicted with rampant alcoholism, marital strife, abuse, violence, suicide, and so on, it is naive to expect that representatives can seamlessly enter the always-stressful world of consensus building across the table from stably employed non-Native participants. The impact on collaborative planning of personal difficulties in the shadow of intergenerational trauma is among the major undiscussable items when First Nations and other governments interact (Dale 2005). And that which remains undiscussable will enduringly and insidiously jeopardize any cross-cultural process.

Involvement in the Central Coast LRMP was primarily limited – as every other Nuxalk forestry-related initiative between 1996 and 2001 was – by the hereditary/elected governance split in the community. During that period, the Nuxalk's elected leadership engaged in a succession of consensus-seeking processes, as well as in more direct and assertive ways of dealing with the assertion of rights over forest lands, but in each case, full resolution stalled:

- In 1998, after racial tensions broke out within Interfor's logging crew, the Nuxalk chief and council and Interfor began a regular and continuing process of consultation to steadily improve the corporate–Nuxalk relationship. After two such meetings the process withered in the face of Nuxalk community opposition to any kind of relationship with a company seen as responsible for the strife and arrests at King Island.
- In 1999, the Nuxalk applied to the BC government for a pilot community forest licence and were one of the very few communities in the province successful in receiving approval. The minister of forests strongly encouraged the Nuxalk to collaborate with local non-Native interests so as to expand the timber allocation. But this and indeed the licence itself were opposed by more traditionalist elements of the community. Without internal consensus, the pilot licence went on hold for much of the next seven years, despite the steady urgings of the local BC Ministry of Forests district manager.
- In 2000, after several culturally modified trees were felled by Interfor contractors, the Nuxalk elected council announced a stop order on the company's operations within its territory. This injunction was not permissible under Canadian law, but Interfor took it very seriously and ceased operations pending an extensive negotiation process. That process yielded a multifaceted and precedent-setting "Accord for Reconciliation."[5] But even when the company pressed during the following years for implementation, the Nuxalk did not fully take advantage of the benefits promised under the accord, again because significant segments of their community wanted no dealings whatever with Interfor.
- In 2000-1, an initiative of coastal First Nations north of Cape Caution called Turning Point gained momentum as a means for developing strategies to enhance managerial authority and economic benefits from forestry. The Nuxalk sent staff to the organizational meetings but eventually decided against participating because all other nations within Turning Point were in the BC treaty process. Especially after the election of a conservative traditionalist majority to council in March 2001, such an association was deemed to be a slippery slope into the rejected treaty process.

In each case, the bitter split within the community and deep-seated fears and resentments due to longer-standing issues of alienation and acculturation precluded the kind of community participation and support essential to participating in and implementing consensus-based processes. Without an informed and supportive constituency, negotiated agreements on public issues are always at risk of falling apart (see Cormick et al. 1996). The tensions in the Nuxalk community meant that information meetings and briefings were poorly attended. As a result, the Central Coast LRMP and the other noted forest-related initiatives were neither understood nor accepted by the community. The deadlock that resulted was unfathomable to those outside the Nuxalk community and seen as little more than stereotyped procrastination and a lack of leadership. But inside the Nuxalk community, it seemed anything but static.

In 2001, the election of a council dominated by the traditionalists meant an immediate cessation of LRMP participation. Thus, mere weeks later when the much-publicized conclusion of Phase 1 of the LRMP was announced (British Columbia 2001), the Nuxalk Nation, whose territory is at the very core of the Central Coast plan area, was absent. Although this did not quiet the publicity machines of the main signatories – the BC government, environmental groups, and forest companies – closer inspection revealed that the Great Bear Rainforest Agreement had far less First Nations support generally than was presumed to exist at that time. It would take another five years before a wider consensus could emerge (British Columbia 2006), and even then one still not enthusiastically accepted by the Nuxalk leadership (Chief Archie Pootlass, pers. comm. June 2007) (see Forsyth, Hoberg, and Bird, this volume).

At the same time that the Nuxalk seemed stalled on what outsiders saw as wonderful forestry-related opportunities, other strategies were being implemented internally to heal the wounds within the community. In 2000, the Nuxalk joined with non-Natives of Bella Coola to participate as one of several North American communities invited to develop a Sustained Dialogue approach (see Saunders 1999) guided by the Kettering Foundation of Dayton, Ohio. A separate participatory research project, organized by the Community Economic Development Centre at Simon Fraser University, also selected the Nuxalk as one of four BC pilot communities for a collaborative forestry project (see Markey et al. 2005). The nation also independently engaged consultants to facilitate the Nuxalk Nation Community Healing and Wellness Development Plan (Bopp and Lane 2000). This plan acknowledged that for healing to occur, the Nuxalk, as well as non-Natives of Bella Coola, would need to find ways to confront the "long history of racism, resentment, misunderstanding and suspicion between the Nuxalk and their non-Aboriginal neighbours" (Bopp and Lane 2000, 57).

In 2001 and 2002, several events were organized that brought Nuxalk and their neighbours together in loosely structured dialogues and events. One of note was a locally organized town hall at which several working groups were formed on the theme of "two cultures, one community." One group focused on the opportunity for a Native – non-Native community forestry initiative. In 2003, the group coalesced into the Bella Coola Nuxalk Resource Society and, for the first time, brought a united front to ongoing negotiations with the BC Ministry of Forests. Perhaps symbolically, the initiative was staffed by a former non-Native employee of Interfor, the same company whose logging on King Island in 1995-97 became such a divisive issue for the Nuxalk and others of the Central Coast. The steady if halting steps to develop local relationships around forestry issues appear to have given the Nuxalk greater comfort in connecting with the larger-scale processes initiated by government and others described above as seemingly golden opportunities. Several years later and after much internal dissension, the Nuxalk finally joined the Turning Point Initiative to explore so-called interim measure possibilities created by the Liberal government since 2001 (see Forsyth, Hoberg, and Bird, this volume).

Conclusion: The Fit of Contemporary Approaches to Resource Conflict Resolution

There have been frequent suggestions that consensus-seeking, collaborative processes of the kind seen in the Central Coast LRMP processes are a "good fit" for Native – non-Native conflicts. One reason often cited is that indigenous communities have always governed by consensus:

> The modern system of alternative dispute resolution has ... roots in Native American culture ... As the Native American communities continue to expand their operations, the possibilities are unlimited for the continued and expanded use of mediation between individual members of a tribe, between one tribe and another, and between a tribe and the "outside world" ... Mediation, "the wave of the future," is firmly rooted in the past." (Garrett 1994, 38, 44-45)

After the harsh and costly struggles surrounding forestry in British Columbia in the 1980s, it is not surprising that most parties want to believe that innovative consensus processes can bring about profound and durable resolutions. Knowing that such conflict and its resolution must centrally engage First Nations, one also hopes that the seeming natural "good fit" between the troubled Native – non-Native context and alternative dispute resolution will dispel the "long and terrible shadow" (Berger 1991) of colonialism. But there is a danger in expecting techniques and philosophies that grew up within nonindigenous, nontraditionalist, and mostly urban Western society to fully remove deep and historical cross-cultural conflict.

The mainstream model in use for consensus processes falls short in several ways that become most apparent when one moves from the high level and sanguine appraisals of processes like the LRMP to a more up-close standpoint. To see these shortcomings, we need to consider the legacy, much of it negative, of what befell Aboriginal peoples more than 150 years ago. This kind of reconciliatory deliberation is not how those who structure and participate in consensus processes usually frame their purposes. Tight deadlines are usually seen as essential to keeping such processes energized, and parties are urged to keep to the here-and-now concrete matters at hand, not the old rhetoric of the there-and-then. In that atmosphere, revisiting in any detail the historic events and factors that most First Nations people see as pivotal may be seen as tangents by others. First Nations perceive the role of history in a much different light:

> It is important for First Nations peoples to tell their stories – and most of those stories are about the past relationship between First Nations and European non-Aboriginals. For First Nations peoples, history defines the present; it is not something to set aside in pursuit of a better tomorrow ... For First Nations peoples history keeps coming up and it probably always will. (Mercredi and Turpel 1993, 13-14)

Against this backdrop of determined remembering, processes such as the LRMP can appear as ahistorical attempts simply to cut deals without sufficiently working through underlying issues. As such, they fall into what M.A. (Peggy) Smith (this volume) would categorize as assimilationist rather than coexistence-based co-management. Mainstream consensus processes may also be anti-cultural, intentionally belittling deeply ingrained cultural communication practices. For example, a core principle in modern alternative dispute resolution is that parties should focus on interests not positions (Fisher and Ury 1981; Susskind and Cruikshank 1987). An interest is seen as an honest expression of what one *really* needs, whereas a position is equated with highly rhetorical and inflexible posturing. But from a First Nations perspective, rhetorical positions may not be at all inappropriate. Indeed, ethnographers have suggested that Northwest Coast cultures' use of rhetoric, including metaphor, hyperbole, and other tropes, are constitutive of their very social interactions:

> This style of discourse action, and thought ... thrives on ambiguity, allusion, and elusion, even silence ... It is striking, I think, that this style is passed on to those who do not speak the language ... Likely because of the traditional

> values so closely associated with this style, younger Haida grow with a sense of the allusive quality of social and political discourse, of the importance of implicit negotiation, of silence and of many of the types of symbolic actions or gestures ... It is this intricate and deeply meaningful style of indigenous communication which is so easily misunderstood by White outsiders. (Boelscher 1989, 201)

As a non-Native who worked for the Nuxalk on forestry-related interactions with government and corporate interests, I was occasionally consulted by the latter on the seeming inability of the First Nation leadership to "get it together." The Nuxalk were perceived as squandering repeated "golden opportunities" to reap the land use benefits of collaboration and modernity. The apparent deadlock that different initiatives encountered was seen from the outside as consistent with long-standing and unflattering views about the Nuxalk community and its leadership (and other First Nations): childlike and wrought with derisible internal squabbling.

Related to the potentially ahistorical and anti-cultural nature of mainstream consensus processes is the need and the expectation many First Nation communities have for intercultural healing. The roots of this need run deep into traumatic experiences with white society: alienated lands, smallpox epidemics, forcedly supplanted religion, the potlatch laws (LaViolette 1973), residential schools (Grant 1996), and disenfranchisement. All these forces have produced Native communities that struggle constantly with the after-effects of pain, addiction, mistrust, and mostly repressed rage. These, to paraphrase Forrester (1999), cannot be left at the door of negotiations rooms, not if durable consensus is to be achieved. To try to do otherwise will seem nothing more than one more assimilationist strategy (see Smith, this volume).

Conversely, recent collaborative steps between the Nuxalk and neighbouring residents in the Bella Coola area on forestry and other issues present a model for moving toward cross-cultural harmony, distinct from higher-level consensus forums such as the treaty-making and LRMP processes. The serendipitous use of an "incubator" of relatively successful local Native–non-Native cooperation has been effective elsewhere in British Columbia. For example, on Haida Gwaii, collaborative planning for the community-based Gwaii Trust appears to have set the stage for numerous subsequent initiatives, including forestry ones, between the Haida and settler communities of the islands (Dale 1999). They are locally driven rather than simply an offshoot of senior government initiatives, rooted in the kind of rapport that grows best among people who see each other every day. Because truly sustained dialogue and durable face-to-face planning are possible for an isolated First Nation only with nearby neighbours, these local processes create a relatively safe space for citizen-level discussion, which Saunders (1999) considers vital in transforming ethnic conflicts. Locally, far more than in dealings with distant governments and corporations, the possibility exists of transcending low-power colonial relationships, and even medium power consultations, in favour of the truly shared, mutually respectful decision making of equals. Freed from the bad optics and, often, the reality of imperious senior government-driven processes, First Nations working with adjacent settler communities can rehearse for the autonomy sought persistently in their long struggles to recover leadership in forest and land governance. If so, and, if seriously committed to bringing the "Indian wars" in British Columbia's forests to a lasting end, senior governments should provide, as needed, the resources for such processes, and otherwise step back and let a deeper peace be locally created.

Epilogue

As noted earlier, the events described and discussed here are from a receding time period in the immediate aftermath of the notorious "wars of the woods." In the years since the ostensible peace was negotiated, implementation of the Central Coast plan has proceeded haltingly, with intermittent and alternating outbursts of praise and condemnation over the conformity of government and industry actions. Those who were on the "inside" of the initially exclusive negotiations of the Joint Solutions Project – primarily forest companies and nonresident environmentalists – continue to acclaim the protection of the area they renamed "the Great Bear Rainforest." Two participants in the process, Smith and Sterritt (2010, 131), narrate a veritable *bildungsroman* about the dawning of "a new era of sustainability," culminating in the conclusion that the alliance, in which they were key players, "demonstrates that complex issues can be resolved ... for the greater good of all British Columbians" (148). Distributional effects and overall effectiveness locally and provincially are not closely considered, however.

In contrast to this unequivocally positive self-assessment, Shaw (2004, 381) concludes, "Many of those most likely to be directly affected by the campaign will have had much less to do with the campaign or decision making process than shareholder campaigners halfway around the world. Far from being an instance of exemplary local alliance, the founding work of Joint Solutions was and still can be seen as an alien and 'green conspiracy.' Inferring a lesson that forcing land use decisions largely without initial First Nation involvement is a model for emulation, seems, at best, debatable.

In the years since the primary gathering of information for this chapter ended, the Nuxalk Nation has moved through a succession of elected councils and also varying dispositions toward engagement in modern forestry and its management. As this book goes to press, the current chief and council have reorganized their natural resources agency and are committed to balancing a role now within the Great Bear Rainforest framework with the independent needs of their community (Chief Councillor Andrew Andy, pers. comm., September 4, 2011). The story, unlike this chapter, is to be continued.

Notes

1 The time frame is restricted mainly to the first phase because that coincides with the author's tenure within the administration of the Nuxalk Nation, an access which is needed for the close-up perspective sought here. This was also the most important period in terms of the internal community conflict that involved deep differences about forestry in the Nuxalk traditional territory.

2 The Nuxalk pulled back from the process for several years (2001-3). The story of this withdrawal is presented in a subsequent section of this chapter.

3 This phrase was far from universally accepted by the formal participants in the preceding four years of land use planning. It had been what Granander and Wigle (2003, 19) aptly refer to as the *nom de guerre* for the environmentalist coalition and was quite offensive to some First Nations and the Central Coast Regional District's local government.

4 Reverend Betty Sangster (pers. comm. September 19, 1998) of the United Church told the author that at the worst of it, the customary planning sessions held for funerals had to be split so that bitter foes did not have to be in the same room as each other.

5 The full title was the Nuxalk/Interfor Accord for Reconciliation Including Culturally Modified Trees and Other Issues of Mutual Significance. It was signed by Chief Councillor Archie Pootlass and Interfor chief forester Ric Slaco, April 2, 2000.

References

BC Claims Task Force. 1991. *Report of the British Columbia Claims Task Force.* Report to the First Nations of British Columbia, the Government of British Columbia, and the Government of Canada. Ottawa: Supply and Services Canada.

BC Commission on Resources and Environment. 1995. *British Columbia's Strategy for Sustainability: Report to the Legislative Assembly 1994-1995.* Victoria: Queen's Printer.

BC Forest Resources Commission. 1991. *Forests for Our Future.* Report of the BC Forest Resources Commission. Victoria: Queen's Printer.

BC Integrated Resource Planning Committee. 1993. Land and Resource Management Planning – A Statement of Principles. Victoria: Queen's Printer.

BC Round Table on the Environment and the Economy. 1991. *Reaching Agreement: Volume I – Consensus Processes in British Columbia.* Victoria: Queen's Printer.

Berger, Thomas R. 1991. *A Long and Terrible Shadow: White Values, Native Rights in the Americas, 1492-1992.* Vancouver: Douglas and McIntyre.

Blomley Nicholas. 1996. "'Shut the Province Down': First Nations' Blockades in British Columbia, 1984-1995." *BC Studies* 111:5-35.

Boelscher, Marianne. 1989. *The Curtain Within: Haida Social and Mythical Discourse.* Vancouver: UBC Press.

Bopp, Michael, and Phil Lane Jr. 2000. "Nuxalk Community Healing and Wellness Development Plan." Unpublished draft report to Nuxalk community. Four Worlds International, Lethbridge, AB.

Brett, Brian. 1997. "Indians Fight Environmentalists." *Vancouver Sun,* June 11, B8.

Briggs, Jean L. 1970. *Never in Anger: Portrait of an Eskimo Family.* Cambridge, MA: Harvard University Press.

British Columbia. 2001. "Coastal Plan Creates Unique Protection Area, Economic Agreement and New Opportunities for First Nations." Press release. Office of the Premier. April 4. http://archive.ilmb.gov.bc.ca/

–. 2006. "Province Announces a New Vision for Coastal BC." Press release. Ministry of Agriculture and Lands/Office of the Premier. February 7. http://www2.news.gov.bc.ca/
Capozza, Korey. 1998. "Nuxalk Reborn." *Albion Monitor,* December 12. http://www.albionmonitor.com/.
Cassidy, Frank, and Norman Dale. 1988. *After Native Claims? The Implications of Comprehensive Claims Settlements for Natural Resources in British Columbia.* Lantzville, BC: Oolichan Books.
Cormick, Gerald, Norman Dale, Paul Emond, S. Glenn Sigurdson, and Barry D. Stuart. 1996. *Building Consensus for a Sustainable Future: Putting Principles into Practice.* Ottawa: National Roundtable on the Environment and Economy.
Dale, Norman. 1996. "Finding the Common Ground in North America's Oldest Public Conflict." *Consensus* 30:4.
–. 1999. "Facilitated Negotiations for Cross-Cultural Community-Based Planning." In *Consensus Building Handbook,* edited by Lawrence E. Susskind, Sarah McKearnan, and Jennifer Thomas-Larmer, 923-50. New York: Plenum Press.
–. 2005. "The Undiscussable in Native Community Economic Development." In *Second Growth: Community Economic Development in Rural British Columbia,* edited by Sean Markey, John Pierce, Kelly Vodden, and Mark Roseland, 186-95. Vancouver: UBC Press.
Deutsch, Morton. 1973. *The Resolution of Conflict: Constructive and Destructive Processes.* New Haven, CT: Yale University Press.
Fisher, Roger, and William Ury. 1981. *Getting to Yes: Negotiating Agreement without Giving In.* Boston: Houghton Mifflin.
Forrester, John. 1999. *The Deliberative Practitioner.* Cambridge, MA: MIT Press.
Garrett, Robert D. 1994. "Mediation in Native America." *Dispute Resolution Journal* 49:38-45.
Granander, Hans, and Michael Wigle. 2003. *Bella Coola: Life in the Heart of the Coast Mountains.* Madeira Park, BC: Harbour Publishing.
Grant, Agnes. 1996. *No End of Grief: Indian Residential Schools in Canada.* Winnipeg: Pemmican Press.
Heiltsuk Tribal Council. n.d. "The Heiltsuk and the LCRMP." www.heiltsuk.com/lrmp.htm.
Jackson, Tony, and John Curry. 2004. "Peace in the Woods: Sustainability and the Democratization of Land Use Planning and Resource Management on Crown Lands in British Columbia." *International Planning Studies* 9(1):27-42.
Knight, Rolf. 1978. *Indians at Work: An Informal History of Native Indian Labour in British Columbia, 1858-1930.* Vancouver: New Star.
LaViolette, Forrest E. 1973. *The Struggle for Survival: Indian Cultures and the Protestant Ethic in British Columbia.* Toronto: University of Toronto Press.
Markey, Sean, John T. Pierce, Kelly Vodden and Mark Roseland. 2005. *Second Growth: Community Economic Development in Rural British Columbia.* Vancouver: UBC Press.
McIlwraith, Thomas F. (1948) 1992. *The Bella Coola Indians.* Toronto: University of Toronto Press.
Mercredi, Ovide, and Mary Ellen Turpel. 1993. *In the Rapids: Navigating the Future of First Nations.* Toronto: Penguin.
Pinkerton, Evelyn. 1983. "Taking the Minister to Court: Changes in Public Opinion about Forest Management and Their Expression in Haida Land Claims." *BC Studies* 57:68-85.
Riddell, Darcy. 2005. "Evolving Approaches to Conservation: Integral Ecology and Canada's Great Bear Rainforest." *World Futures* 61(3):63-78.
Ross, Rupert. 1992. *Dancing with a Ghost: Exploring Indian Reality.* Markham, ON: Reed.
Saunders, Harold H. 1999. *A Public Peace Process: Sustained Dialogue to Transform Racial and Ethnic Conflicts.* New York: St. Martin's Press.
Shaw, Karena. 2004. "The Global/Local Politics of the Great Bear Rainforest." *Environmental Politics* 13(2):373-92.
Smith, Merran, and Art Sterritt. 2010. "Towards a Shared Vision: Lessons Learned from Collaborations between First Nations and Environmental Organizations to Protect the Great Bear Rainforest and Coastal First Nations Communities." In *Alliances: Re/Envisioning*

Indigenous – Non-Indigenous Relationships, edited by Lynne Davis, 131-48. Toronto: University of Toronto Press.

Susskind, Lawrence, and Jeffrey L. Cruikshank. 1987. *Breaking the Impasse: Consensual Approaches to Resolving Public Disputes*. New York: Basic Books.

Tennant, Paul. 1990. *Aboriginal Peoples and Politics: The Indian Land Question in British Columbia, 1849-1989*. Vancouver: UBC Press.

14
Co-Management of Forest Lands
The Cases of Clayoquot Sound and Gwaii Haanas

Holly S. Mabee, D.B. Tindall, George Hoberg, and J.P. Gladu

Co-management of natural resources between First Nations and state governments is becoming a common tool to resolve resource use conflicts and to accommodate Aboriginal rights to the land base. The overarching goals of co-management are to include Aboriginal values and knowledge systems in managing resources, and to promote cooperation and compromise, reducing conflict between Aboriginal communities and government or industrial resource users.

This chapter briefly examines the context for the development of co-management in Canada and the province of British Columbia in particular.[1] We then present two case studies where innovative co-management arrangements have been in place for over a decade: Gwaii Haanas, a protected area being managed by the Government of Canada and the Council of the Haida Nation, and Clayoquot Sound, a region being managed for multiple uses, including logging, by the Central Region Nuu-chah-nulth First Nations and the Government of British Columbia.[2] (Complementary material is provided about these cases in the chapters by Forsyth, Hoberg, and Bird, and by Pechlaner and Tindall.) The challenges and benefits arising from each of these examples will be discussed, highlighting contrasts and similarities. We hope that the lessons learned from these case studies provide insight into the issues that can arise with co-managing protected areas and co-managing areas involving commercial resource extraction. Sharing the experiences of the case studies may also help guide future co-management efforts on forest lands in British Columbia and elsewhere.

The case studies presented here are both examples of shared power consistent with the "collaborative endeavours" theme of the book. However the institutional, cultural, and socioeconomic challenges that exist in both case studies, particularly issues of decision-making authority, differing world views, and capacity gaps, demonstrate the grey area that exists between the shared power perspective and the colonization perspective, where co-management is seen as assimilation (see Smith, this volume).

Context and Rationale for Co-Management in Canada and British Columbia

Co-management regimes between First Nations and non-First Nations parties are being established across Canada. Co-management has arisen in Canada as a response to the ongoing struggle of Aboriginal peoples to regain control over their traditional lands and resources (Pinkerton 1989; Singleton 1998; Castro and Nielson 2001; for a broader perspective on co-management in an array of diverse settings see Wilson, Nielsen, and Degnbol [2003]). For thousands of years, First Nations peoples lived from coast to coast in complex and diverse societies, stewarding their lands according to their own cultural traditions, laws, and governance systems. Forests are critical to the survival of First Nations cultures across much of Canada, as they contribute to subsistence, spiritual, and ceremonial purposes. Because many First Nation communities are situated in remote areas, commercial forestry is one of their best options for local economic development (NAFA 1996; Brubacher 1998; Ross and Smith 2002).

Introduction to the Case Studies

The two case studies chosen are among the most advanced cases of co-management in Canada in terms of power sharing with First Nations. These case studies showcase co-management issues in protected areas (Gwaii Haanas) and forest lands managed for sustainable timber extraction (Clayoquot Sound). Each case represents the first agreement of its kind in British Columbia. Both give a relatively high degree of power and jurisdiction to First Nations partners, and both have been in place for over ten years, enough time to warrant evaluation (Hoberg and Morawski 1997; RCAP 1997; Gardner 2001; Goetze 2005).

For each case study, the history and context leading up to co-management are briefly discussed. The institutional structure of the co-management arrangement is also described. Following the case study descriptions, the main challenges and benefits experienced throughout the co-management arrangements are explained and contrasted. Figure 14.1 shows the geographic locations of the two cases. Figure 14.2 provides a chronology of key events for each of the two cases.

Clayoquot Sound – Description of the Case

Clayoquot Sound is one of the last remaining areas of relatively undisturbed, old-growth forest on Vancouver Island and is home to five Nuu-chah-nulth First Nations: Ahousaht, Hesquiaht, Tla-o-qui-aht, Toquaht, and Ucluelet. There has been long-term conflict over the appropriate use of the natural resources in the region (BC MoF 2000). In the mid-1980s and early 1990s, this conflict attracted national and international attention as environmental nongovernmental organizations (ENGOs) and First Nations began setting

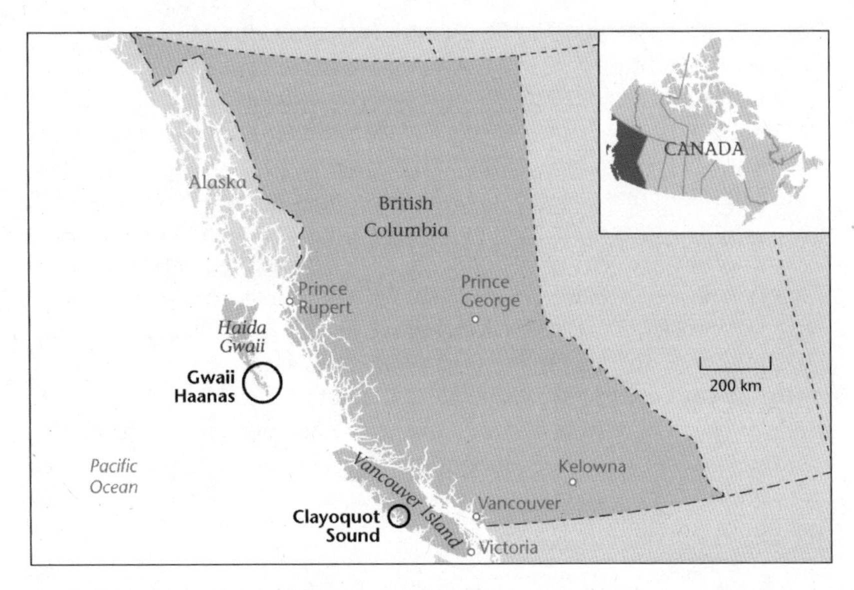

Figure 14.1 **Locations of the case studies**

up blockades on logging roads and leading international boycotts of BC forest products. A critical turning point in the dispute was in 1985, when chiefs of the Ahousaht and Tla-o-qui-aht First Nations obtained an injunction to stop logging on Meares Island until their land claim was resolved. This was the first time the courts had halted resource development on Crown land in British Columbia because of an Aboriginal title claim, and it prompted the provincial government to enter into treaty negotiations (Tennant 1990; Darling 1994; Hoberg and Morawski 1997).

In 1989, the BC government set up the multistakeholder Clayoquot Sound Sustainable Development Task Force in an attempt to resolve the conflict. The task force failed to reach consensus on a land use plan, and the process was returned to Cabinet, which announced the Clayoquot Sound Land Use Decision in 1993 (Iisaak 2002). This decision protected 34 percent of the land base and placed another 21 percent in special management zones but did little to placate the concerns of the environmentalists who wanted to see all of Clayoquot protected. Furthermore, the decision angered the Nuu-chah-nulth, as they felt that they were not adequately consulted in developing the land use plan. Huge protests ensued in the summer of 1993, when over eight hundred people were arrested in the largest incident of its kind in Canadian history (see Shaw 2003; Robinson et al. 2007).

This pressured the province to speed up negotiations with the Nuu-chah-nulth, leading to the Interim Measures Agreement (IMA) signed on March 19, 1994.[3] It was the first IMA in the province and remains the only one of its kind. The agreement provides for local land and resource co-management, and economic development. The IMA has been extended in 1996, 2000, and 2006, providing the Nuu-chah-nulth with some control over and protection of their traditional territories while treaty negotiations proceed (British Columbia et al. 1994, 1996, 2000, 2006, 2008, 2009).

The IMA created a resource co-management body called the Central Region Board (CRB), which has equal representation from the Nuu-chah-nulth and the provincial government and is responsible for reviewing all proposals for development on the land base (LUCO 1996). The CRB consists of five BC government-appointed local representatives, five First Nations members representing each of the Central Region nations, a government co-chair, and a First Nations co-chair. Provincial representatives (excluding the co-chair) are selected from local non-Aboriginal communities and represent various interest groups, and include some local government officials.

The objectives of the CRB are to promote sustainable economic development, to reduce unemployment levels within Aboriginal communities, to assess compliance with forestry standards, to provide a viable sustainable forest industry while increasing local ownership, and to work toward reconciliation between environmentalists, labour, industry, First Nations, and recreational users. The CRB is funded through annual contribution agreements from the province.

Decision making within the CRB is by consensus; if consensus can't be reached, a double-majority vote occurs where a majority of board members and a majority of the First Nations members must approve. In the two cases where the CRB has relied on a vote, the First Nations double majority has influenced the decision only once. If board recommendations are not implemented satisfactorily within thirty days, the board can refer the matter to the parties to the IMA, which then refers the matter to Cabinet. If either party is not satisfied by a Cabinet decision that varies a decision of the CRB, the Central Region Resource Council dispute resolution body is activated, composed of the ministers or their designates and the hereditary chiefs of the First Nations.

Gwaii Haanas – Description of the Case

The Gwaii Haanas National Park Reserve and Haida Heritage Site is located at the southern end of South Moresby Island in Haida Gwaii (the Queen Charlotte Islands), home of the Haida First Nations people for thousands of years (see Figure 14.1).

Because of its valuable timber, the forests of Haida Gwaii have largely been placed under long-term licence by the Province of British Columbia. Fears of unsustainable logging practices led to the formation of the multistakeholder Islands Protection Committee (IPC) in the 1970s (Islands Protection Society 1984; May 1990). The IPC developed the Wilderness Proposal to protect South Moresby after a long and protracted struggle with the forest licensees, the provincial government, and the federal government.

The impetus for the formation of the IPC and the Wilderness Proposal was the Haida's objection to a logging plan on Burnaby Island (Broadhead 1995). The province issued a five-year moratorium on logging on Burnaby Island but continued to issue logging permits elsewhere in the region. The Haida-led protests to logging became large scale in 1985, following on the heels of the Meares Island injunction in Clayoquot Sound won by the Nuu-chah-nulth. With this legal precedent driving their hopes, the Haida blockaded logging on Lyell Island with the cooperation of local and international environmental groups (RCAP 1997; NAFA and Wildlands League 2003). The protest drew national and international media attention (Broadhead 1995).

The negative public opinion toward the government garnered by this protest led the federal government to enter into treaty negotiations with the Haida. With support from the provincial government, the National Park Reserve was established in 1988 with the signing of the South Moresby Agreement. The Council of the Haida Nation then began negotiations with Parks Canada for co-management terms. The Haida had already designated the area as a Haida Heritage Site in 1985 under the authority of the Haida constitution and had been managing it as such (RCAP 1997).

The Gwaii Haanas Agreement (GHA) ratified in 1993 is the only co-management agreement of its kind in Canada, containing parallel statements from the Haida and the federal government on sovereignty, ownership, and jurisdiction, agreeing to disagree on this issue until it is resolved in treaty negotiations or by the courts. The parties agree to the need to protect the ecological and cultural heritage of Gwaii Haanas for future generations, and to state how the parties will work together.

The agreement provided for the creation of the Archipelago Management Board (AMB). Two of the representatives on this board are appointed by the Council of the Haida Nation, and two by Parks Canada. Each party has an alternating co-chair. The board members ensure that the laws and policies of their respective governments are carried out. They are also responsible for all activities associated with planning, management, and operations within Gwaii Haanas. This includes creation of a joint management plan in consultation with the public; overseeing Haida cultural and traditional harvest activities; identification and protection of Haida cultural sites; control of visitor activities; information management and research; and licensing for tourism operators.

The AMB makes decisions by consensus; if consensus cannot be reached, the issue is referred to the Council of the Haida Nation and the Government of Canada, with a mutually agreed-on mediator (GHA sections 5.3 to 5.5). The AMB receives its funding on a two-year basis by a contribution agreement

Figure 14.2

Timeline of key local events leading up to co-management

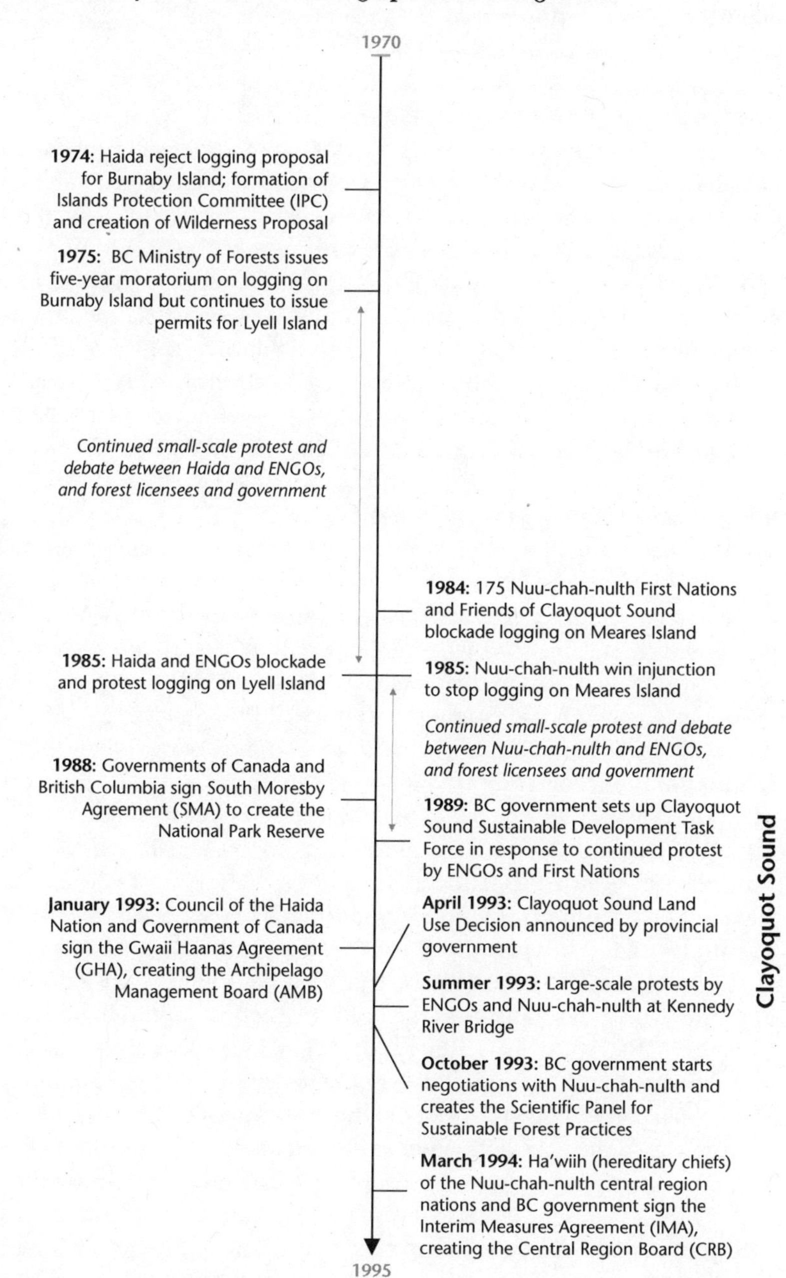

from the federal Ministry of the Environment. The GHA includes preferential hiring policies for Haida Nation members within the park, as well as opportunities for training so that Haida members can be employed at all levels.

Challenges and Benefits of Co-Management

We categorize challenges and benefits as follows: *Institutional* constraints or benefits are associated with the way the co-management agreement itself is set up, or the broader institutional structure in which co-management functions. *Cross-cultural* issues are associated with the differences in Aboriginal and Western world views and cultural norms. *Socioeconomic* aspects arise from colonial policy and resulting capacity gaps between First Nation and non-First Nation communities in Canada. Where applicable, contrasts are drawn between the Clayoquot Sound case and the Gwaii Haanas case. Analysis of the case studies is primarily based on interviews conducted by the authors in 2002 (see NAFA and Wildlands League 2003; Spiro 2003).

Institutional Challenges and Benefits

Institutional challenges are particularly common in co-management situations arising from conflicts over land use (crisis-based co-management). The terms and conditions of co-management partnerships are often left at very vague levels, resulting in difficult implementation. First Nations often differ from state governments in their idea of what the role, authority, and power of co-management boards should be (Gardner 2001; Weitzner and Manseau 2001). Terms such as "equal partners" can be misleading when ultimate statutory authority and property rights reside with government and industry partners (Hawkes 1996; Beckley 1998; Mabee and Hoberg 2006).

In the Clayoquot Sound case, First Nations are not equal partners in decision making because the district manager of the Ministry of Forests retains statutory decision-making authority (Spiro 2003). First Nations are not joint signatories on forestry plans, so the CRB is a de facto advisory body. In contrast, the Council of the Haida Nation does have equal decision-making authority with Parks Canada, as the Haida insisted on joint signatory status (AMB 1996; Hawkes 1996; Gardner 2001; NAFA and Wildlands League 2003). This difference may arise from the lengthy negotiation process for the GHA, where each party came to understand the other's perspectives and priorities, or from the fundamental difference between co-managing protected areas versus resource extraction involving a third-party industrial licensee.

Some First Nations CRB members feel that their nation does not have sufficient power in decision making because multiple communities provide representatives to the board (Spiro 2003). When discussing issues affecting Ahousaht territory, only one Ahousaht representative is present at the table with five provincially appointed representatives. A related issue that suggests that individual government-to-government agreements are preferable is territory overlap. In Clayoquot Sound, decision making in the CRB is often delayed when a development application is in an area where there is an overlapping land claim by two or more First Nations. Each First Nation may have a different land use goal for that area, requiring negotiation among themselves before the CRB can reach a decision. The GHA is a simpler co-management arrangement in this regard, as there is only one First Nation involved.

A key problem with crisis-based co-management arrangements is that they often lack a robust dispute resolution mechanism. This can be particularly problematic where state government partners retain statutory decision-making power, since support for the co-management arrangement may decline once the immediate crisis has passed (Castro and Nielson 2001). The Central Region Resource Council (CRRC) was named as the dispute resolution body in the Clayoquot Sound IMA, but it was unclear as to how this council would actually work. This problem became evident in 2003 when the CRB rejected a forest development plan, which was subsequently approved by the BC Ministry of Forests. The First Nations CRB members used their double-majority vote to reject a plan that did not adequately protect the Tla-o-qui-aht Nations' cultural resources of cedar and medicinal plants. The Ministry of Forests approved the plan because, in their analysis, these resources existed in sufficient amounts elsewhere in the cutblock. However, the outcome was that International Forest Products – the tenure holder – did not log the block, even though its plan had been officially approved, in order to maintain First Nations' support for their operations in Clayoquot Sound. Therefore, the dispute resolution mechanism did not have to come into play (Clogg, Hoberg, and O'Carroll 2004; Mabee and Hoberg 2004). Gwaii Haanas hasn't experienced similar difficulties as yet, as there is no commercial resource extraction involved. Co-management agreements should include clearly defined rules of accountability and dispute resolution processes, although it may be impossible to have a clearly defined, rigid dispute resolution process while the sovereignty question remains unresolved (Hawkes 1996).

The dual statutory authority in Gwaii Haanas leads to problems with enforcement. The laws of both parties apply, but neither party will recognize the ultimate authority of the other. For example, backcountry permits are usually issued by Parks Canada in national parks, but because neither party in Gwaii Haanas would agree to defer to the other, the park now uses "backcountry registration forms" with the Council of the Haida Nation logo on one side and Parks Canada's logo on the other. The problem with this system is that these registration forms lack the force of law (Hawkes 1996). Research is needed on the question of how to deal with enforcement issues where there is joint statutory authority, especially in light of First Nations' calls for joint stewardship over their entire traditional territories in addition to sole authority of selected settlement lands (Penikett 2003).

Co-management board decisions can be influenced by the fact that most if not all of the board's funding comes from provincial and federal governments, which may constrain certain outcomes (Gardner 2001; Natcher and Davis 2007). The annual or biannual nature of this funding means that co-management boards are unable to do any meaningful long-term planning and are vulnerable to changing government priorities (RCAP 1997; Abbott 2001). Over the years since its inception, annual funding to the CRB has decreased, resulting in staff cuts and an inability to effectively work toward stated goals. The funding situation for the AMB is a little better, as it is funded on a two-year basis, but the funding remains insufficient to allow for long-term planning. Concerns about governments not living up to funding commitments is a consistent complaint among First Nations involved with co-management efforts in British Columbia (Ecotrust and CPAWS 1999).

The potential for co-management to create employment for First Nation community members can be constrained by union hiring rules and related legislation. The Nuu-chah-nulth feel that there is not enough recruitment of their members by the local forest industry, especially at higher levels. Local licensees have said that they would like to hire more First Nations employees and contractors but are constrained by powerful unions and legislation that protects existing contractors (Spiro 2003). In contrast, the Gwaii Haanas agreement has built-in detailed provisions for employment of Haida members in the park reserve, and Haida are employed at all levels of management and operations in the absence of powerful industry unions.

The one institutional benefit that was seen in both case studies was improved information management. In Clayoquot Sound and Gwaii Haanas, extensive inventories on the status of ecological and cultural resources, using both scientific and traditional ecological knowledge (TEK) approaches and sources, have been conducted (CSSP 1995; AMB 1996; Spiro 2003). The AMB has created a geographic information system (GIS), which includes and relates all available historical and current data related to resource management (AMB 1996). Having all the relevant data easily accessible will make for better informed and more efficient management planning in the future.

Cross-Cultural Challenges and Benefits

An extremely difficult challenge for co-management is the need to incorporate two very different world views and knowledge systems, often the source of miscommunication and conflicting expectations, into one management regime that respects and acknowledges both systems equally in its processes and products (Gardner 2001; Natcher and Davis 2007; see also the chapters in this volume by Marc Stevenson; Lewis and Sheppard; and Blackstock). In most cases of co-management in Canada, the planning and management process is biased toward Western resource management systems, and integration of First Nations traditional knowledge is often left at an anecdotal, superficial level (Rusnak 1997; Robinson 1998; Houde 2007).

Protected areas can be a particularly divisive aspect of co-management planning, since they are a foreign concept to most Aboriginal cultures. First Nations see the land holistically and consider themselves to be an integral part of the environment, rather than outside the ecosystem or controlling it. First Nations agree with (and indeed have often initiated) the idea that certain areas need to be spared from industrial development activities, but excluding all resource extraction activities (including subsistence hunting or berry picking) in the name of protecting the ecosystem does not make sense to them (Ecotrust and CPAWS 1999; RCAP 1997; Parks Canada 2000; Gardner 2001; Sapic, Runesson, and Smith 2009).

It is important for co-management agreements to ensure First Nations access to resources within their territories as per their Aboriginal and treaty rights, something that is not always sufficiently addressed. In Gwaii Haanas, Haida access to resources for subsistence and ceremonial purposes is ensured in the GHA. In Clayoquot Sound, however, the IMA does not explicitly consider Nuu-chah-nulth access to resources within ecological reserves. Although these reserves were intended to exclude commercial resource extraction rather than First Nations subsistence use, there is uncertainty over their eventual impact (Spiro 2003).

In Clayoquot Sound, one of the most difficult challenges with co-management has been incorporation of TEK into the resource management process. Part of the difficulty is that TEK has a highly spiritual and ceremonial context, and many Nuu-chah-nulth members say that ecological knowledge cannot be separated from this context without losing its meaning (Spiro 2003; Stevenson 2004 and this volume). Therefore, it is difficult for Western resource managers and non-First Nations people involved in co-management to understand TEK.

A major benefit associated with co-management is that the experience allows the two parties to learn about each others' cultures and world views. The informal cross-cultural education that occurs through participation in co-management agreement negotiations, board meetings, and activities is invaluable in creating mutual respect and understanding. In Clayoquot Sound, non-First Nations partners in co-management, as well as other government and forest industry players, have developed at least a basic understanding of and respect for First Nations world views and TEK (Spiro 2003). Similar cross-cultural exchange and learning has occurred in Gwaii Haanas through the activities of the AMB.

Many co-management agreements in Canada have not achieved the level of integration of traditional and scientific knowledge and management systems that First Nations would like to see, often because traditions of resource management are not well understood by federal and provincial government partners. Through the continued interaction and dialogue that co-management processes provide, the two parties can learn to understand each others' world views and cultures (Lertzman 1999). To allow for better integration of world views and knowledge systems, it may be wise to include provisions for a more proactive, formal approach to cross-cultural education in co-management agreements (Duffy, Roseland, and Gunton 1996; Parks Canada 2000; Gardner 2001) – even though some First Nations are confident that TEK will implicitly be incorporated into management decisions as long as holders of TEK are at the table (Christensen and Grant 2007).

There is no doubt that co-management can help to build respect for local indigenous law and place. In both the CRB and the AMB, consensus decision making has been adopted from Nuu-chah-nulth and Haida governance traditions. Thus, co-management is taking some first steps toward filling the need for non-First Nations people and their governments to respect First Nations' governments and their peoples' own laws, regulations, codes, and practices for protecting the earth (Robinson 1998; Parks Canada 2000).

Socioeconomic Challenges and Benefits

The capacity of First Nations to meaningfully participate in co-management is often cited as a challenge for co-management (CSSP 1995; Treseder and Krogman 2000). Indigenous groups may face difficulty because of a lack of experience in the dominant society's negotiating methods and institutions (Castro and Nielson 2001; Natcher and Davis 2007). The technical language and process of resource management are unfamiliar for many First Nations participants. The Clayoquot Sound case study demonstrated that lack of capacity and experience of non-First Nations parties in First Nations governance and resource management systems present an equally significant challenge (Spiro 2003).

Co-management requires a substantial commitment of time and resources from all parties (Singleton 1998; Treseder and Krogman 2000). This level of commitment is difficult to achieve for small community partners. Compared with government parties to co-management agreements, the Nuu-chah-nulth and Haida have few human and financial resources to draw on. If First Nations do not have the capacity to take on co-management responsibilities, the potential benefits of community empowerment will not be fully realized. In addition, First Nations have to deal with many pressing issues within their communities arising from the colonial policy legacy and the tragedy of residential schools (as discussed in the chapters by Frideres and Dale, this volume). The average education level is lower than the national average, especially in the sciences, and very high unemployment levels exist

(Statistics Canada 2001). These issues impact First Nation communities' capacity to participate in forest management and planning.

Nuu-chah-nulth members involved in co-management have found that it is difficult to balance economic development interests with protection of cultural values. Some members do not want to see any commercial logging, as it goes against their spiritual values, whereas others want to see more logging to provide employment and revenue (Spiro 2003; Mabee and Hoberg 2004). This is less of an issue for Gwaii Haanas as a protected area, though similar issues have arisen in regard to the amount and kind of tourism development that the AMB will allow in and around the park. In both Gwaii Haanas and Clayoquot Sound, some First Nations members don't even support the concept of co-management, as they feel that the First Nations are sole owners of the land and they should be managing it themselves (Weitzner and Manseau 2001; Spiro 2003).

Two of the most commonly cited benefits of co-management are relationship building and trust building among actors and parties with a previous history of mistrust and conflict (Pinkerton 1989; Singleton 1998; Treseder and Krogman 2000; Gardner 2001; NAFA and Wildlands League 2003). This holds true in Clayoquot Sound and Gwaii Haanas, where a major benefit of co-management has been the positive relationships that have developed between First Nations and non-First Nations and the realization that they can work together despite their differences (Ecotrust and CPAWS 1999; Spiro 2003; Goetze 2005).

Other benefits often associated with co-management are local economic development and capacity building in the local First Nation communities. Co-management agreements should contain provisions for First Nations employment and training, and for monetary benefits from commercial activity. These benefits may be associated with tourism development in and around protected areas, or with logging or other resource extraction in multiple use areas. The creation of Gwaii Haanas through the GHA has created more jobs for local people, especially Haida members, than were historically available in the logging industry (Broadhead 1995). Furthermore, tourism associated with the park reserve has led to increased business development and revenues in the islands' service sector (ibid.; NAFA and Wildlands League 2003). The Clayoquot Sound IMA also included local economic development provisions, and a significant amount of First Nations business development in the resource sector has occurred.

In addition to benefits associated with the process of co-management, the existence of co-management agreements provides relative stability. If parties share responsibility for management of land and resources equally, the question of which party has ultimate responsibility in a dispute becomes practically irrelevant (Hawkes 1996). Furthermore, well-designed co-management agreements can create a fair and efficient process for resolving any future resource conflicts in the area. The existence of a co-management board reduces the likelihood that future resource conflicts will need to be resolved in court. Both the Haida and the Nuu-chah-nulth are more confident that their lands and cultures will be protected and passed down to future generations (NAFA and Wildlands League 2003; Mabee and Hoberg 2006). This is an advantage for all, since consensus deliberation among board members is likely to result in a win-win solution, whereas court decisions generally produce unpredictable win-lose solutions (Hawkes 1996).

Key Differences between the Case Studies

From looking at the aforementioned challenges and benefits, we can see that co-management issues in Clayoquot Sound are generally more complex than in Gwaii Haanas. (See Table 14.1 for a summary.) The Clayoquot Sound area is managed for multiple uses, including commercial timber extraction, whereas Gwaii Haanas is a protected area. Commercial logging at any scale

Table 14.1

Summary of challenges and benefits of co-management in Clayoquot Sound and Gwaii Haanas

	Challenges	Benefits
Institutional	Unequal decision-making power (CS) Territory overlap and divergent goals (CS) Unclear dispute resolution process (CS) Uncertain legal basis for enforcement (GH) Insecure funding (B)	Improved information access and management (B)
Cross-cultural	Different world views on people and nature relationship (B) Different knowledge systems (B)	Improved cross-cultural understanding (B) Respect for local indigenous law and place (B)
Socioeconomic	Capacity gaps (B) Lack of trust (B) Balancing economic and cultural values (B)	Relationship building (B) Local economic development (B) Stability (B)

Note: The case study each particular benefit or challenge applies to is indicated in brackets: GH = Gwaii Haanas, CS = Clayoquot Sound, B = both.

is in much greater conflict with First Nations traditional land use than protected area management, making it difficult to balance economic development goals and cultural protection needs. Furthermore, co-management for resource extraction or so-called sustainable development is a newer concept in Canada than is co-management of protected areas (Notzke 1995), so it is not surprising that there are more challenges with these kinds of agreements.

When more than one First Nation is involved, issues can arise with conflicting priorities among First Nations parties to the agreement. The Clayoquot Sound IMA has five First Nations as signatories, whereas the GHA includes the Haida Nation alone. In Clayoquot Sound, this has had the effect of slowing decision making, but it could have more severe consequences if First Nations parties have vastly divergent goals.

Lessons Learned and Future Directions for Co-Management in British Columbia and Canada

The case studies presented in this chapter are two of the most advanced examples of co-management in British Columbia and arguably the whole of Canada. Several key criteria make them successful and are important to note for those embarking on new co-management agreements, or for those currently struggling to make failing agreements work.

One of the most important aspects of the Gwaii Haanas Agreement and the Clayoquot Sound IMA is legal accountability on the part of the government. The GHA is legislated by the Parliament of Canada, and the IMA is a binding agreement for the Province of British Columbia. This protects the co-management agreements from changing priorities of governments after elections or because of other external factors.

In contrast, the case of the Wendaban Stewardship Authority in Temagami, Ontario, illustrates the problem with legally informal co-management agreements. A memorandum of understanding was signed in 1990 between the Ontario provincial government and the Teme-Augama Anishnabai, creating the WSA – the local co-management board. After the 1995 Ontario elections, the newly elected premier did not support the agreement and was able to withdraw all provincial funding and commitments with a simple policy change (RCAP 1997). Another important lesson is the need to empower First Nations within the context of co-management agreements. The Haida are generally more satisfied with their role in co-management than are the Nuu-chah-nulth, as the AMB has a full mandate with joint decision-making authority, whereas the CRB is essentially an advisory board, with the province retaining statutory authority – although thus far this hasn't inhibited the First Nations from ensuring that their values are protected on the land base (Clogg, Hoberg, and O'Carroll 2004; Goetze 2005). Complications that may arise with enforcement in models like Gwaii Haanas with joint statutory authority need to be resolved.

An issue where further research is needed is that of community members' participation in co-management processes. Most research on co-management in Canada has focused on interviews with members of co-management boards rather than on local community members. It would be interesting for researchers to conduct a random sample survey of members of First Nation communities in Clayoquot Sound and Haida Gwaii to determine their levels of awareness and involvement in co-management efforts, and whether they feel they are benefiting from the co-management agreement in any tangible way.

Lessons learned from co-management agreements such as the GHA and the Clayoquot Sound IMA will be of great significance to the Province of British Columbia as the treaty process proceeds. Many First Nations are requesting co-jurisdiction over their entire traditional territories as part of treaty settlements and are not willing to settle for sole jurisdiction over a small percentage of their lands (Penikett 2003). The Haida recently gained greater control over their traditional territory through the 2007 Haida Gwaii Strategic Land Use Agreement, which follows ecosystem-based management objectives set by the Haida in accordance with their laws, policies, customs, traditions, and decision-making processes (CHN and British Columbia 2007). It is likely that the relationship and trust building that occurred through the Gwaii Haanas co-management agreement facilitated the process for the new land use plan.

As co-jurisdiction continues to expand in the province, through the treaty process and through legal judgments affirming Aboriginal rights (see BC Supreme Court 2009; Price, Roburn, and MacKinnon 2009), it will be critical to design arrangements that build on the successes and minimize the challenges of previous co-management efforts so as not to further exacerbate problems associated with uncertainty on the land base as it affects resource development (Forsyth, Hoberg, and Bird, this volume; Natcher et al. 2009).

The extension periods for the Clayoquot Sound IMA are getting shorter; in 2008, it was extended for only one year. This makes it difficult for the CRB to commit to any long-term plans (CRB 2009). First Nations parties can benefit more from co-management arrangements that provide adequate and secure funding (Natcher and Davis 2007).

The case studies discussed in this chapter are examples of situations that are moving toward the shared power perspective in this book's typology of perspectives on First Nations and forest lands. From analyzing these two cases, we conclude that there is a need for strong institutional and legal structures in co-management agreements, and that particular attention should be paid to ensure that decision-making powers and dispute resolution mechanisms are understood and interpreted in the same way by all parties. Attention to such design factors at the outset may prevent First Nations from perceiving co-management to be another assimilation tool as implementation proceeds (Mabee and Hoberg 2006; Smith, this volume).

Acknowledgment

We would like to thank Ron Trosper for his helpful comments on an earlier draft of this chapter.

Notes

1 A detailed consideration of what is meant by the term "co-management" was provided earlier in this volume by Smith.

2 Canada and Council of the Haida Nation. 1993. Gwaii Haanas/South Moresby Agreement. Ottawa: Parks Canada. This chapter covers events up to the mid-2000s.

3 An equally important outcome of the struggle from an ENGO (environmental non-governmental organization) perspective was the Clayoquot Sound Scientific Panel (see CSSP [1995] and Spiro [2003] for details).

References

Abbott, K. 2001. "Co-Management in Canada." In collaboration with First Peoples Worldwide. http://www.firstpeoples.org/

AMB (Archipelago Management Board). 1996. *Draft Strategic Management Plan for the Terrestrial Area*. Public Planning Program, newsletter no. 3. Haida Gwaii, BC: Gwaii Haanas Haida Heritage Site and National Park Reserve.

BC MoF (British Columbia, Ministry of Forests). 2000. *Clayoquot Sound Land Use Decision*. www.for.gov.bc.ca/.

BC Supreme Court. 2009. *Ahousaht Indian Band and Nation v. Canada (Attorney General), Reasons for Judgement*. BC Supreme Court 1494.

Beckley, T.M. 1998. "Moving Toward Consensus-Based Forest Management: A Comparison of Industrial, Co-Managed, Community and Small Private Forests in Canada." *Forestry Chronicle* 74(5):736-43.

British Columbia and the Ha'wiih of the Tla-o-qui-aht First Nations, the Ahousaht First Nation, the Hesquiaht First Nation, and the Ucluelet First Nation. 1994. Interim Measures Agreement.

–. 1996, 2000, 2006, 2008, 2009. Clayoquot Sound Interim Measures Extension Agreement: A Bridge to Treaty.

Broadhead, J. 1995. *Gwaii Haanas Transitions Study*. Skidegate, BC: Queen Charlotte Islands Museum Press.

Brubacher, D. 1998. "Aboriginal Forestry Joint Ventures: Elements for an Assessment Framework." *Forestry Chronicle* 74(3): 353-58.

Castro, A.P., and E. Nielson. 2001. "Indigenous People and Co-Management: Implications for Conflict Management." *Environmental Science and Policy* 4(4-5):229-39.

CHN and British Columbia (Council of the Haida Nation and the Province of British Columbia). 2007. Haida Gwaii Strategic Land Use Agreement. Victoria: Ministry of Agriculture and Lands.

Christensen, J., and Grant, M. 2007. "How Political Change Paved the Way for Indigenous Knowledge: The Mackenzie Valley Resource Management Act." *Arctic* 60(2):115-23.

Clogg, J., G. Hoberg, and A. O'Carroll. 2004. *Policy and Institutional Analysis for Implementation of the Ecosystem-Based Management Framework*. Victoria: Coast Information Team.

CRB (Clayoquot Sound Central Region Board). 2009. *Annual Report 2008-2009 and Strategic Plan 2009-2014*. Ucluelet, BC: CRB.

CSSP (Clayoquot Sound Scientific Panel). 1995. *Report 3: First Nations Perspectives Relating to Forest Practices Standards in Clayoquot Sound*. Victoria: Cortex Consultants.

Darling, C.R. 1994. *In Search of Consensus*. Victoria: University of Victoria Institute for Dispute Resolution.

Duffy, D.M., M. Roseland, and T.I. Gunton. 1996. "A Preliminary Assessment of Shared Decision-Making in Land Use and Natural Resource Planning." *Environments* 23(2):1-16.

Ecotrust and CPAWS (Canadian Parks and Wilderness Society). 1999. *First Nations Co-Operative Management of Protected Areas Workshop: Summary Report*. Vancouver: Ecotrust and CPAWS.

Gardner, J. 2001. *First Nations Cooperative Management of Protected Areas in British Columbia: Foundations and Tools*. Vancouver: Canadian Parks and Wilderness Society, BC Chapter and Ecotrust Canada.

Goetze, T. 2005. "Empowered Co-Management: Towards Power-Sharing and Indigenous Rights in Clayoquot Sound, BC." *Anthropologica* 47:247-65.

Hawkes, S. 1996. "The Gwaii Haanas Agreement: From Conflict to Cooperation." *Environments* 23(2):87-100.

Hoberg, G., and E. Morawski. 1997. "Policy Change through Sector Intersection: Forest and Aboriginal Policy in Clayoquot Sound." *Canadian Public Administration* 40(3):387-414.

Houde, N. 2007. "The Six Faces of Traditional Ecological Knowledge: Challenges and Opportunities for Canadian Co-Management Arrangements." *Ecology and Society* 12(2):34.

Iisaak (Iisaak Forest Resources). 2002. *Iisaak Forest Resources*. Ucluelet, BC: Iisaak.

Islands Protection Society. 1984. *Islands at the Edge: Preserving the Queen Charlotte Islands Wilderness*. Vancouver: Douglas and McIntyre.

Lertzman, D.A. 1999. "Planning between Cultural Paradigms: Traditional Knowledge and the Transition to Ecological Sustainability." School of Community and Regional Planning, University of British Columbia.

LUCO (Land Use Coordination Office). 1996. "Clayoquot Sound Interim Measures Agreement Extended for 3 Years." http://www.luco.gov.bc.ca/.

Mabee, H.S., and G. Hoberg. 2004. "Protecting Culturally Significant Areas through Watershed Planning in Clayoquot Sound." *Forestry Chronicle* 80(2):229-40.

–. 2006. "Equal Partners? Assessing Co-Management of Forest Resources in Clayoquot Sound." *Society and Natural Resources* 19(10):875-88.

May, Elizabeth E. 1990. *Paradise Won: The Struggle for South Moresby*. Toronto: McClelland and Stewart.

NAFA (National Aboriginal Forestry Association). 1996. *Aboriginal Forest-Based Ecological Knowledge in Canada*. Ottawa: NAFA.

NAFA (National Aboriginal Forestry Association) and Wildlands League. 2003. *Honouring the Promise – Aboriginal Values in Protected Areas in Canada*. Ottawa: NAFA and Wildlands League.

Natcher, D., and S. Davis. 2007. "Rethinking Devolution: Challenges for Aboriginal Resource Management in the Yukon Territory." *Society and Natural Resources* 20:271-79.

Natcher, D., C. Hickey, M. Nelson, and S. Davis. 2009. "Implications for Tenure Insecurity for Aboriginal Land Use in Canada." *Human Organization* 68(3):245-57.

Notzke, C. 1995. "A New Perspective in Aboriginal Natural Resource Management: Co-Management." *Geoforum* 26(2):187-209.

Parks Canada. 2000. "Section D: Aboriginal Peoples and National Parks; 'Unimpaired for Future Generations?'" In *Protecting Ecological Integrity with Canada's National Parks: Report of the Panel on the Ecological Integrity of Canada's National Parks*. Ottawa: Minister of Public Works and Government Services.

Penikett, T. 2003. "The Haida Don't Let Go Easily." *Canadian Dimension* 37.

Pinkerton, E. 1989. "Introduction: Attaining Better Fisheries Management through Co-Management – Prospects, Problems, and Propositions." In *Co-Operative Management of Local Fisheries*, edited by E. Pinkerton, 3-36. Vancouver: UBC Press.

Price, K., A. Roburn, and A. MacKinnon. 2009. "Ecosystem-Based Management in the Great Bear Rainforest." *Forest Ecology and Management* 258:495-503.

RCAP (Royal Commission on Aboriginal Peoples). 1997. *Report of the Royal Commission on Aboriginal Peoples*. Ottawa: Indian and Northern Affairs Canada.

Robinson, C. 1998. "Local Indigenous Perspectives of Community and Co-Management Arrangements." Paper presented at Crossing Boundaries, seventh annual conference of the International Association for the Study of Common Property, Vancouver, June 10-14.

Robinson, Joanna L., D.B. Tindall, Erin Seldat, and Gabriela Pechlaner. 2007. "Support for First Nations' Land Claims amongst Members of the Wilderness Preservation Movement: The Potential for an Environmental Justice Movement in British Columbia." *Local Environment* 12(6):579-98.

Ross, M., and P. Smith. 2002. *Accommodation of Aboriginal Rights: The Need for an Aboriginal Forest Tenure.* Edmonton: Sustainable Forest Management Network.

Rusnak, G. 1997. *Co-Management of Natural Resources in Canada: A Review of Concepts and Case Studies.* Minga Working Paper no. 2. Working Paper Series of the IDRC Program Initiative: Minga – Managing Natural Resources in Latin America and the Caribbean. Ottawa: International Development Research Centre.

Sapic, T., U. Runesson, and M.A. Smith. 2009. "Views of Aboriginal People in Northern Ontario on Ontario's Approach to Aboriginal Values in Forest Management Planning." *Forestry Chronicle* 85(5):789-801.

Shaw, K. 2003. "Encountering Clayoquot." In *A Political Space: Reading the Global through Clayoquot Sound,* edited by W. Magnusson and K. Shaw, 25-66. Montreal and Kingston: McGill-Queen's University Press.

Singleton, S. 1998. *Constructing Cooperation: The Evolution of Institutions of Co-Management.* Ann Arbor: University of Michigan Press.

Spiro, H. 2003. *An Implementation Analysis of the Clayoquot Sound Scientific Panel Recommendations on First Nations Perspectives*. Vancouver: University of British Columbia.

Statistics Canada. 2001. "Aboriginal Peoples Census." http://www.statcan.ca/.

Stevenson, M.G. 2004. "Decolonizing Co-Management in Northern Canada." *Cultural Survival Quarterly* Spring 28(1).

Tennant, P. 1990. *Aboriginal Peoples and Politics: The Indian Land Question in British Columbia, 1849-1989.* Vancouver: UBC Press.

Treseder, L., and N. Krogman. 2000. *The Effectiveness and Potential of the Caribou – Lower Peace Cooperative Forest Management Board.* Edmonton: Sustainable Forest Management Network, 2000-19.

Weitzner, V., and M. Manseau. 2001. "Taking the Pulse of Collaborative Management in Canada's National Parks and National Park Reserves: Voices from the Field." In *Crossing Boundaries in Park Management: Proceedings of the 11th Conference on Research and Resource Management in Parks and on Public Lands,* edited by David Harmon, 253-59. Hancock, MI: George Wright Society.

Wilson, D.C., J.R. Nielsen, and P. Degnbol, eds. 2003. *The Fisheries Co-Management Experience: Accomplishments, Challenges, and Prospects.* London, UK: Kluwer Academic Publishers.

15

Changing Contexts

Environmentalism, Aboriginal Community and Forest Company Joint Ventures, and the Formation of Iisaak

Gabriela Pechlaner and D.B. Tindall

In the past two decades, forestry has been affected by two powerful yet at times contradictory social forces: a shift of public support toward the non-economic values of forestry and a growing recognition of Aboriginal rights. Although environmental concern fluctuates to some degree over time, in recent decades, publics have increasingly questioned the negative environmental impacts of conventional industrial practices, and environmental organizations have engaged in a growing number of protests against forestry operations (Tindall and Begoray 1993; Wilson 1998; Braun 2002; Tindall 2002). Second, as discussed elsewhere in this volume, in recent years First Nations in Canada have demonstrated a growing ability to assert the legal and moral legitimacy of their rights. In British Columbia, where no significant treaties had been signed prior to the 2000 Nisga'a treaty, this is a considerable factor affecting the stability of forestry operations and investor confidence.

These changing social forces have coincided to create a unique set of circumstances that may lead the way toward a new model of socially sustainable forestry. Both government and industry alike have begun to recognize that the current conflicts over land use are unlikely to resolve themselves. In words, if not always in practice, they acknowledge the need to address the First Nations and environmental claims for the good of economic stability. As a result, there has been a "growing phenomenon" of joint ventures and innovative partnerships with First Nations (NAFA and IOG 2000, 1).

In many cases, industry has been quicker than government to recognize the importance of partnerships with First Nations and to forge ahead with their formation prior to treaty settlement. At the same time, there have been an increasing number of First Nation alliances with environmental groups, often formed in opposition to unsustainable industry practices on disputed territory (Robinson et al. 2007; see Braun [2002] on the cultural dimensions of the wilderness preservation movement and its portrayal of First Nations). The nature of these alliances differs from band to band. Further, they are not limited to the forestry sector. Aquaculture, in particular, is another environmentally controversial rural-based industry with vastly differing alliances between First Nations, industry, and environmentalists.

This chapter is concerned with joint ventures that include all three groups – First Nations, environmentalists, and industry – and that act as an attempt to resolve their competing resource claims.[1] Such agreements are increasingly evident in British Columbia's forestry sector, where land use conflicts have been volatile.[2] Even without the mounting pressure of First Nation claims, the long-term sustainability of resource communities has become a concern for many in resource extraction. If successful, joint ventures that address such issues could resolve some significant land use conflicts prior to the slow-moving treaty process.

The first half of this chapter provides a general discussion of joint ventures: their context, goals, and structure. The second half looks at a specific joint venture from the Clayoquot Sound region in British Columbia: Iisaak Forest Resources (Iisaak). It is difficult to find a better example of the compromised business climate resulting from land use conflicts than that created by the volatile protests that erupted in Clayoquot Sound during the 1980s and 1990s. After the failure of many less-radical appeasement measures, Iisaak was ultimately created in 1993 through an agreement between the Nuu-chah-nulth First Nations and MacMillan Bloedel (later purchased by Weyerhaeuser), and with the support of environmental groups. Of particular interest in this venture is the manner in which First Nations, industry, and environmentalists interacted to develop a new model for forestry that incorporated the concept of social sustainability. Iisaak has existed for a number of years, yet its level of success in realizing this new model of forestry remains unclear. If nothing else, however, the continued development of the venture, and the ongoing support of groups previously in conflict with each other, is an indication of its importance to the transformation of the conventional model of forestry.

This chapter uses a case study design and focuses primarily on the context for the development of Iisaak, and its early existence – from the period of the 1990s up until 2003, when the research for this chapter was conducted. It does not cover in detail developments since 2003. Data were obtained through a review of available documents and through unstructured interviews with a purposive sample of key informants.

The Context for Joint Ventures

Canada has a long tradition as a staples exporting region (Innis 1954a, [1930] 1954b; Watkins 1963; Marchak 1983). During the period covered by this study, nationally, the forestry sector contributed almost 3 percent to the GDP (in 2001), for a total of $28.5 billion (NRC 2003). These figures are much higher if activities conducted in support of the industry are considered.

The forest industry contributed 352,800 direct jobs, primarily centred in Quebec, British Columbia, and Ontario (ibid.). The forestry sector is particularly important to British Columbia. Historically, forest products have been the province's most important export commodities. In 2001, the industry accounted for 90,600 direct employment positions in the province (ibid.). Considering forestry's economic impact, changes in the sector are important. Environmental protests over clear-cut logging in the 1990s were a constant and significant irritant to the sector, particularly in British Columbia. Although the effectiveness of such protests depends in part on the ebb and flow of public support (Tindall 2003), the increasing legal weight of First Nations land claims is another threat to the industry that is likely only to grow.

There are approximately 680 First Nations across Canada, in twelve jurisdictions (Bombay in Hagerman 1998, 369). British Columbia is distinct in that for most of its history the province persistently declined to address First Nations land claims. Effectively, treaties were not addressed until the current slow-moving BC treaty process that began in the 1990s (see Mark Stevenson, this volume). A long sequence of court rulings has now made ignoring these claims increasingly difficult, however. Recent court rulings not only have clarified the means by which they can be addressed but also have resulted in implications for interim resource management prior to settlement (see Trosper and Tindall, this volume). The Government of British Columbia has limited the available Crown land for treaty settlement to 5 percent (Lee and Symington 1997, 349); however, even this is a very significant amount of land, particularly with respect to rural-based resource extraction industries. Although a detailed outline of the number of court decisions that have shaped the current legal context is beyond the scope of this chapter, suffice it to say that, overall, they have progressively strengthened the legal status of claims to rights and to title.

First Nations' Economic Development and the Growing Phenomenon of Joint Ventures

The forestry sector is very important to the provincial and national economies, but it is even more important to those living in rural communities. First Nations are disproportionately rural based, with 80 percent of Canada's First Nation communities located within forested lands (NAFA and IOG 2000, 3). Opportunities for traditional lifestyles and living off the land have declined, yet alternatives have been slow to materialize. Unemployment in rural communities, which is already very high, is even higher for rural reserve communities, sometimes exceeding 90 percent (see Frideres, this volume).

Despite their rural proximity and their increasing interest in forestry, there are several barriers to First Nations' entry into the industry. Forestry has become a highly mechanized sector, with high capital and technological expertise requirements, and high-volume production necessary for break-even operation (Brubacher 1998, 353). Further, historically in British Columbia, there has been a requirement for the ownership of processing facilities as a prerequisite for obtaining timber harvest licences (Brubacher 1998). This is a significant barrier for newly economically developing First Nations. Despite such barriers to entry, however, the rural proximity of First Nations and the changing legal climate around resource access make forestry an important candidate for First Nations' economic development. The formation of partnerships or joint ventures with forestry companies can facilitate the potential for this development.

A joint venture arrangement is one in which the Aboriginal and non-Aboriginal parties are joint owners of the venture, with both parties bringing some form of contribution to the venture. For example, industry helps First Nations overcome barriers to entry into the sector by providing necessary capital, technology infrastructure, and marketing venues; First Nations can provide industry with "access to timber resources, a labour force, regional good will, and improved corporate image and political leverage" (NAFA and IOG 2000, 14). The contributions made by the parties are thus not limited to material contributions. Although certification is one industry motivation to enter into such an arrangement, community acceptance is also a significant contribution to industry stability that First Nations' support can provide. In the current context, "social licence" is a highly valuable commodity for the much-maligned forestry industry.

Because of its joint ownership structure, the joint venture allows the First Nation community to have a much greater degree of control over the nature and shape of economic development. M.A. (Peggy) Smith, technical advisor for the National Aboriginal Forestry Association, characterizes the difference between economic development determined externally – which often employs First Nations primarily in the capacity of loggers – and that determined internally:

> While there is nothing inherently wrong with being a logger, these type of large-scale, high-volume extraction industries have often depleted the resource in a short time, then moved on. Smaller scale, employment intensive, culturally appropriate, higher value added operations would provide long-term sustainable industries. (Smith in Hagerman 1998, 368)

Once large-scale forestry operations have depleted the resource, they follow the inherent logic of capital, and move on. First Nation communities lose under this mobility logic, and to the extent that they can shape economic development in favour of sustainable industries, their communities benefit. Joint ventures offer such potential.

The Growing Phenomenon of Joint Ventures

Over time, Aboriginal–industry partnerships have become increasingly important in forestry. In June 2000, a report from the National Aboriginal Forestry Association and the Institute on Governance identified five types of partnership relationships: joint ventures, cooperative business arrangements, forest services contracting, socioeconomic partnerships, and forest management planning (NAFA and IOG 2000). These differed in function, structure, and range from primarily economic agreements (e.g., Aboriginal harvesting marketed through non-Aboriginal channels) to more socially oriented ones (e.g., partnerships that are foremost community development and secondarily economic). Certification was one motivation for industry to undertake the latter.

As noted, the joint venture structure is ownership shared between both parties. This is significantly different from contracting out services. Examples of joint ventures in British Columbia include Tl'oh Forest Products, which operated an I-beam and finger-joint mill in Fort St. James; Ecolink, a silviculture service and logging operation in Alkali Lake; and West Chilcotin Forest Products, a lumber manufacturing company in Anahim Lake. In total, seventeen Aboriginal–industry partnerships were identified in British Columbia, nine of which were joint ventures (NAFA and IOG 2000). In the region with the second-highest number of such partnerships (Ontario), none was a joint venture. Instead, the emphasis was on socioeconomic partnerships – for example, as a requirement of certification (ibid.). The number of joint ventures in British Columbia is arguably one consequence of the growing concern over unsettled treaties.

The Context for Iisaak

Clayoquot Sound is located on the west side of Vancouver Island. Geographically, its boundaries vary depending on perspective, but the region covers approximately 350,000 hectares and comprises remote islands, fjords, watersheds, and the largest nearly intact temperate rainforest ecosystem on Vancouver Island. This last point was raised extensively during the height of the Clayoquot Sound controversy. As home to endangered ancient rainforest, it is considered by some to be a priceless natural heritage, and clear-cut logging is seen as an affront to this. The Clayoquot Sound region is also home to the Nuu-chah-nulth First Nations. There are fifteen Nuu-chah-nulth, of which three Central Region groups – the Hesquiat, Ahousaht, and Tla-o-qui-aht First Nations – are within the Clayoquot Sound region, as defined by the 1993 Clayoquot Sound Land Use Decision (Clayoquot Sound Research Group 2003). The region also includes the non-Aboriginal community of Tofino. The non-Aboriginal communities of Ucluelet and Port Alberni are in very close proximity and are intricately bound up in the economics, culture, and politics of the region.

In addition to its importance to environmentalists and industry, the region has immediate economic importance for those who live there. With the decline in traditional resource extraction industries, the economic underpinnings of British Columbia's coastal communities have crumbled. Successive fishing fleet reduction plans beginning in 1968 devastated the fishing and fishing-related industries, while forestry suffered at least two decades of economic decline. Even after decades of hardship in the resource sector, the Alberni-Clayoquot regional district still had a high level of income dependency on the primary industries of forestry, fishing, and trapping. Around the time of this study, the district ranked fifth out of twenty-seven regional districts for economic dependency on the primary sector, with 39 percent dependency in 1996. Thirty-six percent of this income dependency fell on the forestry sector (versus 9 percent for British Columbia overall) and 3 percent dependency on fishing and trapping (versus 1 percent for British Columbia) (BC Stats 2003, 3). Over the previous two decades (prior to the formation of Iisaak), the provincial GDP contribution from forestry dropped from 9 percent to 6 percent; employment in the industry dropped from 7 percent to 5 percent (Hallin 2001, chap. 4, 2). With respect to the fishing industry, Ucluelet and Tofino were among the fifteen communities hardest hit by the 1995 fishing fleet reduction plan. Ucluelet suffered a 4 percent decline, and Tofino suffered a 3 percent decline in jobs as a percentage of community employment (GS Gislason and Associates 1998). This loss of high-wage employment is particularly difficult in rural communities with few alternative options and where education levels are typically low. For example, in 1996, 33.7 percent of the population in the Alberni-Clayoquot regional district were without high school completion (BC Stats 2003, 5).

The economic stability of the region is important to note not only as a context to the joint venture's formation but also because one goal of Iisaak was social sustainability, not just environmental preservation. A complete retraction of logging in the area for the sake of environmental preservation would not allow for social sustainability without a viable economic replacement. Although Tofino has a long-standing claim as a tourist destination, it is questionable whether tourism could fill the economic gap for the whole region.

Conflict in Clayoquot Sound

The history of the Clayoquot Sound conflict can be characterized by protests and blockades over logging, followed by extensive studies and the formation of task forces and committees to study the problem, continued logging, and then more protests and blockades, to start the cycle again. The difference between this history and many similar ones is that this familiar cyclical pattern was broken by the force of mass public opposition to clear-cutting in the region (Shaw 2003). Karena Shaw (2003, 27) describes the politically engaged Tofino residents in these early years as united in their effort to challenge "the narrative" of logging such that

> the Sound would be clearcut logged by large multinational logging corporations ... that the profits from this logging would flow to the urban headquarters of these corporations; and that the local inhabitants – Nuu-chah-nulth and non-Natives – would be left to carry on with their local affairs, coping with whatever impacts (positive or negative) the logging had on their livelihoods.

Although the provincial government established the Clayoquot Sound Scientific Panel in a further attempt to resolve the conflict, it also shifted the forum of the conflict through its acceptance of the Nuu-chah-nulth First Nations as one of the first Nations to begin treaty negotiations with (Shaw 2003, 42). By March 1994, this acceptance was accompanied by the signing of a two-year Interim Measures Agreement (IMA) between the provincial government and the five First Nations of the Nuu-chah-nulth Central Region: the Ahousaht, Hesquiaht, Tla-o-qui-aht, Toquaht, and Ucluelet First Nations. The IMA acknowledged that "the Ha'wiih [hereditary chiefs] of the First Nations have the responsibility to conserve and protect their traditional territories and waters for generations to which will follow" (Iisaak 1999, 4); therefore, it granted joint management of these territories until completion of the treaty negotiations (see Mabee et al., this volume).

A New Company

The signing of the IMA represented a fundamental shift in the politics of the sound. Although both MacMillan Bloedel and Interfor were logging in the region in 1992, MacMillan Bloedel was by far the larger operator. After the early mass-protest year of 1993, the campaigns shifted to the international forum, facilitated by Greenpeace International and other environmental organizations. This strategy successfully threatened MacMillan Bloedel's customers with consumer boycotts, resulting in cancellation of a number of pulp and paper contracts (Cashore, Vertinsky, and Raizada 2000, 101). More importantly, it began to threaten the confidence of the heretofore unaccommodating MacMillan Bloedel.

MacMillan Bloedel was already suffering serious economic losses in the sound. In 1995, when the scientific panel presented its reports on sustainable management in Clayoquot Sound, its recommendations were fully accepted by the province, thus further reducing the potential for logging. In 1996, MacMillan Bloedel had losses of $7 million on its Clayoquot Sound operations, harvesting only 52,000 cubic metres of a planned 100,000 (Cashore, Vertinsky, and Raizada 2000, 103). Subsequently, MacMillan Bloedel announced a temporary shutdown of its logging operations. Ultimately, this campaign stimulated a corporate transformation and brought MacMillan Bloedel to an acknowledgement of the need to change (Cashore, Vertinsky, and Raizada 2000).

The idea of a joint venture between MacMillan Bloedel and the Central Region First Nations had been raised already in 1994 in the Interim Measures Agreement, but it had not progressed. In 1996, an extension to the IMA was signed, in part of which committed MacMillan Bloedel and the Central Region Nuu-chah-nulth to begin serious negotiations toward the formation of a joint venture company. The stated goal of the partnership was to be based on a "shared commitment to achieving change" (British Columbia, Aboriginal Affairs Branch 1996, IMEA Schedule II, 1).

In 1997, Ma-Mook Natural Resources was founded in order to represent the collective economic interests of the five Nuu-chah-nulth Central Region First Nations. An agreement was reached between MacMillan Bloedel and Ma-Mook, and Iisaak Forest Resources. Iisaak was signed into existence in November 1998. Iisaak was 51 percent owned by the Nuu-chah-nulth, with MacMillan Bloedel's share capped at 49 percent. Weyerhaeuser subsequently bought out MacMillan Bloedel's interest. Finally, in 2005, the Nuu-cha-nulth First Nations purchased the outstanding interest from Weyerhaeuser, and Iisaak became 100 percent First Nations owned (and thus no longer a joint venture).

The name Iisaak is Nuu-chah-nulth for "respect" and is in keeping with the new company's commitment to the concept of Hishuk-ish ts'awalk, "the Nuu-chah-nulth belief of respecting the limits of what is extracted and the interconnectedness of all things" (Iisaak 1999, 6). The goals of the new company were to "develop and deliver new, innovative ways of managing the resources of Clayoquot Sound which respect cultural, spiritual, recreational, economic and scenic values" (ibid.). Essentially, Iisaak was to embrace a new model of socially sustainable forestry.

The Changing Terrain of Sustainability

Although Clayoquot Sound had its share of pro- and anti-forestry divisions, it was also a location where there was a growing, general, societal recognition that industry's exit was more a question of when rather than whether, and that community members may well be worse off for industry having operated there. The issue of logging in the Meares Island watershed is a case in point (Tennant 1990). In consequence, the environmental call for "sustainable" forestry shifted beyond the traditional ecological concept of limiting harvest extraction rates to not exceed production, to include the idea of social sustainability. This transition is replicated outside the sound. In June 2002, for example, loggers in the Queen Charlotte Islands walked off the job to ally with the Haida against Weyerhaeuser's control of the island's resources. The loggers were afraid that "if they do nothing, they will share the fate of resource towns like Tahsis, Gold River and Youbou: Vancouver Island communities devastated when the main forest industry employer shut down" (Hamilton 2002).

Social sustainability means that the resource must be environmentally and economically sustainable, as well as sustainable for the communities that depend on it. Unfortunately, there is as yet no governmental or academic consensus regarding the definition of social sustainability (CCFM 1997, 105). Further, even within a given group or institution's definition, there is difficulty establishing exactly what it entails empirically (Nadeau, Shindler, and Kakoyannis 1999; Beckley, Parkins, and Stedman 2002; Beckley et al. 2008). A 1997 technical report by the Canadian Council of Forest Ministers (CCFM 1997) identified five elements necessary for a socially sustainable forest management practice: (1) the incorporation of First Nations and treaty rights (i.e., legal obligations), (2) the participation of First Nation communities in sustainable forest management, (3) the sustainability of forest communities, (4) management based on fair and effective decision making, and (5) management based on informed decision making (CCFM 1997, Criterion 6.0). Most definitions of social sustainability emphasize the interdependence of human and ecological values.

Iisaak Forest Resources

It is just such a balance of human and ecological needs that Iisaak proposed to strike, conducting forestry in a manner that was environmentally, culturally, socially, *and* economically sustainable. If successful, it could set a worldwide standard for innovative forest management. The Iisaak business model has three core sections: timber, nontimber, and conservation values. As a forestry company, Iisaak obviously intended to continue logging in the region. However, the company made several commitments toward sustainability that distinguished it from conventional forestry companies.

First and foremost, while emphasizing the need for adaptive management, Iisaak uses the Clayoquot Sound Scientific Panel recommendations as the technical basis for its forestry management approach (Iisaak 1999, 20). Iisaak's ongoing operations include several provisions for increased environmental sustainability: higher forest retention, reduced road construction, and transition to second-growth logging. Iisaak has also committed to not log in the pristine valleys of Clayoquot Sound.[3] Therefore, environmentally, Iisaak represents a switch to conservation-based forestry. Instead of maximizing profit through maximizing extraction (and utilizing economies of scale), Iisaak proposed to profit by emphasizing quality, diversifying to more products than just timber, and stressing value-added production – and by an overall emphasis on the environment as an economic good. Thus, Iisaak represents a transition from volume-based forestry to value-based forestry, with the actual volume of harvest determined by "an application of the principles of sustainable ecosystem management outlined in the Scientific Panel" (Iisaak 1999, 22).

Subsequent to the formation of Iisaak, a memorandum of understanding (MOU) was signed between the company and five of the six key environmental groups that had previously organized against MacMillan Bloedel: Greenpeace Canada, Greenpeace International, Natural Resources Defense Council, Sierra Club of BC, and the Western Canada Wilderness Committee. Under the MOU, these five groups committed to "supporting Iisaak's operations, actively engaging in promotion of markets for products produced by Iisaak, and developing ongoing mechanisms for sustaining cooperation" (Iisaak 1999, 5). The MOU therefore represented not only an industry's commitment to innovative forestry, but it also demonstrated a new role for environmental groups. The sixth group, the locally based Friends of Clayoquot Sound, while supportive of the venture, remained external to the MOU in a watchdog capacity.[4]

Iisaak's goals were not strictly environmental, however, but also included a fundamental shift in attitude toward the social environment in which forestry is practised. First and foremost, Iisaak reversed the traditional role played by First Nations. Before it became solely owned (in 2005) by the First Nations (and was thus still a joint venture), the board of directors of Iisaak consisted of two Weyerhaeuser representatives and three Nuu-cha-nulth representatives, therefore retaining the balance of control in the local (First Nations) communities (Iisaak 1999, 21).[5] Further, consistent with the share structure of the joint venture company, the structure and goals of the new joint venture were to

> create training and employment opportunities for First Nations in forest-related activities that will foster economic initiative and independence in First Nations communities in Clayoquot Sound and help provide sustainable long-term employment for both First Nations and local communities. (British Columbia, Aboriginal Affairs Branch 1996, IMEA Schedule II, 1)

This commitment to increased First Nations participation was accompanied by a commitment to honour First Nations cultural values. Monitoring for culturally sensitive sites is a key component of this. Iisaak would also require permission of any nation within whose territory it would like to conduct its harvesting.

Economic development is considered by some to have slipped in priority below ecological protection in the scientific panel recommendations, yet job creation and increasing community stability remained one of Iisaak's stated objectives. This was not limited to First Nations' involvement – though this remained a priority – but extended to the broader, non-First Nation community. The maintenance of visual quality, for example, allowed for recreation and tourism development opportunities. Thus, the prevailing sentiment in tourism-based Tofino was generally supportive of the venture. Although further developments of this kind were intended, the goal was diversification, rather than a shift from a forestry- to a tourism-based economy. In the hope of attracting new small-manufacturing operations, for example, at the time of our research, Iisaak had committed 30 percent of its timber volume harvest to local value-added businesses (Iisaak 1999, 18).

Iisaak's socially sustainable forestry goals therefore represented an attempt to bridge the divide between many traditionally oppositional groups. Considering the diversity of goals underpinning the new enterprise, innovation and flexibility would be required. Consequently, ongoing monitoring and adaptive management were incorporated as an integral part of the operation. Lastly, while attempting to amalgamate these timber and non-timber forest values into a new model for conservation-based forestry, Iisaak remained a business enterprise and, as such, it intended to be profitable. Iisaak began its operations in Clayoquot Sound in August 2000.

A Lasting Development?

Given the differing motivations that could be behind any party's participation in a joint venture, some consideration needs to be given to whether they might change over time. For industry and First Nations, the motivations are likely to remain stable in the near future. Statistics provided through the course of Iisaak's formation cite 70 percent unemployment in the region's Native communities (Iisaak 1999, 8). Earlier attempts at integration into the developing tourism economy had resulted in conflict with the broader community. Therefore, economic development was key for local First Nations. At the same time, First Nations involvement brought social licence to industry, where it was blatantly obvious that conventional operations were otherwise unlikely to proceed profitably. Further, the public relations impact was a direct threat to the profitability of MacMillan Bloedel itself.

While not meaning to diminish the real business partnerships that existed between MacMillan Bloedel/Weyerhaeuser and the Nuu-chah-nulth, it is fair to suggest that the outcome might have been different without the organized support and international pressure tactics of the environmental movement (Robinson et al. 2007). At the very least, it would have taken longer to achieve. Therefore, given the role of environmental groups not only in impacting industry's bottom line but also in maintaining an ongoing presence in the joint venture, their goals and motivations are important to consider for the future of the venture. Although environmental organizations could have continued to struggle for a complete cessation of logging in the area, such as through the creation of a park, this approach would likely have alienated the First Nations that were in need of some form of economic development. Further, environmentalists had begun to raise concerns that focused attention on site-specific preservation, which can lead to the neglect of other valuable areas. Shaw, for example, discusses how environmentalists reacted to MacMillan Bloedel's cessation of logging in 1997:

> By making Clayoquot their "poster child," the environmentalists realized that they might only have transferred the basic problem of excessive and damaging clear-cut logging to other parts of British Columbia. The problem that had brought them to the barricades had been addressed only superficially, as a conflict over the future of a particular area, rather than structurally, as a context over the future of forestry in British Columbia. (Shaw 2003, 50)

In Canada, out of the 1,027,711 hectares harvested in the year 2000, almost 93 percent, or 924,188 hectares, were clear-cut. In British Columbia, the statistics were similar, with 193,177 hectares clear-cut out of a total of 204,472 hectares harvested, or almost 95 percent (Statistics Canada 2003). Consequently, environmentalists had and continue to have a strong motivation to support a new model that attempts to address their structural concerns with forestry – one that could make Clayoquot Sound "the leading global example of ecologically sensitive harvesting techniques" (British Columbia, Aboriginal Affairs Branch, 1996, IMEA Schedule II, 1).

Social Sustainability: Did It Work?

The most immediate question raised regarding the formation of Iisaak Forest Resources is, did it work? Did the venture resolve the tensions in the First Nations – environmental movement – industry dynamic and achieve a new form of socially sustainable forestry? One way to answer this question would be to undertake a social impact assessment. As noted, however, operationalized indicators of social sustainability are clearly still in development and little data is available, as "the social dimension of sustainable forestry has only [relatively] recently become a priority for government and university researchers" (CCFM 1997, 105). The effectiveness of such an approach is further limited by Iisaak's political context: as an Iisaak representative noted, when conventional forces would be happy to see them fail, there is limited motivation to submit to the potentially damaging process of a quantitative assessment. Lastly, a quantitative assessment can be effective only when assessed in the context of the numerous contributing factors and influences.

The latter is particularly salient given that the softwood lumber dispute pushed many already-stressed coastal forestry companies to the breaking point (and, more recently, many operations have been hurt by the rise of the Canadian dollar, and by the recession in general, the related downturns in the US economy, and the US housing market in particular). Due to a number of internal and external factors, British Columbia's coastal forest industry had reached a twenty-year low at the time of our study, matched only by the recession of the 1980s in mill closures and layoffs (Hamilton 2003, A1). Weyerhaeuser's July 2003 layoffs brought the number of coastal mills that were either "down or with curtailed operation" to thirty-three out of thirty-five (ibid., A12). In this context, indicators such as unemployment rate, household income, and even migration patterns may be less revealing of the impact of a venture than of the external forces acting on it. For a new company such as Iisaak, with little history to act as reference, it is particularly difficult to separate external from internal factors.

These limitations to an indicator approach to sustainability support the growing idea of process approaches (Beckley, Parkins, and Stedman 2002; Beckley et al. 2008). Further, to some extent, the achievements of socially sustainable initiatives are not exclusively related to outcome measurements. Rather, "the social process of decision-making and management can be as important to society as the social outcomes" (Sheppard 2001, 1).

Initial Operations, and a Recent Update

Iisaak defines its overall goal as "building a business case for conservation" (Iisaak, email communication, August 15, 2003). Its goals for forestry management include undertaking "economically viable, socially acceptable, ecologically defensible and culturally relevant practices that are designed to achieve conservation as a primary objective" (ibid.). Without a model for such a socially sustainable forestry company, its early progress was slow. It is a model they are still building. At the same time, Iisaak is subject to the same needs as other forestry companies. It is not subsidized and needs to make a profit in order to continue. Given the difficulties involved in its undertaking, at times, Iisaak has considered its sheer survival its greatest success (ibid.).

Iisaak began logging in Clayoquot Sound in August 2000. In its first year of operation it harvested approximately 10,000 cubic metres of primary harvest and 2,500 cubic metres of salvage; in 2001, Iisaak harvested 1,500 cubic metres of primary harvest and 10,000 cubic metres of salvage; and in 2002, it harvested almost 34,000 cubic metres of primary harvest and 12,000 cubic metres of salvage (Iisaak 2003c). For several years, Iisaak struggled economically.

In 2006, Ecotrust Canada, and its partner, Triumph Timber, signed a contract to manage Iisaak and implement a turnaround strategy for the company. By the end of 2007, the following accomplishments had been realized (Ecotrust et al. 2007):

- About 85,000 cubic metres of timber were harvested in 2007.
- 47 percent of jobs were First Nations.

- 67 percent of jobs were local.
- FSC certification was reinstated.
- The company was financially profitable.

Although it may be too early to assess the long-term likelihood of success, by 2007, Iisaak's operations appeared to be economically viable. Further, during the period since it began operations, Iisaak's forestry operations supported its other (social, environmental, and cultural) goals. As noted, perception remains an important factor. Iisaak has not managed to magically brush away the years of land use conflict, and increasing community consensus remains (and likely will remain) a long-term goal. However, in its few years of operation, Iisaak has managed a number of steps toward its goals.

On February 10, 2000, Clayoquot Sound was designated a UNESCO Biosphere Reserve, as part of the Man and Biosphere Programme. This designation requires a high level of community commitment toward its goals. Rather than assigning a strictly ecological designation, the UNESCO program seeks to strike a balance between human and ecological needs. Research is key to achieving this balance, and one of the benefits of the UNESCO designation is the allocation of seed money ($12 million from the federal government) for research, education, and training projects related to the reserve.

Another key aspect of Iisaak's intended shift from volume- to value-based forestry was to apply for Forest Stewardship Council (FSC) certification. FSC certification indicates that a company is managing its forests in compliance with a number of environmental and social criteria: long-term security for the forest; maintenance of environmental functions; sustained yield forestry production; positive impact on local communities; and the existence of a system for long-term forest management planning, management, and monitoring (Smartwood 2003). This is a market-driven approach, charging consumers a premium for these sustainable forest products. Although there are a number of certification schemes, not all provide for the recognition of indigenous rights, as does FSC certification and Canadian Standards Association (CSA) certification. In July 2002, Iisaak gained FSC certification.[6]

Facilitating the building of skills for conservation-based forestry in the community is one of the goals Iisaak is working toward. Less than 1 percent of Canadian Aboriginal postsecondary students were enrolled in natural resource management programs in 1992 (NRC 2003). Iisaak is also working at increasing the amount of employment it provides. In 2003, Iisaak had thirteen full-time staff, eleven of whom were First Nations.[7] Iisaak also launched a three-year Iisaak Sustainable Forestry Project (ISFP), which had two main objectives: the "iterative analysis of criteria-based indicators, to define and test sustainable forest management practices on the ground" and the "long-term building of First Nations capacity in resource management" (Iisaak 2003a). The project provided for the employment and training of a five-person crew in a number of activities. These include: ecological monitoring, cultural inventories, engineering, and silviculture (Iisaak, email communication, August 15, 2003). Cross-training allowed for these positions to offer year-round employment rather than simply seasonally.

Iisaak has also worked in partnership with the University of British Columbia toward establishing a methodology for measuring carbon sequestration on land covered by its tree farm licence. When the methodology is established, this will be one of the activities that the ISFP crew will be trained to undertake. Once the impact of Iisaak's forest management on carbon sequestration can be established (for example, versus the impact of clear-cutting), the carbon credits can be marketed under the Kyoto protocol. This would provide an environmental source of income for the undercapitalized business.

Another method of increasing profits would be to get as close to the end consumer as possible, and building the Iisaak brand remains an important goal. This requires establishing a consistent supply of timber, and with this supply in hand forging relationships with "primary breakdown mills and value added manufacturing facilities" (Iisaak, email communication, August 15, 2003). Iisaak is working toward this objective.

With respect to its economic diversification goals, Iisaak is interested in supporting various initiatives, such as eco-tourism and environmental services. Whereas Iisaak normally maintains a support and marketing function for such nontimber initiatives, the Ma-Mook portion of the enterprise was 100 percent owner of Clayoquot Sound Wildfoods. This initiative used ingredients found in the Clayoquot Sound temperate rainforest, which were then "harvested, produced, packaged and marketed" by the people of the Nuu-chah-nulth Central Region (Clayoquot Sound Wildfoods 2003).

Continuing on the path of the MOU, Iisaak built on its strategic alliances with environmental organizations. The World Wildlife Fund, for example (not one of the original signatories of the MOU), assisted Iisaak in reaching its FSC certification goals. Strategic alliances were also made with Clayoquot Biosphere Trust and Ecotrust Canada (Iisaak, email communication, August 15, 2003). Iisaak also seeks strategic alliances with other organizations and institutions, such as is evidenced by the UBC carbon sequestration project, in order to "benefit from a broad range of expertise" (Harkin, Wong, and Bull 2003a, 2003b; Iisaak, email communication, August 15, 2003; Wong and Bull 2003).

Iisaak has not been free from criticism, most notably for not providing sufficient socioeconomic benefits. Typically, these critiques come from those with ties to the conventional logging sector, the latter of whom are part of a declining workforce. According to Iisaak, these benefits still exist but are distributed differently because of the different skills required by more ecological harvesting techniques, such as helicopter logging. One informant familiar with the communities and operations in the region told us that support for Iisaak in Ucluelet is typically low compared with the conventional forestry still practised in the sound by Interfor because Iisaak represents a decrease in employment and, further, draws much of its employment from outside the local community.[8] Iisaak also at times has received critique from the environmental community (Garcia 2006; Hume 2006a, 2006b). Obviously, the company still has inroads to make toward community cohesion.

Although it would be impossible to make any transition from the industrial model without employment disruption, the transition away from economic reliance on high-volume resource extraction industries is one that is already being replicated up and down the coast. Ironically, at the same time as it faces these socioeconomic criticisms, Iisaak is under increased scrutiny from the watchdog environmental group Friends of Clayoquot Sound (as well as other environmental groups) for its slowly increasing rate of harvest and its apparent intentions to build new logging roads and pursue new logging operations (FOCS, personal communication, August 18, 2003; FOCS 2010; Mychajlowycz 2009; Pynn 2010, 2011; Hume 2011; Lewis 2011).[9]

Conclusion

In this chapter we have focused on the context that surrounded the formation of Iisaak (a forest company – First Nations joint venture). This context included changes in First Nations rights, changes in public opinion, changes in the forest industry, and pressure created by environmental organizations and related changes in public opinion. Iisaak was developed as a forest company that explicitly set out to do forestry in a different way. In particular, it was tasked with taking First Nations values and environmental values into serious consideration, while also trying to be profitable as a business. Relatedly, for the Nuu-chah-nulth First Nations, it attempted to address social sustainability and environmental sustainability concerns at the same time as it struggled to be economically sustainable. This venture arose out of a relatively unique set of circumstances, yet still may serve as a model for other communities and companies.

Notes

1 "Environmentalists" refers to environmental nongovernmental organizations, and members of the environmental movement.

2 For example, the David Suzuki Foundation supported the General Protocol Agreement on Land Use Planning and Interim Measures (North Coast), and the Great Bear Rainforest Agreement (Central Coast).

3 At the time this manuscript was going to press, there was evidence that Iisaak may have to backtrack on this commitment. In 2009, a report by the Friends of Clayoquot Sound (Mychajlowycz 2009, 13) noted that "both Iisaak and Ma-Mook Coulson have stated that by 2010 they will run out of wood to cut in the partly logged watersheds and will need to begin logging in the remaining intact watersheds in order to maintain their required cut

levels." In its fall 2010 newsletter, and in a posting on its website, FOCS stated that Iisaak was surveying an old-growth valley on Flores Island, and that logging could begin in 2011, which FOCS noted would be a violation of the 1999 MOU signed by Iisaak and environmental organizations. In 2011 both Iisaak and FOCS issued competing media statements about this issue. In April 2011 Iisaak received a permit from the BC government to build logging roads on Flores Island. According to FOCS, Iisaak has since withdrawn the cut permit application.

4 Friends of Clayoquot Sound have continued to serve in a watchdog capacity regarding Iisaak's activities, logging more generally in Clayoquot Sound, as well as other activities such as fish farming and mining. Various updates and reports are provided at www.focs.ca.

5 As noted earlier, MacMillan Bloedel, the original corporate partner, was bought out by Weyerhaeuser during the period this study documents. In 2005, the Nuu-chah-nulth First Nations obtained Weyerhaeuser's interest in Iisaak and became the sole owner of the company (Ecotrust, Iisaak, and Triumph Timber, 2007).

6 In September 2006, Iisaak's FSC certification was temporarily suspended because the company was not meeting FSC's standard for business management performance; it was reinstated in August 2007 (Ecotrust et al. 2007).

7 By 2007, forty-three jobs in Iisaak (including managers, employees, and contractors) were held by First Nation people (Ecotrust et al. 2007).

8 By contrast, recent figures from Ecotrust et al. (2007) state that 67 percent of all managers, employees, and contractors are local residents.

9 However, there is little doubt that FOCS prefers the operation of Iisaak to that of Interfor. On August 13, 2003, FOCS protested against Interfor's refusal to state its intention regarding logging in pristine areas. For recent information and evaluation about Iisaak's performance, and assessments of the forest lands that it manages, see Iisaak (2008), Mychajlowycz (2009), Tripp and Butt (2010).

References

BC Stats. 2003. "Regional District 23 – Alberni-Clayoquot Statistical Profile." Socio-Economic Profiles. April. http://www.bcstats.gov.bc.ca/.

Beckley, Thomas M., Diane Martz, Solange Nadeau, Ellen Wall, and Bill Reimer. 2008. "Multiple Capacities, Multiple Outcomes: Delving Deeper into the Meaning of Community Capacity." *Journal of Rural and Community Development* 3(3):56-75.

Beckley, Thomas, John Parkins, and Richard Stedman. 2002. "Indicators of Forest-Dependent Community Sustainability: The Evolution of Research." *Forestry Chronicle* 78(5):626-36.

Braun, Bruce. 2002. *The Intemperate Rainforest: Nature, Culture and Power on Canada's West Coast.* Minneapolis: University of Minnesota Press.

British Columbia. Aboriginal Affairs Branch. 1996. *Clayoquot Sound Interim Measures Extension Agreement: A Bridge to Treaty.* http://www.aaf.gov.bc.ca/.

Brubacher, D. 1998. "Aboriginal Forestry Joint Ventures: Elements of an Assessment Framework." *Forestry Chronicle* 74(3):353-58.

Cashore, B., I. Vertinsky, and R. Raizada. 2000. "Firms' Responses to External Pressures for Sustainable Forest Management in British Columbia and the Pacific Northwest." In *Sustaining the Forests of the Pacific Coast: Forging Truces in the War in the Woods,* edited by D. Salazar and D. Alper, 80-120. Vancouver: UBC Press.

CCFM (Canadian Council of Forest Ministers). 1997. *Criteria and Indicators of Sustainable Forest Management in Canada*. Technical report. http://www.nrcan.gc.ca/.

Clayoquot Sound Research Group. 2003. "What Is Clayoquot Sound?" Website: *A Political Space: Reading the Global through Clayoquot Sound.* Available at: http://web.uvic.ca/clayoquot/home.html.

Clayoquot Sound Wildfoods. 2003. "About Us." http://clayoquotsoundwildfoods.com.

Ecotrust, Iisaak, and Triumph Timber. 2007. *Iisaak: Wood with Respect. 2007 Report.* Vancouver: Ecotrust, Iisaak, and Triumph Timber. Available online: http://fngovernance.org/resources_docs/Iisaak_Wood_With_Respect.pdf. Last accessed: May 3, 2012. http://www.ecotrustcan.org/.

FOCS (Friends of Clayoquot Sound). 2010. "Fall 2010/Winter 2011 newsletter" *Friends of Clayoquot Sound*. www.focs.ca.

Garcia, Diego A. 2006. "The Clayoquot Betrayal." *Globe and Mail*, August 7, A10.

GS Gislason and Associates. 1998. *Fishing for Money: Challenges and Opportunities in the BC Salmon Fishery.* Final report. Prepared for the BC Job Protection Commission. June 10.

Hagerman, E. 1998. "The National Aboriginal Forestry Association: An Interview with Harry Bombay and Peggy Smith." *Forestry Chronicle* 74(3):367-69.

Hallin, L. 2001. "A Guide to the BC Economy and Labour Market." BC Stats, Human Resources Development Canada, and British Columbia Ministry of Advanced Education, contributing partners. http://www.guidetobceconomy.org/.

Hamilton, G. 2002. "QC Loggers Walk Off Job, Join Haida." *Vancouver Sun*, June 4. Available online at: Creative Resistance, http://www.creativeresistance.ca/.

–. 2003. "Another 2,200 Layoffs Shut Down Coast Logging: Weyerhaeuser Feels Bite of Slumping Wood Markets." *Vancouver Sun*, final edition, June 25, A1, A12.

Harkin, Z., A. Wong, and G.Q. Bull. 2003a. *Principles of Carbon Accounting and Forest Carbon Inventory: Iisaak Forest Resources Ltd. – Report No. 2.* Vancouver: Department of Forest Resources Management, University of British Columbia.

–. 2003b. *Planning and Field Measurement Procedures for a Forest Carbon Inventory: Iisaak Forest Resources Ltd. – Report No. 3.* Vancouver: Department of Forest Resources Management, University of British Columbia.

Hume, Mark. 2006a. "Clayoquot Opened to Logging: Furious Environmental Groups Threaten Resumption of BC's 'War of the Woods.'" *Globe and Mail*, Vancouver edition, August 2, A1, A5.

2006b. "Green Coalition Shocked by Clayoquot Forest Plan." *Globe and Mail*, August 3, A4.

–. 2011. "The Sound and the Fury." *Globe and Mail*, April 11, S1.

Iisaak (Iisaak Forest Resources). 1999. *Iisaak*. Ucluelet, BC: Iisaak.

–. 2003a. *The Iisaak Sustainable Forestry Project: Summary Description.* Ucluelet, BC: Iisaak.

Iisaak Forest Resources Ltd. [Iisaak]. 2003b. Email Communication with Iisaak Representative. August 15, 2003.

–. 2003c. "Iisaak Harvest for 2000, 2001, and 2002." Unpublished document. Ucluelet, BC: Iisaak.

–. 2008. "2008 Annual Monitoring, Sustainable Forest Management and Adaptive Management Report."PDF available online at http://www.iisaak.com/documents/rad2008_%20Monitoring_SFM_AdMan_Report.pdf Ucluelet, BC: Iisaak. www.iisaak.com.

Innis, Harold A. 1954a. *The Cod Fisheries.* Toronto: University of Toronto Press.

–. (1930) 1954b. *The Fur Trade in Canada.* Toronto: University of Toronto Press.

Lee, C.A., and P. Symington. 1997. "Land Claims Process and Its Potential Impact on Wood Supply." *Forestry Chronicle* 73(3):349-52.

Lewis, Dan. 2011. "Flores Island Flashpoint." *Friends of Clayoquot Sound*, Fall 2011/Winter 2012:6.

Marchak, M. Patricia. 1983. *Green Gold: The Forest Industry in British Columbia.* Vancouver: UBC Press.

Mychajlowycz, Maryjka. 2009. "Overview of Logging in Clayoquot Sound: 2000-2009." Tofino, BC: Friends of Clayoquot Sound, www.focs.ca/logging/.

Nadeau, Solange, Bruce Shindler, and Christina Kakoyannis. 1999. "Forest Communities: New Frameworks for Assessing Sustainability." *Forestry Chronicle* 75(5):747-54.

NAFA and IOG (National Aboriginal Forestry Association and Institute on Governance). 2000. *Aboriginal-Forest Sector Partnerships: Lessons for Future Collaboration.* Ottawa: National Aboriginal Forestry Association. www.nafaforestry.org/.

NRC (National Resources Canada). 2003. "Statistics on Natural Resources: Statistics and Facts on Forestry." July. Natural Resources Canada, http://canadaonline.about.com/.

Pynn, Larry. 2010. "Green Groups Blast Plan to Log Old-Growth Forest on Flores Island." *Vancouver Sun*, December 4, A7.

–. 2011. "First Nations Logging Company Gets Flores Island Road Permit: Fears Raised That Area's Old-Growth Forest May Be at Risk." *Vancouver Sun*, April 6, A4.

Robinson, Joanna L., D.B. Tindall, Erin Seldat, and Gabriela Pechlaner. 2007. "Support for First Nations' Land Claims amongst Members of the Wilderness Preservation Movement: The Potential for an Environmental Justice Movement in British Columbia." *Local Environment* 12(6):579-98.

Shaw, K. 2003. "Encountering Clayoquot." In *A Political Space: Reading the Global through Clayoquot Sound,* edited by W. Magnusson and K. Shaw, 25-66. Montreal and Kingston: McGill-Queen's University Press.

Sheppard, S. 2001. "Would You Know a Socially Sustainable Forest If You Saw One? Why a Results-Based Approach May Not Be Enough." UBC Faculty of Forestry Jubilee Lecture Series. November 6. http://forestry.ubc.ca/.

Smartwood. 2003. "Certification Guidelines and Applications." www.smartwood.org.

Statistics Canada. 2003. "Forest Land Harvested and Clearcut." January 8. http://www.statcan.ca/.

Tennant, Paul. 1990. *Aboriginal Peoples and Politics: The Indian Land Question in British Columbia, 1849-1989.* Vancouver: UBC Press.

Tindall, D.B. 2002. "Social Networks, Identification, and Participation in an Environmental Movement: Low-Medium Cost Activism within the British Columbia Wilderness Preservation Movement." *Canadian Review of Sociology and Anthropology* 39(4):413-52.

–. 2003. "Social Values and the Contingent Nature of Public Opinion, Attitudes and Preferences about Forests." *Forestry Chronicle* 79(3):692-705.

Tindall, D., and Begoray, N. 1993. "Old Growth Defenders: The Battle for the Carmanah Valley." In *Environmental Stewardship: Studies in Active Earth Keeping,* edited by S. Lerner, 296-322. Waterloo, ON: University of Waterloo Geography Series.

Tripp, Tania, and Gordon Butt. 2010. "High Conservation Forest Assessment Clayoquot Sound, Western Vancouver Island, British Columbia, Canada: British Columbia Forest Stewardship Council Standards – Principle 9." Nanaimo: Madrone Environmental Services. www.iisaak.com.

Watkins, Mel. 1963. "A Staple Theory of Economic Growth." *Canadian Journal of Economics and Political Science* 29:141-58.

Wilson, Jeremy. 1998. *Talk and Log: Wilderness Politics in British Columbia.* Vancouver: UBC Press.

Wong, A., and G.Q. Bull. 2003. *Resource Assessment and Data Gap Analysis for Forest Carbon Inventory and Monitoring in Clayoquot Sound: Iisaak Forest Resources Ltd. – Report No. 1.* Vancouver: Department of Forest Resources Management, University of British Columbia.

16

Unheard Voices

Aboriginal Content in Professional Forestry Curriculum

Trena Allen and Naomi Krogman

Forestry education is changing in Canada in response to an increasing recognition of multiple values the forest holds for various groups of people. Among these changes is the increasing pressure from Aboriginal people, the courts, nongovernmental organizations, forest certification processes, and government policies to equitably involve Aboriginal people in forest management. Although these changes are encouraging, at best, forestry education in mainstream Canadian forestry programs is teaching students about "shared power," as Smith refers to in her chapter, but providing very little context for future foresters to imagine and negotiate their work in a context where First Nations, as co-managers, are setting their own levels for annual allowable cut and deciding where to cut and how to integrate Aboriginal traditional knowledge and values within contemporary forest management.

Frideres pointed out in his chapter that nearly half of the Aboriginal population in Canada lives in the boreal forest, yet it is only in the past decade that most accredited forestry schools in Canada have provided any curriculum on Aboriginal rights to, and interests in, the forested areas in which Aboriginal people may live, illustrating the distance forestry curriculum is from being truly decolonized. Currently, we suspect most forestry graduates from accredited forestry programs in Canada have had little or no exposure to how Aboriginal treaty rights and title are interpreted in their province, the status of land claims in their province, and other issues involving sectoral agreements and self-government. More strikingly, methods of consultation with Aboriginal communities, nontimber values and traditional forest use, and exposure to Aboriginal communities in field training are largely absent from forestry education.

Evidence of some movement to a new orthodoxy of Aboriginal forestry includes novel programs to recruit and retain Aboriginal students through the First Nations Forestry Program and the BEAHR (Building Environmental

Aboriginal Human Resources) Circle to Success. Some of the forestry schools have also developed advisory boards for groups outside the university to suggest improvements to forestry programs, and most notably, the University of British Columbia's First Nations Council of Advisors has substantial influence on changes to the university's forestry curriculum and program priorities. Greater efforts are in place to increase Aboriginal enrolment in the forestry technician programs, as found at the Nicola Valley Institute of Technology, an Aboriginal-controlled postsecondary institution, to train Aboriginal students who may not have the prerequisite high school math and science courses that are required for entry into other forestry programs. We are also encouraged by the ways in which students learn about "shared power" and "coexistence" within numerous courses and field experiences recently implemented in forestry schools, discussed later in the chapter.

There is considerable room for improvement in the role of Aboriginal content in forestry education. It is important to emphasize up front that the creation and implementation of forestry management practices continue to be the responsibility and duty of professional foresters. Professional foresters or registered professional foresters (RPFs) are held accountable to practise forestry in line with all professional forestry legislation in Canada (Bombay 2000) and thus are in key positions to implement forestry practices that represent the values of Aboriginal people and non-Aboriginal peoples' interests on the land.

Smith (2002) contends that by 2002, there were over 30 Aboriginal RPFs; according to the National Aboriginal Forestry Association a decade ago, there were at least 8,500 non-Aboriginal RPFs (Bombay 2000). These estimates suggest that Aboriginal foresters represented less than 0.4 percent of all practising professional foresters in Canada, and we are not sure if this situation has improved much given the downturn in the forest industry. Forest managers need to understand Aboriginal rights, values, and land management priorities to avoid protests (see Wilkes and Ibrahim, this volume), litigation, and misunderstandings when acting on land used by neighbouring Aboriginal communities, yet very little attention is given to forestry-related Aboriginal issues in forestry accredited programs (McKay 2001; Smith 2002). Few foresters employed by industry and government are equipped to address Aboriginal-related issues (Smith 1995; Bombay 1996), and this lack of understanding, communication, and negotiation with Aboriginal people may result in intensified conflicts.

In general, there is a growing demand for Aboriginal expertise in natural resource management. Environment sector employment statistics show that of the thirteen hundred Canadian communities where 20 percent or more of the population are of Aboriginal ancestry, eleven hundred such communities are within two hundred kilometres of mining or natural resource industries. The environment sector has experienced a growth of 4.6 percent annually and is expected to experience dramatic levels of retirement within all institutional levels (CCHREI and AHRDCC 2002). Nevertheless, Aboriginal people continue to experience low levels of employment in these areas. One reference suggests 2 percent of the First Nations and Métis population are employed in the forestry sector (GFW 2000). Many other Aboriginal communities are negotiating employment agreements with forest companies as new forest management plans are created (Natcher 2001). Wellstead and Stedman (2008) report that Aboriginal people are more often employed at the periphery of the forest sector and not at the core, where forest management decisions are made.

We argue that forestry education should provide a better foundation of Aboriginal content so that foresters can adapt their forest management practices to the movement toward the "new orthodoxy" of Aboriginal sovereignty, alongside curriculum that is focused on interim improvements, through certification requirements, co-management arrangements, employment agreements, and consultation among government, industry, and Aboriginal groups. While an increasing number of other professionals participate in forest management plans, foresters continue to be key social change agents. Foresters are in a pivotal role for addressing Aboriginal issues in forest management plans and to improve working relationships with Aboriginal communities.

What follows is a brief overview of the historical context of Aboriginal involvement in forest management, focusing on key laws and policies, and trends in industrial forestry. We then cover the needs of Aboriginal communities for forestry professionals and the training expectations within the professional designation of forestry, in terms of forestry curriculum, recruitment, and representation and retention of Aboriginal students in forestry programs. Our conclusion addresses the challenges for forestry education to embrace all three perspectives, as outlined by Smith in this volume, with the commitment of continual improvement of curriculum to fit the social context of Aboriginal progress in Canada, so that foresters are leaders, rather than trailers, in cooperation with Canada's First Peoples.

Historical Relationships That Have Shaped Forestry Education

Increases in industrial forestry operations have taken place alongside an increase in controversy over Aboriginal land rights. In large parts of the country, Aboriginal people have called on the Crown to honour historic treaties between the Crown and Aboriginal peoples. In British Columbia, parts of the territories, Quebec, and Labrador, other Aboriginal groups are undertaking land claims or treaty negotiations (Ross and Smith 2002). The tradition of treaty making in what was eventually to become Canada is well established and goes as far back as 1701 with the signing of Peace and Friendship treaties (Atlantic Canada) and the Great Peace of Montreal. Efforts to open up the Canadian west and northwest to settlement and development resulted in a new wave of signed treaties that started in 1850, whereby an additional three hundred treaties were signed over the next 150 years. The James Bay and Northern Quebec Agreement, signed in 1975, was the first comprehensive claim or modern treaty. The most recent comprehensive claim is the Labrador Inuit Land Claims Agreement, which was signed in 2004. These modern treaties provide certainty to the federal and provincial or territorial governments regarding resources and lands, and grant the Aboriginal beneficiaries of the agreement Aboriginal title to or exclusive rights over the land.

In 1930, the federal government transferred ownership and control of Crown land to the provinces under the National Resources Transfer Agreement (NRTA). Embodied within the NRTA was legislation that extinguished the rights of Aboriginal peoples to harvest forest resources. Since that time, Aboriginal groups have contested the transfer of ownership and control of Crown lands, which they argue occurred without their consent. Provincial governments were allowed to impose provincially regulated restrictions on traditional harvesting activities but were not bound to the same Aboriginal fiduciary responsibilities as the federal government. In 1982, treaty and Aboriginal rights became constitutionally protected under the Constitution Act (section 35). Since then, the Supreme Court of Canada and lower courts have seen numerous cases that have further defined the nature and scope of Aboriginal and treaty rights, and governments' obligation to Aboriginal peoples.

Despite constitutional recognition and affirmation of treaty and Aboriginal rights, Supreme Court of Canada decisions, and many Aboriginal communities rejecting the NRTA, Aboriginal groups continue to experience difficulty in exercising their rights and title to their surrounding forested land base. Aboriginal rights are inherent rights to perform traditional activities on the land, such as hunting, fishing, or harvesting ceremonial plants, that occurred on the land at time of contact with European explorers. Treaty rights are benefits granted pursuant to treaties signed by the Crown and Aboriginal people. Treaties were signed "to define, among other things, the respective rights of Aboriginal people and governments to use and enjoy lands that Aboriginal people traditionally occupied" (INAC 2000, 1). Treaty rights may include reserve lands; gratuities; and hunting, trapping, and fishing rights within treaty areas (ibid.). Forestry education at accredited forestry schools in Canada has focused on harvesting a volume of timber from across a large land base, or from large-area, long-term tenures. The presence of overlapping Aboriginal title, Aboriginal treaty rights, and Aboriginal rights has generally not been integrated into forestry instruction as part of responsible forest management of Crown land.

Despite the general unawareness among forestry professionals about Aboriginal and treaty rights, there has been substantial attention to the constitutional basis for the duty to consult with Aboriginal people before resource harvests are underway. The 1982 Constitution Act requires "government to give Aboriginal and treaty rights priority, and to infringe upon them only to the extent necessary to achieve a compelling and substantial objective" (Sharvit, Robinson, and Ross 1999, 5). In the important 1997 *Delgamuukw* case, the British Columbia Supreme Court further affirmed that "consultation must be in good faith, and with the intention of substantially addressing the concerns of Aboriginal peoples whose lands are at issue."[1] Thus, governments that consult with Aboriginal groups must seek to minimize the impact on Aboriginal interests (Sharvit, Robinson, and Ross 1999; Usher 2000) by demonstrating continued protection of prior established interests of the Aboriginal people (now entrenched in section 35 of the Constitution Act), and to provide compensation for title infringement. Most recently, in *Mikisew Cree* (2005), the Supreme Court of Canada upheld a lower court decision and reiterated that the Crown governments have a duty to consult and accommodate First Nations when making land and resource use decisions, even when those decisions negatively affect treaty rights. Failure to adequately consult in advance of interference with existing treaty rights, as in this case, undermines the process of recognition and reconciliation of those treaty rights.[2]

Sharvit, Robinson, and Ross (1999) contend that, ultimately, reasonable consultation has to be seen as reasonable by the courts. Reasonable consultation, they argue, means that notification of meetings and opportunities for Aboriginal groups to participate must be widely displayed in the community of interest. Governments must make every effort to provide a process whereby affected Aboriginal groups are informed about proposed developments and asked for their concerns before the development plans are set in stone, which may require several meetings.

Under court rulings, only government bodies have the obligation to consult potentially affected Aboriginal groups, "not resource industries exercising rights to extract or develop resources on traditional Aboriginal lands" (Sharvit, Robinson, and Ross 1999, 8). Nevertheless, Aboriginal groups may enter into discussion with resource industry groups if the government makes it a condition for obtaining permits or licences (Sharvit, Robinson, and Ross 1999), or where government assistance is not forthcoming and companies are told to "work it out" with Aboriginal people. In many cases, complex details regarding the likely impacts of future resource extraction on Aboriginal groups can be discovered only when discussions occur between the responsible industry and the affected Aboriginal group, yet the Crown is obliged to be involved. We are aware of one court ruling that concluded that consultation between a timber company and an Aboriginal group was not sufficient because it did not meet the requirements of a proper consultation process, given government representatives were absent at all industry–First Nation discussions (Sharvit, Robinson, and Ross 1999).[3]

Court rulings have significant implications for future relationships between Aboriginal groups, federal and provincial levels of government, and resource development industries operating in the boreal forest. An understanding of treaty and Aboriginal rights, and the constitutional fiduciary obligation to meaningfully consult and accommodate, may be fundamental for future foresters in their daily work on contested lands. In addition, Canadian forestry operations are being scrutinized on a host of other criteria that have emerged in both the national and international context of increased environmental concerns and goals for sustainable development.

National and International Agreements and Policies

International and national organizations have increasingly recognized Aboriginal values, beliefs, and priorities for sustainable forest management (Smith 1998; Berkes 1999). At the provincial levels, for example, delegated provincial staff positions are specifically allocated to negotiate with Aboriginal communities on natural resource management practices in British Columbia, Ontario, and Alberta. The provincial governments of British Columbia and Ontario have drafted provincial legislation and regulations to incorporate Aboriginal involvement into the forest management planning process (Smith 1995). However, these initiatives, in addition to written policies, fall short in the face of more urgent issues. In fact, Aboriginal communities have voiced frustration in negotiating with provincial governments in a system that weighs financial and human resources too heavily on the government side while Aboriginal communities face excessive demands for "consultation" with few resources (Smith 1995, 7).

In fact, there are many challenges for Aboriginal communities to keep pace with the requests for consultation with companies wishing to develop in their traditional territory. Among these challenges are the shortfalls in their own experts on forest management; the communication barriers and limited opportunities to explain to foresters Aboriginal knowledge and concerns about forest management; foresters' sensitivity to culturally and contextually appropriate levels of effort to engage Aboriginal people in communities of varying capacity to respond; and for governments to address Aboriginal legal obligations, in addition to providing the services and programs in Aboriginal communities that can lead to parity in education, expertise, and so on with government and industry representatives. Each relationship among government, industry, and Aboriginal communities is unique, and all parties must have the patience and competence to work toward mutual goals.

A detailed report of Aboriginal involvement in forest industry developments by the National Aboriginal Forestry Association and the Institute on Governance (NAFA and IOG 2000) provides examples of partnerships from across the country. Growth and success of such partnerships involving Aboriginal communities, ranging from joint ventures to socioeconomic partnerships, require the implementation of carefully constructed building blocks, which include

> an understanding and awareness of each culture and acknowledgement of differences, well thought-out and involved partner selection and clarification of each partner's role, clear communication leading to open and informed decisions, expectations of relationship and benefits based on long-term perspective, and not hurrying any stage of the development process. (NAFA and IOG 2000, 83)

Industrial Activity in the Boreal Forest

Approximately 80 percent of Aboriginal peoples live in and around these forest regions. The suitable commercial timber land within boreal forests, approximately 244.6 million hectares, is presently being accessed for resource extraction purposes, such as industrial forestry, mining, and oil and gas developments. Large-area, long-term forest tenures, increasing oil and gas activity, and greater access to remote parts of the forest have substantially affected traditional uses of the forest by Aboriginal peoples (Schneider 2002). The cut requirements in long-term forest tenures focus on extraction of maximum volumes of timber and do not take into adequate account Aboriginal land uses, especially those associated with nontimber forest uses (Ross and Smith 2002).

Intensified land uses within forested regions of western Canada have contributed to the confrontation among government, industry, and First Nations about how to protect certain forested areas. Heightened pressure to protect British Columbia's coastal forests, for example, led to the cooperative effort between the Haisla First Nation, BC government, and West Fraser Mills to protect the Kitlope Valley, six hundred kilometres north of Vancouver. As part of this agreement, West Fraser Timber Company gave up plans to log 317,000 hectares of the Haisla's traditional territory. The result of this agreement is that one of the world's largest intact tracts of coastal temperate rainforest is protected (GFW 2000). This example illustrates the ways in which natural resource companies, governments, and Aboriginal communities can work together to make substantial changes to forest management, and suggests that the range of opportunities for such collaboration will be important for new foresters to understand.

Certification

In addition to global, national, and provincial Aboriginal involvement initiatives, certification systems such as the Canadian Standards Association (CSA) and Forest Stewardship Council (FSC) have created separate measures of Aboriginal involvement in the planning, development, and implementation phases of forest management practices (Smith 1998). Despite the voluntary efforts of the CSA and FSC in sustainable forest management to recognize Aboriginal rights, actual outcomes remain unknown. We suspect the recent drop in the economy may discourage companies from investing in CSA or FSC certification, because these two certifications in particular require evaluations of the involvement of Aboriginal people in the industry and land use planning as part of their certification processes, potentially requiring forest companies to spend more time and money to meet various FSC or CSA certification criteria. Smith (1998) suggests that Aboriginal groups, to protect their interests on the land, need to become educated in the various certification processes and forestry-related legislation to secure their place in the planning, implementation, monitoring, and improvement of the forest management process. Correspondingly, foresters may need exposure to collaborative efforts with Aboriginal people to meet CSA and FSC standards.

Changes Proposed to Forestry Education

At an Aboriginal Professional Development in the Forest Sector workshop in 2001, attended by Aboriginal forestry professionals from across Canada, participants concluded that Aboriginal forestry interests are poorly represented in Canadian forestry schools (McKay 2001) and that this was reflected in the reluctance of professional foresters to proactively address the interests of Aboriginal people in forest management. During that workshop, a group of Aboriginal forest practitioners formed a working group to design an Aboriginal core forestry education curriculum to fill the gap in Aboriginal forestry courses. In general, Harry Bombay, executive director of the National Aboriginal Forestry Association, called on forestry education to prepare all foresters to be able to address

> the importance of traditional knowledge to sustainability, the maintenance or revitalization of forest-based traditions and practices, and the role modern technology can play in developing co-ordinated, multiple use, natural resource management plans which are consistent with community needs, and aspirations with respect to the pace of development. (Bombay 2000, 4)

Integration of traditional Aboriginal values into commercial forest management plans is a difficult task, and to date, traditional ecological knowledge (TEK) has made very little impact on forest management practices and, in general, has received cursory treatment in forestry education programs. In part, this is because of the emphasis in forestry curriculum on competency in traditional forest topics (i.e., biophysical science competencies), leaving very little room for new curriculum relating to the social context in which foresters will work.

Professional Education

Competence

Professional forestry education in Canada places a great deal of emphasis on competency and advancing novice students, through their acquisition of forestry-related knowledge and skills, to the status of forestry professionals. In Canada, professional forestry education concentrates on forestry-related technical skills and natural science knowledge, thus requiring a core forestry curriculum. To demonstrate competency and become a registered professional forester, a person progresses through several stages: (1) completion of an accredited four-year forestry program, (2) enrolment in a professional forester association and work within the field of forestry for twenty-four months under the tutelage of a mentoring professional forester, and (3) successful completion of the professional forester registration entrance examination. Successful completion of core curriculum requirements is necessary for registration in any of the six provincial professional forestry-related associations. Forestry schools in Canada have until recently focused on the generalist approach to forestry education, where forestry graduates are expected to reach proficient and expert levels via experience in a work setting and continuing education. Finally, professional forester associations require a combination of mandatory, and voluntary, continuing education and self-assessment to assure postgraduate competency in professional forestry (ABCPF 2003).

Several scholars have noted that the strong foundation in basic and applied natural sciences as well as the technical skills required in the previous Canadian Forestry Accreditation Board (CFAB) accredited four-year forestry programs left little room for adequate integration of basic social science and humanities courses into the curricula requirements (Tindall 2001; Lukai 2002; Klenk and Brown 2007). In 2009, a revised CFAB-accredited standard came into effect. The new standard allows for specialization into forestry streams, has condensed core requirements, and expanded opportunities for electives. It also has increased the social sciences and humanities portion. However, very little of the continuing education courses available in Canada include Aboriginal content, and foresters are less likely to take courses on Aboriginal consultation, treaty and Aboriginal rights, and TEK than other continuing education courses (Krogman et al. 2007).

Canadian Professional Forester Undergraduate Education

Forestry education in Canada has been guided by the CFAB, jointly created by the Canadian Institute of Forestry, Association of University Forestry Schools of Canada, and the Canadian Federation of Professional Foresters Association in 1989. Its role is to build forestry programs that are responsive to the varied needs of forestry stakeholders of the twenty-first century (Apsey et al. 2000). The CFAB reviews Canadian university forestry baccalaureate programs to ensure they meet or exceed educational standards for registration in the provincial professional forester and forest engineer associations, and performs other facilitative functions to maintain a high standard of university education (Apsey et al. 2000, 39). The CFAB, as a result of a formal review process for the continuing accreditation of forestry schools, has recently revised its accreditation standard, a standard that had not seen any change since the 1970s (Krogman et al. 2003).

Forestry programs are accredited for a maximum of six years, which involves a forestry school site visit by an accreditation team to check that the program is meeting CFAB curriculum standards. The accreditation team comprises a broad range of industry, government, and academic experts, but there is no formal appointment of a reviewer to evaluate the social science or Aboriginal content of forestry curriculum. Forestry schools place high importance on achieving the prestigious CFAB accreditation and thus likely force attention on the strict guidelines of CFAB course requirements.

Krogman et al. (2003) found in a focus group study with faculty members and those involved in forestry curriculum development at the University of British Columbia, University of Alberta, and Lakehead University, in Thunder Bay, Ontario, that members from each of the schools identified the difficulty to update curriculum in the context of rapid societal shifts that affect forestry. Forestry faculty members also asserted they had a limited budget, and limited time available, to educate excellent "generalist" foresters (those foresters who have a suite of basic skills in forestry vs. specialist foresters on silviculture, land use planning, etc.) *and* provide specialty courses for emerging areas of importance. Respondents expressed widespread support for more courses that hold social and Aboriginal content, though there was little consensus on what courses might be cut to accommodate such courses in an accredited forestry school's curriculum. Forestry instructors in our focus group were reportedly encouraged to include social and Aboriginal issues and values in their required course offerings but have little guidance, training, or additional resources to update their courses (Krogman et al. 2003; Klenk and Brown 2007).

Professional Forestry Schools

The eight forestry schools in Canada have adopted unique approaches to integrate Aboriginal issues into their accredited curriculum. Their individual methods are a result of site-specific capacities, such as school location, size, leadership, and the breadth of faculty. Most forestry schools have made some steps to implement new program design and created courses about Aboriginal and social issues in forest management; developed partnerships with Aboriginal communities; and funded Aboriginal and social science forestry faculty positions to improve undergraduate forestry education.

For example, the Faculty of Natural Resources Management at Lakehead University created a tenure-track position in "Aboriginal Forestry" (Smith 2002). The University of New Brunswick's Faculty of Forestry and Environmental Management, the only five-year undergraduate forestry program in the country, has redesigned its program around an outcome-based reflective practitioner philosophy where students must work hands on to produce the best outcome (Naysmith and Crichlow 1995). Godbout (1997) reports that Université Laval Faculté de foresterie de geographie et de géomatique implemented a teaching initiative centred on indigenous forestry issues (Smith 2002). In 2001, the University of British Columbia's Faculty of Forestry introduced the First Nations Strategy, developed by the First Nations Council of Advisors, to the curriculum as one of many changes to upgrade its forestry programs.[4] The Ecosystem Science and Management Program in the College of Science and Management at the University of Northern British Columbia (UNBC) has developed a research and educational partnership with the Tl'azt'en First Nation through creation of the thirteen-thousand-hectare John Prince Research Forest (Smith 2002). UNBC was also the first school to be accredited under the revised CFAB accreditation standard, in July 2009. The inclusion of social and Aboriginal issues and values is slowly occurring across the country.

Retention and Recruitment of Aboriginal Students

A number of initiatives at the federal government, university, and forestry-school levels have been created to increase the presence and retention of Aboriginal students at forestry schools, at technical institutions, and within the environment sector. Two programs of particular interest at the national level are the First Nations Forestry Program (FNFP), funded by Natural Resources Canada, and Aboriginal Affairs and Northern Development Canada (previously named Indian and Northern Affairs Canada), and the Building Environmental Aboriginal Human Resources (BEAHR) program, funded by the Government of Canada's Sector Council Program. To capitalize on future employment opportunities and increase Aboriginal involvement in forest management, the FNFP provides funding for partnerships with the forest industry, for business development training, and for institutional capacity building in sustainable forest management. The BEAHR program focuses on meaningful participation and increased employment of First Nations, Métis, and Inuit people in the environment sector; supports environmental excellence in such communities; and runs programs to provide support to students who pursue training in an environmental field. It is not clear to what extent university administrators and forestry educators promote these programs and provide complementary recruitment scholarships and mentoring programs to encourage more Aboriginal students to choose forestry as a profession.

Community Contributions to University Programming

One impetus to change can be through the continuous participation of Aboriginal communities and forestry practitioners as advisors to forestry programming. Forestry faculty members in conjunction with interested Aboriginal and public groups have devised committees and boards that acknowledge the importance of communication between forestry schools and outside groups. For example, in 1995, the University of British Columbia's Faculty of Forestry established the First Nations Council of Advisors to work in tandem with the newly appointed First Nations forestry co-coordinator to build a First Nations Strategy "through which" key stakeholders can determine the perceived issues surrounding First Nations and forest land use. Results from the First Nations Strategy interview process provided the council with a clearer picture of the important Aboriginal issues that have challenged forest managers across the province, such as the impact of provincial court rulings on Aboriginal rights and title (UBC Faculty of Forestry 2001). The detailed strategy adopted by the Forestry Faculty in the spring of 2001 provided the faculty with guiding principles, goals, and objectives necessary to meet proposed targets on an annual basis. Similarly, Lakehead University's Faculty of Natural Resources Management established a public advisory board that provides input on current and future forestry curriculum.

In August 2000, the Faculty of Natural Resources Management, as mentioned earlier, created an "Aboriginal Forestry" tenure-track position (Smith 2002). The faculty has worked closely with (a newly established) forest technician program at Confederation College, also in Thunder Bay. The college has devised an outreach program to increase Aboriginal enrolment in the forestry technician program. Similarly, in the Interior of British Columbia, the Nicola Valley Institute of Technology, located in Merritt, offers a forestry technician program aimed at attracting Aboriginal students, Council of Forest Industries sponsorship, and summer employment opportunities. After completing the one-year program, students can continue their postsecondary forestry technician training to eventually complete the two-year technician-forestry option program. Students may continue their education with the Faculty of Forestry at the University of British Columbia and use their forestry technician courses as transfer credits toward the four-year professional forestry degree program.

Pedagogy

Of particular debate within the field of forestry and wildlife management is the role of traditional ecological knowledge as a valid alternative way of knowing. (See the chapters in this volume by Marc Stevenson and Blackstock) Berkes (1999) derived a working definition of traditional ecological knowledge (TEK) based on the combination of its most significant components, where it is a

> cumulative body of knowledge, practice, and belief, evolving by adaptive processes and handed down through generation by cultural transmission, about the relationship of living beings (including humans) with one another and with their environment. (Berkes 1999, 8)

Knowledge associated with TEK and indigenous knowledge has gained recognition, acceptance, and relevance for agency scientists, policy makers, and academics. In fact, the respect, preservation, promotion, and use of TEK are called for in the United Nations Conference on Environment and Development – the Convention on Biological Diversity (Berkes, Colding, and Folke 2000; Kimmerer 2002). The former Sustainable Forest Management Network of Centre of Excellence, for example, funded dozens of research projects on the potentialities for TEK to inform, and in some cases lead, forest management. Marc Stevenson points out in this volume the richness of opportunity to incorporate traditional knowledge into sustainable forest management. Michael Blackstock discusses similar issues with regard to water resources and forest lands.

Despite these strides, the integration of TEK into forest management remains a challenge for most Aboriginal communities, forest industries, and government agencies. More importantly, there are few courses in forestry programs that take the time required to grapple with the philosophy of science. The notion that there are alternative ways of knowing can be a foreign supposition for students whose education has focused on deductive reasoning, human control of the landscape, and isolation of causal variables. In forestry education, an important contrast in knowledge systems is the emphasis on quantifying various features of specific resources, such as tree or animal abundance, ages, conditions, and regeneration rates, whereas in many traditional knowledge systems, more emphasis is placed on knowing the factors that influence the behaviours, availability, distributions, movements, and interrelationships of resources (Stevenson and Webb 2003). Stevenson and Webb (2003, 84) contend that these qualitatively different approaches have resulted in the use of TEK in a selective and isolated way.

Under most cooperative management arrangements with a mandate to incorporate TEK, the common practice has been to take whatever aspects of this knowledge managers and natural resource scientists find useful and merge them with Western scientific knowledge to inform environmental resource management objectives, and ultimately, planning and decision making (Stevenson and Webb 2003, 84).

There are large variations in the degree to which different disciplines acknowledge other knowledge systems, and we argue that the forestry curriculum could be improved by incorporating far more fundamental training on ontology, epistemology, and the nexus between science and its utility alongside other knowledge and decision-making systems. Rather than teaching knowledge as a set of facts, it could be taught that knowledge is what many people can agree on as "accurately representing reality" at a certain time, thus recognizing that knowledge is couched within a broader sociocultural context, and is treated with suspicion in some cultural contexts.

Forestry educators have expressed discomfort around incorporating TEK into their forestry courses given their unfamiliarity with it, the suspicious status it holds in many scientific circles, and the lack of references for the integration of TEK into forest management (Krogman et al. 2003; Krogman et al. 2007). Increasingly, practitioners and scholars are developing helpful resources to address the potential use of TEK in forest management (for example, see Karjala, Sherry, and Dewhurst 2003; Orlowski and Menzies 2004).

Discussion

The lack of content on Aboriginal issues in forestry is found in other university curriculum on natural resource management, parks management, rural development, and conservation biology. Yet, the societal implications for omitting Aboriginal interests in forestry curriculum are significant given the special licence registered professional foresters have with which to sign off on forest management plans, the expansion of forestry operations throughout the boreal forest, and the growing tension between numerous Aboriginal communities and forestry companies operating in an Aboriginal community's traditional territory.

Aboriginal curriculum should address the role of treaties across the provinces, given that negotiations between government, industry, and Aboriginal communities in British Columbia, for example, where most First Nations have not signed treaties, might be entirely different from those in Alberta, where most Aboriginal communities have signed treaties. Foresters would be wise to understand the importance of recent court decisions on the interpretation of Aboriginal rights regarding traditional use – for example, in regard to the potential infringement on Aboriginal rights. It is becoming necessary for many foresters to be involved with Crown fiduciary responsibilities and to consult with Aboriginal groups to minimize development impacts on Aboriginal interests. Ignorance about the court interpretations of Aboriginal rights makes forest companies and provincial governments more susceptible to confrontation in the form of resistance and lawsuits. Such ignorance could also make it all the harder on practising foresters to adjust to shifts in land tenure laws that provide greater forest land-base access to Aboriginal peoples. At a minimum, forestry students should understand Aboriginal constitutional rights to perform traditional activities on the land – such as hunting, fishing, or harvesting ceremonial plants that occurred at the time of contact with European explorers (Natcher 2000) – and be exposed to traditional land use and occupancy studies, and cultural heritage inventories as ways to document Aboriginal forest values.

Forestry students are more likely to respect Aboriginal treaty rights and individual rights if they learn about the ways in which Aboriginal people have used and depended on the forest in the past, and the central role the forest plays in Aboriginal culture. This deeper sense of respect for Aboriginal ways of knowing and ways of doing goes beyond how-to courses on Aboriginal involvement in forest management. An understanding at this deeper level means building in course time and cross-cultural experiences that foster students' contemplation on the role of traditional knowledge, the role of nontimber forest values, and the role of cultural continuity in Aboriginal community well-being. Future foresters will likely be more responsive to international pressures, manifested in government targets and certification requirements, to honour indigenous people's rights and interests in forest management if they understand and ultimately value the reasoning behind such pressures. Holmes Rolston III, one of the leading environmental philosophers in North America, argues that through repeated exposure, as people learn more about the spectrum and basis of values in a landscape, they internalize those values (Rolston 1988). The higher calling of education is to go beyond the provision of skills and knowledge to cultivate a sense of commitment to and respect for the First Peoples who live in Canada's boreal forest.

Some recent trends are encouraging. Canadian professional undergraduate forestry education has undergone a revision, away from a generalist model. Students will be exposed to additional electives, and social science and humanities courses. The Canadian Forestry Accreditation Board has recognized the rigidity of a set of course requirements that allows so few additional courses on other important forestry-related topics, though the accreditation requirements for forestry schools has not changed to reflect Aboriginal priorities for forest management. Many forestry faculty members feel uncomfortable adding Aboriginal issues to their courses, so Aboriginal content is sought from new forestry faculty who have specialized in this area through their graduate school training and research. Given that all forestry schools in Canada have experienced a decline in student enrolment, it is a relatively difficult time to justify the hiring of new faculty to focus on Aboriginal content in forestry courses. Nevertheless, numerous additional classroom and field courses that address Aboriginal issues continue to be developed across Canada's eight forestry schools.

Conclusion

We argue in this chapter that very little forestry curriculum in Canada appears to teach forestry from the vantage point of co-management as controlled by First Nations or Aboriginal sovereignty. If this were the case, there would be far greater emphasis on curriculum that addresses treaty rights and Aboriginal rights, provides examples of First Nation-owned or -managed forestry companies, details the challenges and successes of First Nations' forestry training programs, explains options for compensation for infringement on treaty rights, and discusses possibilities for alternative tenures and mill requirements that recognize Aboriginal interests and traditional knowledge, and the graduated progress many Aboriginal groups would need to have to manage a mix of commercial timber and other nontimber interests.

Thus, we believe a fruitful next step is to develop curriculum that recognizes "shared power" that would cover forest co-management, processes that respect Aboriginal and treaty rights to establish protected areas, integrated resource management plans, water and watershed management practices (Graham and Fortier 2006), and other land-based planning efforts that cannot reasonably, in the eyes of most Aboriginal peoples, be separated from forest management. This curriculum would focus on more government-to-government relations and the pathways for coexistence among private landowners, provincial landowners, treaty holders, and Aboriginal traditional territory interests.

To date, we argue that most of the curriculum is on what M.A. (Peggy) Smith has referred to as the "old orthodoxy," which focuses on consultation so that forestry can go ahead, and hiring practices to include more Aboriginal people. Improvements to curriculum have been made, but the content offered about Aboriginal interests is generally relegated to the optional category, so many forestry students obtain very little training in this area.

In May 2003, Ovide Mercredi, former national chief of the Assembly of First Nations, gave a speech at the Community-University Research: Partnerships, Policy, and Progress conference in Saskatoon. In his opening keynote address he argued that increased accessibility has not led to a comparable change in the foundations and assumptions of university curricula, research, teaching practices, and gathering of knowledge (Community University Expo 2003). Although forestry programs may become more accessible to

Aboriginal people, the content of the curriculum must also change to authentically address the value differences between status quo industrial forestry and collaborative forest management with Aboriginal people. Educational change in forestry programs is part and parcel of practical and cultural changes in forest practices. Foresters have their own culture, which was for many years sheltered from criticism by women, environmentalists, people of colour, and most importantly here, Aboriginal people. The unheard voices in forestry have been plentiful, but increasingly, the professional forester must be one who seeks to understand these voices, has dialogue with these voices, and strives to manage forests in the interests of a multicultural public, especially in accordance with the values of the First Peoples of Canada.

Notes

1 Reason for judgment in *Delgamuukw v. British Columbia,* [1997] 3 S.C.R. 1010, 1997, note 6 at 1112.

2 *Mikisew Cree First Nation v. Canada (Minister of Canadian Heritage),* [2005] 3 S.C.R. 388, 2005 S.C.C. 69.

3 The aforementioned court ruling is *Blueberry River Indian Band v. British Columbia (Ministry of Employment and Investment),* 1997.

4 The University of British Columbia adopted the First Nations Strategy in the summer of 2001.

References

ABCPF (Association of British Columbia Professional Foresters). 2003. *Continuing Competency for Registered Members: A Discussion Paper.* Vancouver: ABCPF.

Apsey, M., D. Laishley, V. Nordin, and G. Paul. 2000. "The Perpetual Forest: Using Lessons from the Past to Sustain Canada's Forests in the Future." *Forestry Chronicle* 76(1):37-42.

Berkes, Fikret. 1999. *Sacred Ecology: Traditional Ecological Knowledge and Resource Management.* Philadelphia: Taylor and Francis.

Berkes, Fikret, Johan Colding, and Carl Folke. 2000. "Rediscovery of Traditional Ecological Knowledge as Adaptive Management." *Ecological Applications* 10(5):1251-62.

Bombay, Harry. 1996. *Aboriginal Forest-Based Ecological Knowledge in Canada.* Ottawa: National Aboriginal Forestry Association.

–. 2000. *Aboriginal Foresters: Key to Maintaining Relationship with the Land.* Ottawa: National Aboriginal Forestry Association.

CCHREI and AHRDCC (Canadian Council for Human Resources in the Environment Industry and Aboriginal Human Resource Development Council of Canada). 2002. *Building Environmental Aboriginal Human Resources: BEAHR, Moving Forward.* Calgary: CCHREI and AHRDCC.

Community University Expo. 2003. "A Day of Enlightenment: When Universities and Funders Collaborate with Communities on Research, Education and Training That Communities Need, Want, and Lead." Paper presented at the Community University Expo international conference, Saskatoon, May 8.

GFW (Global Forest Watch). 2000. *Canada's Forests at a Crossroads: An Assessment in the Year 2000.* Global Forest Watch Canada, http://pdf.wri.org/.

Godbout, C. 1997. "Forestry Education in Canada." *Forestry Chronicle* 73(3):341-46.

Graham, John, and Evelyn Fortier. 2006. "Building Governance Capacity: The Case of Potable Water in First Nations Communities." Paper presented at the Aboriginal Policy Research Conference, Ottawa, March 23. http://www.iog.ca/.

INAC (Indian and Northern Affairs Canada). 2000. *Treaties with Aboriginal People in Canada.* Information sheet. Ottawa: Indian and Northern Affairs Canada.

Karjala, Melaine, Erin Sherry, and Stephen Dewhurst. 2003. *The Aboriginal Forest Planning Process: A Guidebook for Identifying Community-Level Criteria and Indicators.* Prince George, BC: Ecosystem Science and Management Program, College of Science and Management, University of Northern British Columbia.

Kimmerer, Robin. 2002. "Weaving Traditional Ecological Knowledge into Biological Education: A Call to Action." *BioScience* 52(5):432-38.

Klenk, Nicole L., and Peter Brown. 2007. "What Are Forests For?: The Place for Ethics in the Forestry Curriculum." *Journal of Forestry* 105(2):61-66.

Krogman, Naomi, Trena Allen, Peggy Smith, and Kendra Isaac. 2007. *Professional Forestry Certification in the New Millennium: Opportunities and Constraints for Forestry Curriculum Change.* Edmonton: Sustainable Forest Management Network.

Krogman, Naomi, Peggy Smith, Trena Allen, and Ken VanEvery. 2003. "Changes in Forestry Education to Integrate Social and Aboriginal Forestry: Focus Group Findings from Three Canadian Forestry Schools." Canadian Aboriginal Science and Technology Society conference, Saskatoon, September 18-20.

Lukai, N. 2002. "Undergraduate Programs Offered by the University Schools of Forestry." *Forestry Chronicle* 78(2):240-44.

McKay, Rebecca. 2001. "Workshop Report: Aboriginal Professional Development in the Forest Sector." Vancouver: UBC Faculty of Forestry. January 11-12. Unpublished document.

NAFA and IOG (National Aboriginal Forestry Association and the Institute on Governance). 2000. *Aboriginal-Forest Sector Partnerships: Lessons for Future Collaboration.* Ottawa: NAFA.

Natcher, David. 2000. "Institutionalized Adaptation: Aboriginal Involvement in Land and Resource Management." *Canadian Journal of Native Studies* 20:263-82.

–. 2001. "Land Use Research and the Duty to Consult: A Misrepresentation of the Aboriginal Landscape." *Land Use Policy* 18:113-22.

Naysmith, J.K., and J. Crichlow. 1995. "Educating the 21st Century Forester: Report on a Symposium of the University Association of Forestry Schools of Canada." *Forestry Chronicle* 71(3):345-52.

Orlowski, Paul, and Charles Menzies. 2004. "Educating about Aboriginal Involvement with Forestry: The Tsimshian Experience – Yesterday, Today and Tomorrow." *Canadian Journal of Native Education* 28(1/2):65-69.

Rolston, Holmes. 1988. *Environmental Ethics: Duties to and Values in the Natural World.* Philadelphia: Temple University Press.

Ross, Monique M., and Peggy Smith. 2002. *Accommodation of Aboriginal Rights: The Need for Aboriginal Forest Tenure.* Edmonton: Sustainable Forest Management Network.

Schneider, Richard R. 2002. *Alternative Futures: Alberta's Boreal Forest at the Crossroads.* Edmonton: Federation of Alberta Naturalists.

Sharvit, Cheryl, Michael Robinson, and Monique M. Ross. 1999. "Resource Developments on Traditional Lands: The Duty to Consult." CIRL (Canadian Institute of Resource Law) Occasional Paper no. 6. February.

Smith, P. 1995. *Aboriginal Participation in Forest Management: Not Just Another Stakeholder.* Ottawa: National Aboriginal Forestry Association.

–. 1998. "Aboriginal and Treaty Rights and the Aboriginal Participation: Essential Elements of Sustainable Forestry Management." *Forestry Chronicle* 74(3):327-33.

–. 2002. "Aboriginal Peoples and Issues in Forestry Education in Canada: Breaking New Ground." *Forestry Chronicle* 78(2):250-54.

Stevenson, Marc, and Jim Webb. 2003. "Just Another Stakeholder? First Nations and Sustainable Forest Management in Canada's Boreal Forest." In *Toward Sustainable Management of the Boreal Forest,* edited by P.J. Burton, C. Messier, D.W. Smith, and W.L. Adamowicz, 65-112. Ottawa: NRC Research Press.

Tindall, D.B. 2001. "Social Science and Forestry Curricula: Some Survey Results." *Forestry Chronicle* 77(1):121-26.

UBC (University of British Columbia) Faculty of Forestry. 2001. "UBC Faculty of Forestry First Nations Strategy." Unpublished document.

Usher, Peter J. 2000. "Traditional Ecological Knowledge in Environmental Assessment and Management." *Arctic* 53(2):183-93.

Wellstead, A.M., and R.C. Stedman. 2008. "Intersection and Integration of First Nations in the Canadian Forestry Sector: Implications for Economic Development." *Journal of Aboriginal Economic Development* 6(1):30-43.

17

In Search of Certainty

A Decade of Shifting Strategies for Accommodating First Nations in Forest Policy, 2001-11

Jason Forsyth, George Hoberg, and Laura Bird

This chapter examines how the Government of British Columbia has sought to use forest policy to accommodate First Nations since the BC Liberal Party took office in 2001. Dramatic changes in policy have occurred in its two and a half consecutive terms in office, despite the premiership remaining constant under Gordon Campbell until 2011. Before entering office, the new premier Gordon Campbell, Attorney General Geoff Plant, and Forest Minister Michael de Jong were plaintiffs in a lawsuit challenging the constitutionality of the only modern-day treaty signed in British Columbia, the Nisga'a Final Agreement. Once in office, the Liberal Party acted on its campaign promise to initiate what became an exceptionally controversial referendum on the treaty process.

The first shift in policy away from the antagonistic approach to First Nations came in the form of forest and range agreements with First Nations, which granted access to timber harvesting and a share of stumpage revenues. Despite over two hundred of these agreements being signed, the focus on using economic tools to accommodate First Nations did not provide a durable solution to the challenge of reconciling the Crown – First Nation relationship. In their second term in office, the BC Liberals entered into a "New Relationship" with First Nations, pledging to establish a form of shared decision making and revenue sharing throughout the province. Since then, the province has concluded unprecedented strategic agreements in the north and central coast and Haida Gwaii. These represent a fundamental and conceptual shift from using economic tools to accommodate First Nations toward using governance tools, including shared decision making. This chapter reviews how both approaches to accommodation have been implemented by the province.

Origins

The origins of the Campbell government's approach to First Nations were strongly influenced by its 2001 election platform.[1] The "New Era" forestry agenda for First Nations set out two specific goals (BC Liberal Party 2001). The first was to protect private property rights in treaty negotiations relating to the politics of the Nisga'a Final Agreement; this is not discussed further here. The second was to work to expedite interim measures with First Nations in order to create greater certainty in the forestry sector. Interim measures agreements had previously been recommended by the BC Claims Task Force in 1990 to provide opportunities to advance First Nations' interests pending treaty settlement. This represented the cornerstone of the BC Liberals' forest policy for First Nations, and aimed to address the impacts of the BC Court of Appeal's *Haida* decision while treaties were still being negotiated. Although also a politically motivated action, this commitment stemmed more from the challenge that enhanced First Nations' title and rights posed for provincial governments, particularly for natural resource management. Supreme Court cases such as *Delgamuukw* and *Haida* continued to put significant pressure on provincial governments to establish mechanisms to consult and accommodate First Nations over resource management. The BC Liberal commitment to expedite interim measures agreements aimed to mitigate such pressures.

The Changes in Policy

The BC Liberals moved quickly in making decisions and implementing new policy. Their overwhelming majority in the legislature (seventy-seven of seventy-nine seats) allowed decisions pertaining to First Nations issues to be carried through with little debate and negligible consultation with First Nations (John 2004). Primary of these early policies were the creation of the Forest (First Nations Development) Amendment Act 2002 (Bill 41), development of the Provincial Policy for Consultation with First Nations, Ministry of Forests Consultation Guidelines, and the Forest Revitalization Plan.

The Forest (First Nations Development) Amendment Act 2002 represented the first of several legislative changes that would have significant impact for First Nations. It provides government with the option to directly award small-scale timber tenures to a First Nation in exchange for the First Nation entering into a treaty-related economic or interim measures agreement with the province. Under sections 47.3 or 43.5 of the Forest Act, the minister of forests now has the discretion to invite a First Nation to apply for small- to medium-scale forest tenure provided that the First Nation "implement or further an agreement between the First Nation and the government" (British Columbia 2002a). These agreements are referred to as "direct award agreements" or "forest and range agreements" and are considered interim measures or economic measures under the amended act.

Shortly after creating the legislative tool (Bill 41) for granting First Nations increased access to timber tenures, the BC Liberal government unveiled a comprehensive provincial policy for consultation with First Nations. This policy describes how provincial ministries, agencies, and Crown corporations must consider the "interests" of First Nations in the allocation, management, and development of Crown land and resources. The policy defines Aboriginal interests as potentially existing and recognizes that consultation must occur with First Nations prior to the province making land and resource decisions (British Columbia 2002b). Individual ministries such as the Ministry of Forests (MoF) subsequently developed their own consultation policies, to be used in conjunction with the provincial policy (BC MoFR 2003b). These consultation policies were developed in response to court rulings, particularly the BC Court of Appeal's *Haida* decision, and instruct the consultation clauses in the forestry interim measures agreements.

With the Forest (First Nations Development) Amendment Act 2002 and the provincial consultation policy for First Nations in place, the 2003 Speech from the Throne spelled out the BC Liberals' intention to make substantial changes to forest policy. The portion of the speech entitled "Opening Up: Recognition and Reconciliation with First Nations" talked of learning "from our mistakes" and declared that "for too long we have been stuck in a rut of our own making, talking past each other and heading in opposite directions" (British Columbia 2003, 15). The Throne Speech explained that

> significant reforms will be introduced this year to ensure that more access to logging and forest opportunities is available to First Nations. Your government will take another bold step to forge a new era of reconciliation with First Nations. (British Columbia 2003, 19)

More specifically, the speech noted that,

> starting this year, funding will be earmarked in the budget for revenue sharing arrangements with First Nations that wish to help revitalize the forest industry in their traditional territories. The distribution of that revenue will be negotiated with First Nations in exchange for legal certainty that allows all regions and all British Columbians to more fairly prosper from their resource industries. (British Columbia 2003, 19)

Such "bold steps" and "significant reforms" the BC Liberals spoke of in the Throne Speech followed a month later in the form of the Forest Revitalization Plan. The comprehensive plan included economic policy actions aimed to help restore the vitality of British Columbia's struggling forest industry. A central feature of the plan was the redistribution of existing forest tenures, totalling approximately 20 percent of the province's allowable annual cut (AAC). The take-back process earmarked about 8 percent of the AAC for First Nations that enter into accommodation agreements with the province and promised that those First Nations would receive a further portion of forest revenues using a per capita formula (BC MoFR 2003a, 14).

Following the release of the Forest Revitalization Plan, the BC Liberal government introduced a flurry of legislative changes in order to implement the plan. In total, five Forest Act amendments were tabled and passed in May 2003.[2] These amendments represent some of the most significant changes to BC forest policy in decades, with far-reaching implications for First Nations (see Clogg [2003] for a full review of the changes and implications for First Nations). However, despite the significance of the changes and the government's desire not to talk "past each other," no formal consultation on these changes was ever held with First Nations. Instead, First Nations across British Columbia, including First Nations associations such as the Union of BC Indian Chiefs, the First Nations Summit, and the Northwest Tribal Treaty Nations, released formal objections, citing infringements on their rights and title. They called on the minister to postpone the proposed legislative changes in order to meaningfully consult with First Nations (Clogg 2003; FNS 2003).

With the enabling legislation for the Forest Revitalization Plan in place, the BC Liberal government tried to quell the mounting dissatisfaction expressed by First Nations by hosting a series of regional workshops to discuss the policy changes. Participating First Nations cited mistrust with the provincial government and questioned the rationale in having First Nations participate in forest policy discussions only after the decisions had been made. This was a particularly sensitive point considering much of the Forest Revitalization Plan was completed in a collaborative behind-the-scenes process organized by provincial government and forest industry representatives (BC MoF and COFI 2002).

First Nations had further concerns about the equity of a per capita revenue-sharing formula in the forest and range agreements (this is elaborated on later), and the quality of the timber to be made available to First Nations (BC MoFR 2003d). Although limited discussions with First Nations did occur during these workshops, they did not lead to any changes in implementing the new policies.

Implementing the New Approach

Despite outstanding opposition from First Nations, the BC Liberals moved forward with conducting interim measures forestry agreements with individual First Nations. The scope, content, and implications of these agreements vary, but they can generally be classified into two main types: direct award agreements and forest and range agreements (FRAs, later called "forest and range opportunity agreements" or FROs). Although there are challenges associated with both agreements, it is the FRAs that essentially represented the province's attempt at accommodation and provoked larger backlash from First Nations.

Direct Award Agreements

The direct award agreements issued to date have almost exclusively fallen under three main agreement types: interim measures agreements, First Nation wildfires agreements, and mountain pine beetle agreements. Direct award agreements provide an invitation for a First Nation to apply for a forest tenure without competition from other bidders. Although the policy allows for long-term, large-volume, replaceable tenures (BC MoFR 2002), the agreements have provided nonreplaceable forest licences ranging in terms from one to ten years, with an average of five years. The average annual timber allocation for each First Nation has been approximately 43,500 to 48,500 cubic metres per year, with the timber made available primarily through mountain pine beetle uplifts in the Interior and licensee AAC undercuts on the coast (BC MoF 2005). First Nation wildfire agreements (FNWA) offer an invitation to apply for a three-year nonreplaceable salvage licence on areas subject to recent wildfires, though there has not been an FNWA signed since 2003. These new tenure allocations are not part of the Forest Revitalization Plan's tenure take-back and are dependent on timber volume being available (BC MoFR 2002).

Forest and Range (Opportunity) Agreements

As outlined in the 2003 Throne Speech and subsequent Forest Revitalization Plan, one of the key BC Liberal government commitments was to build on the direct award agreements by opening up new tenures for First Nations and including revenue sharing as part of the agreements. The first of these new interim measures, called forest and range agreements (FRAs, later called "forest and range opportunity agreements" or FROs), was signed in October 2003. Some 129 FRAs/FROs have been signed under the program, with the last one completed in March 2009.

The first iteration of these agreements, the FRAs, were generally for a five-year term and outlined economic benefits (revenue sharing and timber volume) in exchange for specific commitments to consultation and accommodation. The revenue-sharing component of the FRA was determined on a per capita basis, under the formula of approximately $500 per registered First Nation member per year. The timber volume component of the majority of the FRAs invited the First Nation to apply for a five-year nonreplaceable forest licence. The volumes made available for each First Nation varied but was based on a fixed-population formula. The volume made available to First Nations was in the range of 30 cubic metres per person, with an upper target of 54 cubic metres per person if other top-up volumes were available, such as undercut or beetle uplifts (BCSC 2005).

The early FRAs included consistent clauses in each agreement citing that these were conditional on clear commitments from the First Nation to participate in an operational and resource management consultation process and not to unduly impede forest resource developments within its traditional territories (BC MoF 2005). In other words, being granted an FRA was conditional on the First Nation agreeing that the government had met its duty to consult and accommodate, and on it forfeiting its rights to challenge the activities otherwise taking place in its traditional territory.

It is not surprising that the challenges to the province's unilateral legislative changes and approaches to accommodation continued and grew more visible. The conditions attached to the FRAs were proving contentious and unable to quell the dissatisfaction about adequate consultation and accommodation throughout the province. First Nations across British Columbia continued to challenge the province's direction through multiple channels and newly formed coalitions. Two First Nations-led demonstrations were among the most significant responses of that time. In May 2004, the First Nation organizers mobilized thousands of Elders, youth, community members, and leaders in a week-long caravan through the province, culminating in a rally in Victoria. It was a direct response to the state of consultation and accommodation, including the provisions of the forestry agreements (UBCIC 2004). The rally was one of the first and largest demonstrations to be initiated by First Nations and served as a significant indicator of the level of dissatisfaction felt across the province.

A second significant response by First Nations was the Islands Spirit Rising campaign by the Haida Nation and its supporters. The campaign was sparked in response to the BC government's decision not to consult the Haida on the proposed tenure transfer of Tree Farm Licence 39 from Weyerhaeuser to Brascan. The Haida maintained that this decision was counter to the 2004 Supreme Court of Canada ruling that governments must consult in good faith and endeavour to seek workable accommodations with respect to granting tenures and management of the land in question. The Haida were also angered by the Ministry of Forests and Range's approval of logging plans in areas proposed for cultural protection under a joint land use planning process being conducted by the Haida and the province (CHN 2005a). As a result, the Haida and their supporters set up two separate blockades that shut down harvesting operations on the island and assumed control over a significant amount of recently harvested timber (Ramsay 2005). This development, in combination with a pending provincial election, put a significant amount of pressure on the province to find a solution. The minister of forests, at the same time, maintained that he had no obligation to consult with the Haida, stating that "if they're not happy [the Haida], the solution is to go back to the court" (CBC 2005). Despite the minister's stance on the dispute, the government did strike a deal with the Haida on protected and forestry operating areas and was able to bring an end to the blockades (CHN 2005b).

Indeed, the province's outward intention in establishing the direct awards and FRAs was the creation of investor confidence and appeasement of court-directed obligations for consultation and accommodation (BC MoFR 2003c; Parfitt 2007). One of the most objectionable components of the early FRA documents, however, was their requirement that the signing First Nation forfeit the right to challenge activities in its traditional territory on penalty of having the agreement suspended or cancelled by the minister. Through subsequent rounds of negotiations with First Nation leaders and the provincial government, those provisions were removed from the agreements by early 2006 (Parfitt 2007).

Apart from the texts surrounding consultation and accommodation and the right to challenge future activities occurring in traditional territories, First Nations challenged the underlying approach to establishing the forest and range agreements, namely through a per capita formula. As mentioned earlier, both the revenue-sharing and volume components of these agreements are established based on the number of registered members in each First Nation. Such a formula indicates a perception by the provincial government of First Nations as largely generic – forgetting, among other things, that the impacts of timber harvesting occurring in each First Nation's traditional territory vary greatly across the province. This formula attempts to accommodate First Nations equally, without accounting at all for the scale of impact actually occurring in a given territory (Parfitt 2007).

The Huu-ay-aht First Nation successfully challenged the FRA framework and the per capita formula in a BC Supreme Court case against the minister of forests. The Huu-ay-aht were in the process of renegotiating a broader agreement in 2003 when the Ministry of Forests ceased negotiations and tabled the standardized FRA template. The Huu-ay-aht repeatedly expressed a desire to continue with the Interim Measures Agreement framework they established with the ministry in 1998 and extended in 2001. This framework created a joint forest council to resolve forest management issues, supported joint forestry planning, and contained economic and forest tenure opportunities. The ministry maintained that it no longer had the mandate or structure to renew such an agreement and that, in its place, the FRA framework would suffice (BCSC 2005). The Huu-ay-aht rejected this position and subsequently petitioned the court to address the Crown's duty to consult in good faith and endeavour to seek workable economic accommodations. The Huu-ay-aht also directly challenged the province's approach to applying a population-based formula when determining accommodation arrangement.

In her ruling, Madam Justice Dillon was damning of the Ministry of Forests' conduct in applying the FRA policy. Justice Dillon found that "the conduct of the Crown from February 2004 through to the end of negotiations was intransigent. Although the government gave the appearance of willingness to consider the [Huu-ay-aht's] responses, it fundamentally failed to do so" (BCSC 2005, 58). Justice Dillon declared that the "FRA policy does not meet the Crown's constitutional obligation to consult the [Huu-ay-aht First Nation]" and that "the Crown failed to follow its own process for consultation as set out in the Provincial Policy for Consultation with First Nations and the Ministry Policy" (BCSC 2005, 56). Justice Dillon also ruled that the population-based formula to determine accommodation does not constitute good-faith consultation and accommodation, does not fulfill the administrative obligations of the Crown to provide such accommodation, and has no rational connection with the legislative objectives of the FRA program (BCSC 2005). Despite the ruling, the province continued to use the per capita formula to calculate subsequent FRAs and, later, FROs (Parfitt 2007).

The New Relationship

The forest policies of the BC Liberals pertaining to First Nations in their 2001-5 mandate amounted to the most significant changes in First Nations-related forest policy in the history of the province. These new policies signalled a major shift in the province's commitment to fulfilling its legal obligations to First Nations by providing revenue and access to forest tenure for economic development. For the province, these agreements provided hope of establishing some investment certainty in the forest industry. But even before the BC Liberals were elected to a second term in 2005, it had already become apparent that the degree of change was not sufficient to address First Nations' concerns and create a climate of certainty and stability sought by the provincial government and industry. The BC Liberals had aggravated the mistrust growing in First Nation communities through their narrow economic conception of accommodation and by sidestepping consultations with First Nations prior to introducing such far-reaching changes to forest policy.

A marked breakthrough was made in 2005 when the Government of British Columbia and the First Nations Leadership Council announced they were entering into a "New Relationship" based on respect, recognition, and accommodation of Aboriginal title and rights, respect for each other's respective laws and responsibilities, and the reconciliation of Aboriginal and Crown titles and jurisdictions (British Columbia 2005, 1).[3] In stark contrast to its earlier policies, the province pledged with First Nations to establish processes and institutions for shared decision making regarding land use, consultation and accommodation, conflict resolution, and revenue and benefit sharing. As a whole, the province was formally agreeing to recognize First Nations as governments, and to engage in a government-to-government relationship from that point forward.

The vision of the New Relationship represents a fundamental departure from the narrow economic-based approach to accommodation taken under the BC Liberals' first term to one that delivers First Nations a role in governance of their traditional territories. It specifically recognizes that the province will be unable to meet its strategic objectives unless First Nations are able to meet their own goals related to self-determination and social and economic status.

The application of the New Relationship has so far come with mixed results. In the years following the announcement, there has been little obvious advancement of the action plans on a provincial scale, and the BC business community has vocalized frustration at a lack of clarity provided in terms of undertaking consultation and accommodation in conducting business (BCBC et al. 2007). In the 2009 Speech from the Throne, the Government of British Columbia pledged to implement the principles of the New Relationship through a "Recognition and Reconciliation Act" that would inform shared decision-making frameworks across British Columbia and would "take priority over all other provincial statues" (British Columbia 2009, 1). But First Nations were not satisfied that the act would ensure acceptable accommodation of their indigenous rights and title and declared it "dead" only months later (FNS, UBCIC, and BCAFN 2009).

On the other hand, reconciliation efforts have been made in two regions of the province that house the traditional territories of more than two dozen First Nations. In response to local and international pressure to implement an ecologically based management approach to British Columbia's north and central coast and Haida Gwaii, the province announced as early as 2001 that it would develop the region's land use plans on a government-to-government basis. In December 2009, unprecedented reconciliation protocols for both regions were announced that formalized Crown–First Nation shared decision-making processes.

Crown – First Nation Collaborative Governance

North and Central Coast (Great Bear Rainforest)

The province's north-central coast is now commonly known as the Great Bear Rainforest. It is a 6.4-million hectare region of coastal temperate rainforest that became the centre of a successful international market campaign by environmental groups in the late 1990s that were seeking its protection from logging. As a result, the Government of British Columbia and the recently formed Coastal First Nations (CFN) announced in 2001 that they were entering into a General Protocol Agreement on Land Use Planning and Interim Measures (British Columbia and CFN 2001).[4] With the support of environmental and industry groups, the agreement pledged that the province and First Nations would develop a land use plan for the north and central coast through government-to-government negotiations with individual First

Nations. A central feature of the agreement was the establishment of ecosystem-based management (EBM), the guiding principles of which include collaborative processes and the accommodation of Aboriginal rights and title (CCLRMP Table 2004).

The reconciled land use plan for the region was announced in 2006. It included the Land and Resource Protocol Agreement between the province and CFN, and a Land Use Planning Agreement-in-Principle between the province and KNT First Nations,[5] as well as strategic land use planning agreements between the province and many of the individual First Nations. Together with the province's Central Coast and North Coast Land and Resource Management Plans, these agreements spell out the Crown's and First Nations' management, social, and economic objectives, as agreed to through the government-to-government negotiations. The negotiations also concluded land use zones for the region that provide for protection of culturally and ecologically significant areas while allowing the rest to be available for resource extraction in line with EBM. The Protocol Agreement also established land and resource forums through which the province and First Nations cooperatively manage the region's land and resources, and implement cooperative economic initiatives and policies (British Columbia and CFN 2006).

To further define the shape of the collaborative governance arrangement, the province and CFN finalized the Reconciliation Protocol in December 2009. The agreement is moreover intended as "a bridging step to further reconciliation of ... aboriginal title, rights and interests with provincial title, rights and interests" (British Columbia and CFN 2009, Preamble). The protocol includes an engagement framework that outlines a shared decision-making process for land and resources. The spirit of the agreements is to operate by consensus, but it is important to recognize that in neither the Reconciliation Protocol nor the informing agreements is a final arbitrator distinguished in the dispute resolution processes. Ultimately, the north and central coast management decisions have been established, and are to continue, on a government-to-government basis – a considerable advancement from the economic-based approach to accommodation under the FRA process.

Haida Gwaii

The Council of the Haida Nation has been a member of the Coastal First Nations since the inception of the Turning Point Initiative in 2000 and a signatory to the 2001 General Protocol Agreement but has since developed independent agreements for its traditional territory of Haida Gwaii. The 2001 agreement put in place a strategic land use planning process involving a community planning forum that was jointly chaired by the province and

Haida Nation (British Columbia and CHN 2007), but the relationship between the province and the Haida Nation has been contentious this decade, as already covered in this chapter.

The agreement made in 2005 to end the blockades set forth several commitments with the aim of resolving conflicts and building new approaches to resource management on Haida Gwaii. By 2007, the province and the Haida Nation had developed the Strategic Land Use Agreement, which, much like on the north and central coast, designated protected, special value and operating land use zones, and EBM objectives for the remainder of the archipelago (British Columbia and CHN 2007). The agreement also designated a co-governance structure for its implementation that pledges collaborative planning and management for all land use zones. As on the coast, the governance structure was developed into a Reconciliation Protocol in December 2009 that specifically spells out a framework for consensus-based decision making over lands and resources (British Columbia and CHN 2009). In May 2010, the Haida Reconciliation Protocol was passed into law through Bill 18 – the Haida Gwaii Reconciliation Act. The act takes a further step toward co-management by giving the five-member Haida Gwaii Management Council, the management body established under the Reconciliation Protocol, ascendancy over some ministerial decisions and even the calculation of the allowable annual cut. The act also formally drops the name "Queen Charlotte Islands," the colonial moniker given to the archipelago, returning it formally to its original name, Haida Gwaii. It is a step rich in the symbolism of a new relationship.

Second Round of Forestry Agreements

In addition to the advances in collaborative land use planning and natural resource governance in the northwest, the provincial government began working to redesign the successor to the FRA program. With several of the original five-year-term FRAs expiring, the provincial government began to make public the next generation of forestry-specific interim measure agreements in 2010. The new agreement framework considered the legal context of the *Huu-ay-aht* ruling and documented recommendations by First Nations and a provincial government-led Forestry Roundtable report (BC MoFML 2009; BC MoFLNRO 2010).

The new agreement structure includes two standalone agreements: the Forest Consultation and Revenue Sharing Agreement (FCRSA) and the Forest Tenure Opportunity Agreement. The FCRSA outlines new consultation and revenue-sharing procedures; the Forest Tenure Opportunity Agreement provides access to a new area-based tenure designed specifically for First Nations.

The major change with the FCRSA was the incorporation of a transition from the population-based accommodation model under the FRA to an activity-based revenue-sharing formula based on a percentage of the forestry-related provincial revenues derived in a First Nations-asserted territory. Although this change speaks to the revenue-sharing part of the *Huu-ay-aht* ruling, it fails to address the degree of infringement a particular forestry activity is having within the territory (Mandell Pinder 2010). Furthermore, this approach does not consider parts of the territory that have already been harvested. Another major change with the FCRSA is the approach to conducting consultation and acknowledgement of accommodation. Consultation is based on a matrix format, siting different levels of consultation for different decisions. The acknowledgement of accommodation section has also been modified to provide the province with an agreement that the First Nation has been adequately consulted and accommodated for any forestry-related infringement, despite the results of the consultation process (ibid.).

Although some aspects of the FCRSA are considered a step backwards from an acknowledgement of accommodation of Aboriginal rights and title perspective, the new Forest Tenure Opportunity Agreement is seen as a very positive development for First Nations (Mandell Pinder 2010). The new forest tenure that will be available through the agreement was brought in to force through Bill 13 – Forest and Range (First Nations Woodland Licence) Statutes Amendment Act in June 2010 (British Columbia 2010). The new tenure addresses most of the recommendations made by First Nations and the Forestry Roundtable report (MoFML 2009). Specifically, the new tenure is long-term, is area-based, and includes expanded stewardship responsibilities such as management planning and determination of the AAC. In addition, the tenure waives annual rent charges and provides flexibility on payments of silviculture security deposits (MoFLNRO 2010).

Conclusion

As this chapter has shown, the provincial conception of accommodation has shifted considerably over the past decade. From narrow economic concessions under forestry-specific interim measure agreements to the inclusion of First Nations in governance measures and new forms of Aboriginal forest tenure, the province has employed a variety of policy tools and fundamentally shifted the forest economy of the province.

Despite the high participation of First Nations in the interim measure agreements, the province has a long way to go to build their confidence in how First Nations forest policy is developed. Without meaningful consultation with First Nations in the policy formulation stage, it is unlikely there will be genuine support during implementation. Furthermore, the positive governance tools employed in the northwest of the province have yet to be considered in other regions. This leaves the question open as to whether these positive steps are a sincere evolution of policy or simply a reaction to legal challenges and pressure applied by market campaigns led by environmental organizations.

Overall, the BC Liberal government's search for certainty in the forest sector has had a mixed success. Although placing First Nations' interests front and centre in the New Relationship, the broader aspects of the agreement have yet to materialize, and there is an impasse on the larger structural issue of providing a legislative basis for Aboriginal rights and title, as indicated by the failure of the Recognition and Reconciliation Act. Progress remains slow. But there has been progress. In several key areas of the province, the BC Liberal government has moved from a position of explicit legal challenges of Aboriginal rights and title to legal agreements that share decision-making authority with First Nations on a government-to-government basis.

Notes

1 The forest policy components of the platform were, in turn, influenced by the Council of Forest Industries' "A Blueprint for Competitiveness" (COFI 1999) and the previous (New Democratic Party) government's 2000 "Shaping Our Future: BC Forest Policy Review." Both spoke to the need to utilize interim measures agreements.

2 These amendments were Bill 27, Forests Statutes Amendment Act, 2003; Bill 28, Forestry Revitalization Act, 2003; Bill 29, Forest (Revitalization) Amendment Act, 2003; Bill 44, Forests Statutes Amendment Act (No. 2), 2003; Bill 45, Forest (Revitalization) Amendment Act (No. 2), 2003.

3 The First Nations Leadership Council represents the First Nations Summit, the Union of BC Indian Chiefs, and the BC Assembly of First Nations. Under the structure of the Leadership Council sits several industry-specific councils, including the BC First Nations Forestry Council.

4 CFN (originally Turning Point) had been formed by First Nations of the coast and Haida Gwaii, with the intention of mutual support in engagement with the province. The traditional territories of the Coastal First Nations span almost the entirety of the planning region. To facilitate the involvement of other First Nations in the region, the province also entered into separate agreements with the First Nations whose traditional territories cover the remaining lands and Haida Gwaii.

5 The KNT First Nations are members of the Kwakiutl District Council, Musgamagw Tsawataineuk Tribal Council, and Tlowitsis Nation, whose traditional territories cover the most southern portion of the planning area, extending onto Vancouver Island.

References

BCBC (BC Business Council), Council of Tourism Associations of British Columbia, BC Chamber of Commerce, Council of Forest Industries, Mining Association of British Columbia, Coast Forest Products Association, BC Roadbuilders and Heavy Construction Association, BC Utilities Advisory Council, Association for Mineral Exploration BC, and BC Salmon Farmers Association. 2007. *First Nation Consultation and Accommodation: A Business Perspective*. Submission to the New Relationship Management Committee. January 19. http://www.bcbc.com/.

BC Liberal Party. 2001. *A New Era for British Columbia: A Vision for Hope and Prosperity for the Next Decade and Beyond*. BC Liberal Party.

BC MoF (Ministry of Forests). 2005. Aboriginal Affairs Branch. Agreements with First Nations Page. Visited April 15, 2005 at http://www.for.gov.bc.ca/haa/FN_Agreements.htm.

BC MoF and COFI (BC Ministry of Forests and Council of Forest Industries). 2002. Forest Policy Discussions. Revised Draft August 9, 2002.

BC MoFLNRO (BC Ministry of Forests, Lands and Natural Resource Operations). 2010. "First Nation Opportunities in the Forest Sector." Minister Pat Bell's presentation to the First Nations Forestry Council, North Vancouver, June 8. http://www.aboriginal-law.com/.
BC MoFML (BC Ministry of Forests, Mines and Lands). 2009. *The Report of the Working Roundtable on Forestry: Moving Towards a High Value, Globally Competitive, Sustainable Forest Industry*. Victoria: MoF.
BC MoFR (BC Ministry of Forests and Range). 2002. *Ministry of Forests Interim Policy: First Nations Access to Timber Tenures through Sections 43.5 and 47.3 of the Forest Act*. Victoria. http://www.for.gov.bc.ca/.
–. 2003a. *The Forest Revitalization Plan: BC Heartlands Economic Strategy – Forests*. Victoria. http://www.for.gov.bc.ca/.
–. 2003b. *Ministry of Forests Consultation Guidelines 2003*. Victoria: MoFR. www.for.gov.bc.ca/.
–. 2003c. *Strategic Policy Approaches to Accommodation: Ministry of Forests – Final Draft*. July 21. Victoria.
–. 2003d. Workshop notes from the Minister of Forests Coastal First Nations Workshop, Nanaimo, September 11. www.for.gov.bc.ca/haa/Docs/Coastal_Draft_Minutes_version2.pdf.
BCSC (British Columbia Supreme Court). 2005. *Huu-ay-aht First Nation et al. v. The Minister of Forests et al*. B.C.S.C. 697, 56-59.
British Columbia. 2002a. Bill 41 – Forest (First Nations Development) Amendment Act, 2002. Enactments 2,4,5,6, and 8.
–. 2002b. *Provincial Policy for Consultation with First Nations*. October 2002. Victoria. Published by the Government of British Columbia, Victoria, BC. http: faculty.law.ubc.ca/mccue/pdf/2002%20consultation_policy_fn.pdf.
–. 2003. Speech from the Throne. The Honourable Iona Campagnolo, lieutenant governor at the opening of the fourth session, thirty-seventh Parliament of the Province of British Columbia, February 11.
–. 2005. *The New Relationship*. Published by the Government of British Columbia, Victoria BC. http://www.newrelationship.gov.bc.ca/.
–. 2009. *Discussion Paper on Instructions for Implementing the New Relationship*. February 19. http://www.gov.bc.ca/.
–. 2010. Bill 13 – Forest and Range (First Nations Woodland Licence) Statutes Amendment Act. http://www.leg.bc.ca/.
British Columbia and CFN (Coastal First Nations). 2001. General Protocol Agreement on Land Use Planning and Interim Measures. www.coastforestconservationinitiative.com/
–. 2006. Land and Resource Protocol Agreement. March 26. archive.ilmb.gov.bc.ca/slrp/lrmp/nanaimo/central_north_coast/docs/Turning_Point_Protocol_Agreement_Signed.pdf.
–. 2009. Reconciliation Protocol. December. http://www.newrelationship.gov.bc.ca/agreements_and_leg/reconciliation.html.
British Columbia and CHN (Council of the Haida Nation). 2007. Haida Gwaii Strategic Land Use Agreement. September 13. http://www.ilmb.gov.bc.ca/slrp/docs/Haida_SLUPA_Dec_07.pdf.
–. 2009. Kunst'aa Guu – Kunst'aayah Reconciliation Protocol. December 11. http://www.newrelationship.gov.bc.ca/shared/downloads/haida_reconciliation_protocol.pdf.
BCCA (British Columbia Court of Appeal). 2002a. *Haida First Nation v. BC (Ministry of Forests) and Weyerhaeuser*. B.C.C.A. 462.
–. 2002b. *Haida First Nation v. BC (Ministry of Forests) and Weyerhaeuser*. B.C.C.A. 147.
CBC. 2005. "Haida set up Blockades on Queen Charlottes." CBC News March 23. www.cbc.ca/news/.
CCLRMP Table. 2004. *Central Coast LRMP Completion Table – Report of Consensus Recommendations to the Provincial Government and First Nations*. http://archive.ilmb.gov.bc.ca/.
CHN (Council of the Haida Nation). 2005a. *Islands Spirit Rising* bulletin no. 1. March 19. http://www.haidanation.ca/islands/Bulletin1.html.

–. 2005b. "Understanding Arising from April 22 2005 Discussions between the Province and the Council of Haida Nation." http://www.haidanation.ca/.

Clogg, Jessica. 2003. *Provincial Forestry Revitalization Plan – Forest Act Amendments: Impacts and Implications for BC First Nations.* Discussion paper. August. Vancouver: West Coast Environmental Law. http://wcel.org/.

COFI (Council of Forest Industries). 1999. *A Blueprint for Competitiveness: Five Ideas for Improving Public Policy Affecting the BC Forest Industry and the People, Businesses and Communities That Depend on It.* Vancouver: COFI.

FNS, UBCIC, and BCAFN (First Nations Summit, Union of BC Indian Chiefs, and BC Assembly of First Nations). 2009. "All Chiefs Assembly United in Rejection of Proposed Legislation." August 29. www.fns.bc.ca/.

John, Grand Chief Edward. 2004. "Increasing First Nations' Participation in BC's Forest Industry." Presentation by Grand Chief Edward John, Member of the First Nations Summit Political Executive. Hyatt Regency Hotel, Vancouver, June 7, 2004.

Mandell Pinder. 2010. *Review of the Forest and Range Consultation and Revenue Sharing Agreement Template.* Prepared for the Union of BC Indian Chiefs. October. Vancouver: Mandell Pinder.

Parfitt, Ben. 2007. *True Partners: Charting a New Deal for BC, First Nations and the Forests We Share.* Resource Economics Project. January. Vancouver: Canadian Centre for Policy Alternatives. http://www.policyalternatives.ca/.

Ramsay, H. 2005. "Haida Protesters Blockade Two Log-Sort Facilities, They Challenge Alleged Violation of Their Rights." Special to the *Vancouver Sun,* March 23.

Supreme Court of Canada. 2004. *Haida Nation v. British Columbia (Minister of Forests).* 3 S.C.R. 511.

Supreme Court Reports. *Delgamuukw v. British Columbia (Attorney General)* [1997] 3 S.C.R. 1010 FNS (First Nations Summit).

UBCIC (Union of BC Indian Chiefs). 2004. "Storm Clouds of Economic Uncertainty Gather over British Columbia: Natives Predict Provincial-Wide Blockade." Press release. May 21. www.ubcic.bc.ca/.

18

Consultation and Accommodation

Making Losses Visible

Ronald L. Trosper and D.B. Tindall

This book is about the struggles Aboriginal people have had in resisting the policies of the dominant society in Canada to assimilate them to Canada's prevailing model of life. In recent years, particularly after the addition of an amendment to the Canadian constitution to recognize preexisting Aboriginal rights, the leverage to resist assimilation has increased. We argue that consultation and accommodation help to highlight the losses that First Nations have incurred, and hold the potential – through making these losses more visible – of reducing the losses.[1]

To argue that some progress is possible may be startling, given the appearance of little change. As the example of Clayoquot Sound shows, provincial authorities strain to retain ultimate decision-making authority. Even when co-management authority seems to have been granted, the Nuu-chah-nulth have had to accept the structure of a traditional for-profit company, Iisaak, in order to obtain some portion of control over activities in their territory, although the rules for harvest were changed by the Clayoquot Sound Scientific Panel, with significant Nuu-chah-nulth participation.

Iisaak operates under standard provisions of a forest licence. Passelac-Ross and Smith present some alternatives to the standard provisions in their chapter, but these alternative provisions would represent a radical change from current timber policy; the proposal in 2010 by the Government of British Columbia to create a specific Aboriginal timber tenure has replicated most of the main provisions of other forest licences. The emphasis remains on timber production, with all final decisions regarding forest management continuing to be held by the Ministry of Forests and Range. This emphasis remains in the new First Nations Woodland Licence, which retains emphasis on timber harvest. In addition, in order to obtain the new licence, a First Nation is required to sign an interim measures agreement.

The chapters of this book show that Aboriginal peoples in Canada, and in British Columbia in particular, have been excluded from access to resources that they need for their livelihoods and have been pressured to assimilate.

What are the consequences of the loss of access and the requirements to change behaviour? A recent article by Nancy Turner and other authors lists the large variety of losses that have occurred. They point out that many of the losses are invisible to people other than those who have experienced the losses:

> If a family is forced from its home, the loss is direct and obvious, and compensation typically is expected. However, if the loss is not obvious to others, is not readily measured, is not represented in a manner recognized as legitimate, or is the result of a series of compounding impacts that are not easily connected to an original action, the consequences can be invisible even though they prove devastating. (Turner et al. 2008, art. 7)

The article goes on to suggest six processes that can contribute to the visibility of invisible losses, including "focusing on what matters to the people affected, describing what matters in meaningful ways, making a place for these concerns in decision making, evaluating future losses and gains from a historical baseline, recognizing culturally derived values as relevant, and creating better alternatives for decision making" (Turner et al. 2008). Most of these processes involve having Aboriginal people participate in a more equal manner in the formulation and subsequent implementation of land use policy.

Since the *Delgamuukw* decision in 1998, the courts of Canada and British Columbia have started to insist that government agencies contemplating actions involving lands subject to claims of Aboriginal rights, including Aboriginal title, must consult and accommodate the interests of the First Nations that are asserting the existence of the rights, even if no court has ruled on the existence of the rights or no treaty has been agreed to. Provincial government departments, particularly the Ministry of Forests and Range in British Columbia, have not listened to court decisions and are dragging their collective feet in complying with the requirements to consult. Because of this reluctance, First Nations have gone to court to have their interests taken into account. Examination of the resulting decisions shows that compliance with the requirements to consult and accommodate has the potential to create the very processes that Turner and colleagues (2008) suggest can contribute to making losses visible.

The courts have also been reluctant to enforce the requirement to act; in a great many of the cases, the parties are ordered to continue to negotiate. Actions have been allowed to continue while the case is argued. The result is that when the First Nation plaintiffs win, the action has already been taken and the losses imposed. One cannot argue, therefore, that the requirement to consult and accommodate has succeeded as yet in making the invisible losses visible on a broad basis. This chapter argues, instead, that the requirements to consult and accommodate have the *potential* to make losses visible. With the losses stated more clearly, through consultation, the possibility exists for accommodation measures to reduce the losses to some extent. To establish this point, one needs first to review the Supreme Court of Canada's requirements for consultation and accommodation.

Haida and *Taku*

After a period of uncertainty, in 2004, the Canadian Supreme Court provided details about its requirements for consultation and accommodation in the pair of decisions *Haida v. BC (Minister of Forests)*[2] and *Taku River Tlingit First Nation v. British Columbia (Project Assessment Director)*.[3] Both cases involved duties owed by the Crown to Aboriginal groups whose Aboriginal rights to land had not yet been determined. Given the very slow pace of treaty negotiations (see the chapter in this volume by Mark Stevenson), development of lands could remove the value of those lands to Aboriginal people once the treaty rights were settled. The need to consult and possibly accommodate would be triggered, the court ruled, on the Crown learning that a potential claim existed. The court ruled that the level of consultation, which could range between the mere provision of notice of an action to "deep consultation," would depend on the interaction of two factors: the strength of the Aboriginal group's claim and the risk of harm from the action. Once consultation was required, the courts were to judge the reasonableness of the actions of the Crown, given the level of consultation as determined by the strength of the claim and the risk of harm. In the Haida situation, the required level of consultation was high and had to be addressed at the strategic level of decision making. Proposals to consult at operational levels were not enough. The Taku River Tlingit's case also required high levels of consultation; the court ruled that British Columbia's procedures in place at the time for environmental assessment provided a mechanism for consultation and this had led to adequate accommodation concerning a road needed to reopen a mine.

First Nations were not allowed to "veto" a plan, which meant that they had to engage in discussion and be reasonable about the changes they would require to be made to protect their claim and reduce the risk of harm. The language regarding "no veto" could be interpreted to mean that the First Nation could not require any changes in what the Crown was proposing to do; but both cases indicated that the Crown had to take seriously the matters raised by the Aboriginal group, and that the Crown needed to change its plans to accommodate the Aboriginal concerns.

The Narrow Scope of Aboriginal Rights

In claiming that there are grounds for progress in the requirements to consult and accommodate, this chapter could be considered to be making a special effort to find something positive from what some argue is a bleak legal environment. Canada's Supreme Court has offered very little support for Aboriginal or indigenous sovereignty. If Aboriginal peoples were able to exercise autonomy and control over at least some of the lands they claim, they could implement their own visions of land management. But in order to obtain access to lands they claim, indigenous people in Canada have to establish the existence of one or both of two types of authority over the land: Aboriginal title and/or Aboriginal rights. Both of these ideas are defined by Canadian courts, and in the process the courts have created quite limited concepts. Aboriginal title, although partially a proprietary interest in land, does not carry with it jurisdiction to exclude the Crown, particularly the federal Crown, from regulating the use of the land. In addition, the court has indicated it will be open to arguments that justify infringement on the rights of Aboriginal people holding Aboriginal title. (Given the limited extent of the recognition of Aboriginal title, the meaning of justified infringement remains undeveloped.) Similarly, Aboriginal rights can be infringed on, and the standards for that infringement have been spelled out more clearly. Aboriginal rights are not a proprietary interest in land or resources; they define a right to use resources.

In arguing for a positive influence of the rules for consultation and accommodation, we need first to recognize that the positive outcomes would occur within a legal structure that offers very little room for recognition of Aboriginal interpretations of their interests. Gordon Christie summarizes the situation:

> When the Crown is obliged to consult with an Aboriginal nation, it is not about how this Aboriginal collectivity might see itself in relation to its land, and about how that vision might inform visions about how people in general will interact with the land in question – rather, the Crown is obliged to consult about how its visions of land use will be implemented.
>
> There is never any question in the Court's mind that the Crown has complete power to determine the broad parameters within which questions will be answered about how Aboriginal lands are to be used. As the fundamental sovereign power, the Crown decides what land "means," to what uses lands may be put, and how people (including Aboriginal peoples) will live in relation to lands and resources ... The duty to consult fits neatly within this paradigm of overarching Crown power. (Christie 2006, 154)

Thus, large issues of land use are not on the table during consultation and accommodation. Yet, even though the Crown's vision of land use is accepted, important constraints exist. Later in his article examining consultation and accommodation, Christie writes:

> The Crown must be directing its mind toward substantially addressing valid Aboriginal concerns when contemplating action that might negatively impact on these concerns. These are legally binding directives, flowing out of the constitutionalization of Aboriginal rights. (Christie 2006, 173)

Christie goes on to note that even though the *Haida* decision, which mainly controls current interpretation of consultation requirements, provides that Aboriginal claimants have no absolute veto power over Crown plans for land, that same case also "dictates that the Crown is *seriously and powerfully constrained* in how it acts vis-à-vis the Aboriginal interests at stake" (Christie 2006, 178). Thus, even though Aboriginal sovereignty is not recognized as co-existing with Crown sovereignty, the Crown is constrained to some degree in its actions when those actions affect even unresolved Aboriginal issues.

Invisible Losses and Process to Address Them

With its emphasis on economic development, Canada's dominant society tends to focus on the visible matters of economic growth: income based on timber harvest, returns from mining, the growth of cities, and so forth. Many boosters of economic development based on resource extraction count the employment generated by those who supply the resource extraction activities – the indirect employment. Some find that the estimates for indirect employment are somewhat inflated. People can debate the proper magnitude of indirect job creation, but the idea is clear enough: the jobs of the people in the logging or sawmill companies are just part of the impact of such activities. The same thing can happen in reverse: when Aboriginal people are pushed off their fishing grounds, the fishers lose income, but so also do other members of the community.

The idea of invisible losses builds on the idea of indirect job loss. Many other indirect impacts occur. Loss of an economic base creates a loss of influence and a reduction in self-determination. When a community's land base is taken from it, other uses of the land besides the large resource extraction become impossible. This includes access to plants for medicine, but it also includes access to plants needed in ceremonies. If a community used wildlife for food, a direct and visible loss is the loss of access to the food sources, but accompanying the visible loss is an indirect loss, that of the hunting lifestyle. If identity is tied to stewardship of the land, as it is for Aboriginal people, then loss of the ability to care for the land means a reduction in valued identity as well.

Many invisible losses are created by differences in world views and visions between two cultural groups. (See the chapters in this volume by Marc Stevenson, Blackstock, and Lewis and Sheppard.) Each can see losses of things it values; matters not valuable to one group are not actually "visible" to that group. For instance, if an Aboriginal community has structured its kinship relations such that they are tightly connected to its relationship with land, then removal of the connections to the land also disturbs the kinship system.

Such disturbances can create disorder and confusion. In British Columbia, inheritance proceeded through female lines – a nephew would inherit from his mother's oldest brother, for instance. Dominant law assumes inheritance through male descent. Suppose a man has managed to obtain a trapline in his own territory, governed by provincial rules. If provincial law provides that his son should inherit the rights to the trapline, this creates disorder if the man's own society says his nephew (his sister's son) should inherit.

Processes to Address Invisible Losses

Given the main cause of the invisibility, differences in world views and cultures, the processes that can reveal such invisible losses all involve improving communication so that the losses can be revealed. As mentioned above, Turner and colleagues (2008) suggest six closely related processes that can allow a group to reveal such losses:

> Focusing on what matters to the people affected, describing what matters in meaningful ways, making a place for these concerns in decision making, evaluating future losses and gains from a historical baseline, recognizing culturally derived values as relevant, and creating better alternatives for decision making.

The first of these, focusing on what matters, obviously supports revealing invisible losses, which are what matter. Describing what matters in meaningful ways, the second process, means allowing stories and oral narratives to complement quantitative measures of changes. The third process, making a place for concerns in decision making, requires ways for what matters to be valued; for instance, by introducing metrics to measure importance that stem from Aboriginal culture. Multi-attribute analysis of alternatives can allow a broader set of metrics to be used.

The fourth process, evaluating losses from a historical baseline, is particularly important because the passage of time weakens memories of how things were. Often current conditions are taken as a baseline, when those conditions already embody important losses. To recognize culturally derived values as relevant, analysts may have to broaden their concept of what are "facts" to allow opinions about how things should be to bear on decisions. Finally, the creation of better alternatives means changing existing plans to reduce impacts of invisible losses if those reductions can be achieved with little cost to the main goal of an alternative. For instance, the timing of the harvest of trees, by being moved a few months, may vastly reduce the impact on migrating ungulates with no reduction in the volume harvested or the costs of harvest.

Courts' Implementation of Consultation and Accommodation, and the Recognition of Invisible Losses

The resistance of government officials in British Columbia to comply with the *Haida* decision has created a series of cases in which courts below the Supreme Court have interpreted the meaning of the *Haida* and *Taku* decisions. These decisions offer a way to examine the extent to which the principles of consultation and accommodation provide some leverage for Aboriginal people to move policies in a way that will reduce the losses they experience.

Discussion of the cases is organized by type of process that can reveal invisible losses: (1) focusing on what matters, (2) describing what matters meaningfully, (3) making a place for concerns in decision making, (4) using a historical baseline, (5) recognizing the relevance of culturally derived values, and (6) creating better alternatives. At least one case illustrates each of these processes. In addition, the discussion of each process and its case(s) indicates the kind of invisible loss that might be revealed in the case as it works its way to final solution.

Focus on What Matters

The first of the suggested processes to reveal invisible losses is to focus on what matters to the people affected. In order to reveal invisible losses, Aboriginal people need to have a way to draw attention to the most important things that affect their well-being. The case of *Wii'litswx v. British Columbia*[4] provides an example. The Gitanyow hereditary chiefs objected to the replacement of forest licences in their territory. They wished to have the replacement licences modified to indicate to licensees the location of their traditional wilp territories. Among the Gitanyow, "wilp" is translated as "house"; it refers to the corporate group led by a hereditary chief in control of the house's territory. With the territories described, a forest company would be able to know which of the hereditary chiefs was responsible for areas in which the company wanted to operate. The Ministry of Forests and Range refused to provide that information in the licences that were renewed. The judge ruled that provision of such information was required: "The replacement of the FLs [forest licences] clearly had the potential to detrimentally affect Gitanyow's Aboriginal rights. In particular, the harvesting of timber from Gitanyow traditional territory without reference to wilp boundaries could result in the effective destruction of individual wilps in terms of both territorial and social considerations."[5] The judge thus ruled that the failure to recognize wilp territories created a risk of harm.

The *Wii'litswx* case creates recognition of the invisible losses of identity and self-determination. Recognition of house (wilp) territories strengthens both of these for the Gitanyow.[6] The judge ruled that dismissing the recognition of wilp territories as impractical "fell well below the Crown's obligation to recognize and acknowledge the distinctive features of Gitanyow's aboriginal society" (paragraph 247). In doing so, the judge recognized the importance of wilps in the structure of Gitanyow society. Since the Elders in charge of each wilp constitute the group of traditional leaders who make political decisions for the community, recognition of their role supported self-determination.

Describe What Matters in Meaningful Ways

The *Haida* decision itself offers an example of how identification of the risk of harm provides a way for Aboriginal people to describe what matters to them. For the Haida, the loss of monumental cedar was a great concern; they need cedar trees older than eight hundred years in order to provide material for totem poles and dugout canoes. The Supreme Court explicitly recognized the risk of losing all of the remaining old cedar. The Haida were able to describe the importance of monumental cedar as a component of identification of the great risk of harm that continued forest operations presented.

Paragraph 72 of the *Haida* decision reads:

> The evidence before the chambers judge indicated that red cedar has long been integral to Haida culture. The chambers judge considered that there was a "reasonable probability" that the Haida would be able to establish infringement of an Aboriginal right to harvest red cedar "by proof that old-growth cedar has been and will continue to be logged on Block 6, and that it is of limited supply" (para. 48). The prospect of continued logging of a resource in limited supply points to the potential impact on an Aboriginal right of the decision to replace T.F.L. 39.

The value of the cedar was for cultural purposes, not as lumber.

Make a Place for Invisible Concerns in Decision Making

Many decision-making processes depend on monetary evaluation of the costs and benefits so that trade-offs among objectives can be made correctly. Matters that do not have monetary value are not eligible for consideration in the decision-making process. In the case of *Klahoose First Nation v. Sunshine Coast Forest District (District Manager)*,[7] the Klahoose objected to the approval of a Forest Stewardship Plan (FSP) for a small portion of the Toba River watershed in their traditional territory. The Klahoose wanted to have information regarding the forest company's plans for the entire watershed rather than the portion covered by the FSP. The province argued that it did not need to reveal more information than required by its own statutes, and that in addition, future impacts could be considered when future operational proposals were made. The Klahoose wanted to understand the impacts on the entire watershed and to use the information held by the province and the company for evaluation of cumulative impacts of forest operations. The judge ruled that constitutional requirements for consultation took precedence over BC statutes, and required that the Crown share information with the First Nation. Last-minute modifications to the FSP, made when the Crown started to recognize it had to make changes, were dismissed because the changes were not discussed with the Klahoose. The judge clearly stated that the Klahoose would need to be involved:

> The consideration of the application to amend the FSP would, I expect, involve an appropriate sharing of information, including information that may not be statutorily required in relation to an FSP, such as operational and access information. It would also involve Klahoose directly in the decision-making process concerning any accommodation of Klahoose's rights. *(Klahoose First Nation v. Sunshine Coast Forest District [District Manager])*[8]

This requirement that the Klahoose be involved in the decisions allows the Klahoose to bring all of their concerns into consideration. This broad ability will allow the Klahoose to address whichever invisible losses they wish to have considered.

Using a Historical Baseline for Evaluation

One key step in the process of consultation and accommodation is consideration of the trigger of the duty to consult. The trigger depends on the strength of the claim, which in turn requires evidence on Aboriginal use of the land and resources in 1846, when Britain asserted sovereignty over the territory that became British Columbia. Although in some places the evidence of resource use in 1846 is weak, in other areas it is quite strong. Because explorers to the coast of Vancouver Island provided written evidence of what they found, the Nuu-chah-nulth have been able to establish their claim based both on their oral history and on the eyewitness accounts of the explorers. In *Ahousaht First Nation v. Canada*, this historical baseline allowed the Nuu-chah-nulth to establish that they harvested and traded fish in great amounts. As a result, they established a right today to engage in commercial use of fish. The Ahousaht case is one that determined Aboriginal rights rather than dealing with consultation and accommodation. It serves as an example of what a strong claim can deliver in a case about rights, thus showing what rights might be jeopardized by ongoing Crown utilization of land and resources without accommodation.

A case that shows the potential in consultation and accommodation is the case of *Carrier Sekani v. British Columbia (Utilities Commission)*.[9] The Carrier Sekani challenged the Energy Purchase Agreement made in 2007 between BC Hydro, the province's electric utility, and Alcan, which operates a dam in Carrier Sekani territory. The BC Utilities Commission decided that it was not responsible for considering whether the Carrier Sekani had been consulted and accommodated. The appeal court held otherwise and required that the BC Utilities Commission have a hearing to consider the evidence of whether the ongoing operation of the dam in question impacted the Carrier Sekani. The judge providing reasons for the decision stated, "I consider [the decision not to review] an unreasonable disposition for, amongst other reasons, the fact that BC Hydro, as a Crown corporation, was taking commercial advantage of an assumed infringement on a massive scale, without consultation."[10] The "massive scale" arises from using a historical baseline to measure the extent of the impact on the fishery of the river impacted by the dam. The extent to which the Utilities Commission will in the end take into account the original condition of the river impacted by the dam is unknown as of the writing of this chapter. The point is that the ability to place the baseline into consideration will be part of the hearing on adequate consultation and accommodation.

As a result of this ruling, it will be possible for the Carrier Sekani to address the visible loss of fish, as well as the invisible losses that accompanied the loss of fish.

Recognizing Culturally Derived Values as Relevant

This process is similar to the inclusion of invisible concerns, and the Hupacasath case also illustrates the recognition of cultural values in the landscape. The decisions in two Hupacasath cases show that nonmonetary values may have to be recognized within the consultation process. The Hupacasath, whose traditional territory is near Port Alberni on Vancouver Island, objected when the company holding the forest licence was allowed to remove its fee lands from its tree farm licence. The original licence had granted the company access to Crown timber partly in exchange for including the fee land in the tree farm. Because the fee land contained sites of cultural importance to the Hupacasath, they objected that some kind of accommodation of the loss of the cultural sites had to be considered. The Crown during negotiations consistently offered funds in compensation for the losses, but the Hupacasath wanted compensation that took into account the loss of the sacred sites. The judge agreed that their position was reasonable and appointed a mediator to assist with the negotiations. The contributions of the cultural sites to Hupacasath identity stood some chance of being recognized because of the court's willingness to consider their concerns to be reasonable.

In ruling that the Hupacasath had negotiated in good faith, Madame Justice Lynn Smith commented:

> I find that the consultation record shows the HFN [Hupacasath First Nation] consistently raising the issue of access to sacred sites and resources important to the preservation of their culture. The fact that, in the face of the Crown's consistent refusal to discuss ways of dealing with those issues on the Removed Lands, they considered other options that would draw on the remaining Crown lands or on economic compensation, in my view shows realism rather than the opposite.[11]

That these concerns might actually be considered was assisted by the decision in this second case: that a mediator be appointed to assist the parties in reaching agreement on what accommodation should occur.

The invisible losses that might be revealed by addressing the loss of sacred sites are culture and lifestyle losses, as well as loss of identity and loss of order in the world. Sacred sites are important for cultural practices, of course, and cultural practices contribute to identity.

Creating Better Alternatives

A process that allows invisible losses to be addressed is discussion and communication among the parties so that better alternatives to the original proposal can be developed. In *Taku River Tlingit,* the companion case to *Haida,* the Canadian Supreme Court ruled that processes of environmental review had allowed the construction of a road to a mine to be altered sufficiently to address the concerns of the Taku River Tlingit, who had participated on the committee that was considering alternative road designs. Although the Tlingit may have desired to prevent the road entirely, that option was precluded by the "no veto" concept of consultation and accommodation.

In 2002, the BC government changed the environmental assessment process, removing the requirement that First Nations participate, removing consideration of cultural impacts from projects, and removing the requirement that a committee be established to evaluate the project. In *Kwikwetlem First Nation v. British Columbia (Utilities Commission),*[12] involving the BC Transmission Corporation, the court ruled that the consultation and accommodation process could not be plausibly assigned to the environmental assessment process because that process no longer complied with the conditions of the *Taku* case (paras. 53 and 69). The inference is that although the environmental assessment process no longer requires the participation given to the Taku River Tlingit, the requirement to consult and accommodate still requires that level of participation. Thus, original alternatives have to be modified to take into account Aboriginal issues.

The case of the university golf course and Musqueam illustrates the ability of the requirement to consult and accommodate to create better alternatives through a negotiation process to settle the lawsuit. The Court of Appeal suspended the sale of the golf course to the University of British Columbia to allow the parties to consult and to achieve a better result. The Musqueam had also obtained a similar result in a case involving creation of a casino on land that they claimed. Both cases were settled in 2008, with both a monetary payment and designation of particular lands to be transferred to the Musqueam. They acquired title to the golf course, subject to the existing lease; golf will continue to be played there until 2083. In this case, the negotiation outcome will allow the Musqueam directly to address their invisible losses through use of the land that they have now acquired.

Conclusion

The limited requirements for consultation and accommodation put forth by the Supreme Court of Canada provide leverage for Aboriginal peoples to obtain some attention to their issues in dealing with the plans of various Crowns to develop lands to which Aboriginal people have claims or to which they have obtained treaty rights. Although Aboriginal groups have no "veto" over the general direction of development, they do have the ability to modify the Crown's plans in some ways. The extent of modification allowed is not clear as yet.

In most of the cases reviewed in this concluding chapter, the courts merely ruled that the provincial Crown in British Columbia, in resisting making any changes to its proposed forestry activities, had not complied with what the Supreme Court desired. In the Klahoose case, the court said that cumulative effects were important and that data and proposed plans for an entire affected watershed had to be shared with the Klahoose Nation. In the two hydroelectric cases, the appeal court ruled that the BC Utilities Commission had to evaluate the extent of consultation and accommodation that was required; Crown corporations such as BC Hydro and BC Transmission could not make plans without modifications being considered.

In the Hupacasath case, concerns over spiritually important lands, as well as concerns about wildlife populations, had to be taken into account. While it was not specifically discussed above, a recent important case of relevance to our argument is West Moberly. The West Moberly case showed that in the presence of treaty rights, the Crown had to provide a way to prevent important caribou herds from becoming extinct within an Aboriginal traditional territory. The Gitanyow case, *Wii'litswx v. British Columbia,* showed that one court felt a refusal to recognize traditional territorial systems of land governance in forest licences provided obvious evidence of a failure to reach a minimal level of accommodation that was required.

The requirements of consultation and accommodation have started to change the Crown's plans. On Haida Gwaii, for instance, following the *Haida* decision, the provincial government and the Haida have agreed to new plans. The allowable cut has been reduced somewhat, areas of the islands have been given total protection from harvest, and local participation in timber harvest will be encouraged. But, emphasizing Gordon Christie's point that Aboriginal people are asked to join in the broad scope of provincial policy, the decision also allows the Haida people to acquire a timber harvesting licence. This means that the Haida will be harvesting timber under provincial supervision should they acquire the licence.

All of these movements may seem small. But they all share an important characteristic: allowing Aboriginal people to communicate, to identify the losses that are important to them. The cumulative impact of this small change could turn out to be important.

Notes

1 In this chapter we primarily use the term "First Nations" because this is the term that has been primarily used in the court cases, legislation, and policies that we discuss here, and because the cases we explore are based in British Columbia, where this term is widely used.
2 2004 S.C.C. 73.
3 2004 S.C.C. 74.
4 2008 B.C.S.C. 1139.
5 2008 B.C.S.C. 1139, paragraph 223.
6 The Gitskan organize their system of land tenure around house territories; each territory is under the ultimate authority of a head titleholder of the house, who is a *wilp*. When the Government of British Columbia recognizes house territories, it potentially returns authority to those heads of houses, who can be consulted regarding forest operations when a timber company proposes to harvest in a house territory.
7 2008 B.C.S.C. 1642.
8 2008 B.C.S.C. 1642, paragraph 152.
9 2009 B.C.C.A. 67.
10 2009 B.C.C.A. 67, paragraph 13.
11 *Ke-Kin-Is-Uqs v. British Columbia (Minister of Forests),* 2008 B.C.S.C. 1505, paragraph 159.
12 2009 B.C.C.A. 68.

References

Christie, Gordon. 2006. "Developing Case Law: The Future of Consultation and Accommodation." *UBC Law Review* 39(1):139-84.

Supreme Court of Canada (SCC). 2004. *Haida Nation v. British Columbia (Minister of Forests),* S.C.C. 73, [2004] 3 S.C.R. 511.

Turner, Nancy J., Robin Gregory, Cheryl Brooks, Lee Failing, and Terre Satterfield. 2008. "From Invisibility to Transparency: Identifying the Implications." *Ecology and Society* 13(2). http://www.ecologyandsociety.org/vol13/iss2/art7.

Contributors

Trena Allen is a member of the Peter Ballantyne Cree Nation in Saskatchewan. Trena is currently completing an MSc in rural and environmental sociology in the Department of Resource Economics and Environmental Sociology at the University of Alberta. Her thesis is on professional forester career challenges and perceptions of accredited professional forester education. Trena is a Registered Professional Forrester (RPF) in good standing with the Ontario Professional Foresters Association.

Laura Bird completed her MSc in the Faculty of Forestry at the University of British Columbia in December 2011. Her thesis examined Crown–First Nations governance frameworks in British Columbia.

Michael Blackstock is an independent scholar of Gitxsan and Euro-Canadian descent with an MA in First Nations Studies from the University of Northern British Columbia. He is a member of the House of Geel, Fireweed Clan, and his Gitxsan name is *Ama Goodim Gyet*. Michael is a Registered Professional Forester (RPF), Chartered Mediator (CMed), poet, and visual artist.

Keith Thor Carlson is a professor in the History Department at the University of Saskatchewan, where he is also Director of the Interdisciplinary Centre for Culture and Creativity. Keith's research and writing have focused on the history of the Coast Salish people of the Fraser Valley. He has authored, edited, or co-edited seven books, including most recently *The Power of Place, the Problem of Time: Aboriginal Identity and Historical Consciousness in the Cauldron of Colonialism* (University of Toronto Press, 2010) and *Orality and Literacy: Reflections Across Disciplines* (University of Toronto Press, 2011).

Brian Chisholm is a senior instructor in the Anthropology Department at the University of British Columbia. He received a PhD in Archaeology from Simon Fraser University in 1987. For over thirty years he has been doing research on prehistoric diet through the use of stable isotope analysis, working in British Columbia, France, Japan, China, Mexico, and the American Southwest.

Ken Coates is Canada Research Chair in Regional Innovation in the Johnson-Shoyama Graduate School of Public Policy at the University of Saskatchewan. He was formerly Professor of History and Dean at the University of Waterloo and the University of Saskatchewan, respectively. His work focuses on indigenous history, Aboriginal rights, and land-claims processes. His major publications include *Best Left as Indians: Native-White Relations in the Yukon, The Marshall Decision and Aboriginal Rights in the Maritimes,* and *A Global History of Indigenous Peoples: Struggle and Survival.*

Norman Dale is a professional mediator and ecologist who currently works with the Lheidli-T'enneh First Nation in the Central Interior of British Columbia. He operates a firm, Rapport Consensus Building, in Prince George and has worked with the Kwakwaka'wakw Nations, the Haida, and other communities of Haida Gwaii and the Wuikinuxv-Kitasoo-Nuxalk Tribal Council. Norman is the co-author of *After Native Claims?* and other articles and books on natural resources and consensus building.

Jason Forsyth is a natural resources and business planner with the Tsleil-Waututh Nation in North Vancouver. His research interests include natural resource policy, co-management, and conflict resolution between Crown and First Nation governments.

James S. Frideres received his PhD from Washington State University. He is currently a professor of Sociology at the University of Calgary and was the Director of the International Indigenous Studies program for many years. His book (with Rene Gadacz) *Aboriginal People in Canada* is now in its ninth edition.

J.P. Gladu is the owner and president of Aboriginal Strategy Group, a consulting company, and serves as senior advisor for business development to Bingwi Neyaashi Anishinaabek First Nation. He has produced a number of publications related to First Nations issues including forest certification, Native values collection, biofuel opportunities, First Nation community land use plans, criteria and indicators for sustainable forestry, and cedar product development. Throughout his career, J.P. has also been proud to be involved in leading-edge issues with First Nations in the Boreal Forest region of Canada.

George Hoberg is a professor in the Department of Forest Resources Management at the University of British Columbia. His research and teaching focus is on governance and policy for the sustainable management of natural resources.

Tamara Ibrahim holds degrees in sociology from the University of British Columbia and the University of Cambridge. She is interested in the social and political determinants of health. She currently works as a research assistant in the Faculty of Medicine at McGill University.

Naomi Krogman, environmental sociologist, is a professor in the Department of Resource Economics and Environmental Sociology. She is currently the Academic

Director for the Office of Sustainability at the University of Alberta, where she is focusing on curriculum development and interdisciplinary understandings of, and university commitments to, a broader view of sustainability (one that truly incorporates social connections and indicators). Her recent research is on sustainable consumption and wetland policy implementation and, to some extent, sustainable agriculture.

Dr. John Lewis is an associate professor in the School of Planning at the University of Waterloo. Prior to joining the University of Waterloo, John was a member of the Collaborative for Advanced Landscape Planning at the University of British Columbia, where he completed his MSc and PhD. His research interests include First Nations resource planning and land management, landscape planning, collaborative planning, and computer-based environmental visualization.

Holly Mabee studied co-management of forest resources from 2000 to 2005 in the Faculty of Forestry at the University of British Columbia. She has since moved to Kingston, Ontario, where she is raising a family, studying yoga, and working in community development with a focus on urban agriculture.

Andrew R. Mason is a principal and senior archaeologist with Golder Associates Ltd. in Vancouver and an expert member and vice president (North America) of the ICOMOS International Committee on Archaeological Heritage Management, an organization that provides advice to UNESCO on archaeological matters. Andrew's research interests include global heritage policy and legislation.

Monique Passelac-Ross has been a Research Associate at the Canadian Institute of Resources Law (CIRL) since 1989. Her research interests include forestry law, natural resources law and policy, Aboriginal law, and environmental law. Most of her research projects have addressed the interface between resource development – notably forestry, oil and gas, and oil sands – and Aboriginal and treaty rights, with a focus on Alberta. She is the author of several publications dealing with these topics.

Gabriela Pechlaner works on society–natural resource interactions, with a particular emphasis on socially contentious technologies, such as those relating to aquaculture or agricultural biotechnologies. Her research has appeared in *Anthropologica, Rural Sociology,* and the *Canadian Journal of Sociology,* among others. Her book, *Corporate Crops: Biotechnology, Agriculture, and the Struggle for Control,* is forthcoming from the University of Texas Press. Gabriela is a faculty member in the Sociology Department at the University of the Fraser Valley, in Abbotsford, British Columbia.

Pamela Perreault is Anishinabae (Ojibway) and a member of Garden River First Nation in Ontario, Canada. She is currently completing her PhD dissertation on the impacts of Aboriginal capacity-building in the Canadian forestry sector. Most of Pamela's academic and professional work experience has focused on the intention or conceptualization of "building Aboriginal capacity" for forest management or other natural resource-related sectors. Pamela remains a dedicated member of the National Aboriginal Forestry Association (NAFA), the Canadian Forestry

Institute (CFI), and the Canadian Aboriginal Science and Engineering Technology Society (CASTS) and continues to encourage Aboriginal youth to pursue and practise their stewardship responsibilities.

Stephen Sheppard teaches sustainable landscape planning, aesthetics, and visualization in the Faculty of Forestry and Landscape Architecture programme at UBC. He directs the Collaborative for Advanced Landscape Planning (CALP), a research group using perception-testing and interactive 3D visualization tools to support public awareness-building, policy change, and collaborative planning on climate change and sustainability issues. Current research interests include climate change planning, outreach, and community engagement; visioning methods and visualization of climate change causes, impacts, and mitigation/adaptation; low-carbon future scenarios visualized in the CIRS Decision Theatre; community energy planning, renewables, and energy literacy; public perceptions, aesthetics, and sustainability; and social aspects of forestry.

Dr. M.A. (Peggy) Smith is an associate professor in Lakehead University's Faculty of Natural Resources Management and a Registered Professional Forester in Ontario. Her research interests focus on the social impacts of resource management, including indigenous peoples' involvement in resource management, community forestry and public participation, northern development, and forest certification. Peggy continues her longstanding affiliation as Senior Advisor with the National Aboriginal Forestry Association, an indigenous-controlled non-profit organization with the goal of increasing indigenous participation in the forest sector.

Marc G. Stevenson, PhD Anthropology (University of Alberta), is an applied social anthropologist currently providing research and policy advice to First Nation and non-Aboriginal clients in western and northern Canada on a broad range of issues at the interface of resource development and Aboriginal rights and interests. Marc has worked with Canada's Inuit, First Nations, and Métis communities on relevant cultural, social, economic, and environmental issues for the last thirty years. He is known for his work in traditional land use, social and cultural impact assessment, traditional knowledge, and co-management discourses.

Mark L'Hirondelle Stevenson is a Métis lawyer whose family comes from the historic Métis community of Lac Ste. Anne in northern Alberta. Mark works for First Nations and Métis organizations in matters related to conflict management, constitutional development, governance, natural resource law, and economic development. He currently works with a number of First Nations as legal counsel and Chief Negotiator under the British Columbia Treaty Process. Mark is a member of the Board of Directors with the Canadian Civil Liberties Association and practises law in Victoria, BC.

D.B. Tindall is an associate professor in the Department of Forest Resources Management, and the Department of Sociology, at the University of British Columbia, where he studies contention over environmental issues such as forestry, wilderness preservation, fisheries, and climate change. A major focus of his research has been environmental movements in British Columbia, and Canada, and their relations with other groups, such as Aboriginal groups and forestry communities. His research has examined various aspects of environmentalism including values, attitudes, and opinions; activism and conservation behavior; media coverage of environmental issues; gender issues; and social networks. His current research focuses on sociological aspects of contention over climate change in Canada, including perceptions about climate change, views about climate justice, and social processes affecting policies for dealing with climate change. As an adult, D.B. Tindall learned that an ancestral branch of his family was Cree.

Ronald L. Trosper has been Head of the American Indian Studies Program at the University of Arizona since July 2011. Ronald's latest work has been in the areas of indigenous economic theory and traditional ecological knowledge. He examined the institutions that provided stability for the peoples of the Northwest Coast in his book, *Resilience, Reciprocity and Ecological Economics: Northwest Coast Sustainability* (Routledge, 2009). He co-edited a book on traditional forest-related knowledge, *Traditional Forest Knowledge: Sustaining Communities, Ecosystems and Biocultural Diversity,* edited by John Parrotta and Ronald Trosper (Springer, 2012).

Rima Wilkes is Associate Professor of Sociology at the University of British Columbia. Her research focuses on collective actions by indigenous peoples in Canada and has appeared in the *Canadian Review of Sociology, American Indian Culture and Research Journal,* and the *American Behavioral Scientist.* She is currently researching how mass media photographs represent these collective action events.

Index